T0302331

Introduction to Modern Scientific Programming and Numerical Methods

The ability to use computers to solve mathematical relationships is a fundamental skill for anyone planning for a career in science or engineering. For this reason, numerical analysis is part of the core curriculum for just about every undergraduate physics and engineering department. But for most physics and engineering students, practical programming is a self-taught process.

This book introduces the reader not only to the mathematical foundation but also to the programming paradigms encountered in modern hybrid software-hardware scientific computing. After completing the text, the reader will be well-versed in the use of different numerical techniques, programming languages, and hardware architectures, and will be able to select the appropriate software and hardware tool for their analysis.

It can serve as a textbook for undergraduate courses on numerical analysis and scientific computing courses within engineering and physical sciences departments. It will also be a valuable guidebook for researchers with experimental backgrounds interested in working with numerical simulations, or to any new personnel working in scientific computing or data analysis.

Key Features:
- Includes examples of solving numerical problems in multiple programming languages, including MATLAB®, Python, Fortran, C++, Arduino, Javascript, and Verilog.
- Provides an introduction to modern high performance computing technologies including multithreading, distributed computing, GPUs, microcontrollers, FPGAs, and web "cloud computing".
- Contains an overview of numerical techniques not found in other introductory texts including particle methods, finite volume and finite element methods, Vlasov solvers, molecular dynamics.

Lubos Brieda holds a Ph.D. in Aerospace and Mechanical Aerospace Engineering from the George Washington University in Washington, D.C., USA and a M.Sc. in Aerospace and Ocean Engineering from Virginia Tech in Blacksburg, VA, USA. He is the president of Particle in Cell Consulting, LLC, Westlake Village, CA, USA, while also serving as a part time lecturer in the Department of Astronautical Engineering at University of Southern California, Los Angeles, CA. Dr. Brieda is the author of numerous simulation codes utilized by the aerospace and plasma modeling communities. Additionally, he maintains an online blog found at particleincell.com/blog focusing on scientific computing and since 2014 he has been teaching online courses on plasma simulations through his website.

Joseph Wang is Professor of Astronautics and Aerospace and Mechanical Engineering at University of Southern California. Prof. Wang received his Ph.D in Aeronautics and Astronautics from Massachusetts Institute of Technology in 1991. Prof. Wang conducts research in computational physics, space technology, and space and planetary science. He and his students have developed many computer simulation models utilizing particle-in-cell, Vlasov, and molecular dynamics simulation for large-scale first-principle based simulations in these areas. He has more than 300 publications.

Robert Martin received his B.S. in Aerospace Engineering from Iowa State University and his M.S. and Ph.D in Computational Science, Mathematics, and Engineering focused on modeling non-equilibrium gas and plasma at the University of California, San Diego. He joined the Air Force Research Laboratory (AFRL) In-Space Propulsion branch in 2011 leading development of a new multiscale/multiphysics modeling framework for spacecraft plasma. He joined the Army Research Office to run the Modeling of Complex Systems Program in 2021.

Introduction to Modern Scientific Programming and Numerical Methods

Lubos Brieda, Joseph Wang and Robert Martin

CRC Press
Taylor & Francis Group
Boca Raton London New York

CRC Press is an imprint of the
Taylor & Francis Group, an **informa** business

MATLAB® is a registered trademark of The MathWorks, Inc. For product information, please contact:

The MathWorks, Inc. 3 Apple Hill Drive Natick, MA 01760-2098 USA Tel: 508-647-7000 Fax: 508-647-7001 Email: info@mathworks.com Web: www.mathworks.com

Designed cover image: Adeline Wang

First edition published 2025
by CRC Press
2385 NW Executive Center Drive, Suite 320, Boca Raton FL 33431

and by CRC Press
4 Park Square, Milton Park, Abingdon, Oxon, OX14 4RN

CRC Press is an imprint of Taylor & Francis Group, LLC

ISBN: 978-0-367-67191-4 (hbk)
ISBN: 978-0-367-67660-5 (pbk)
ISBN: 978-1-003-13223-3 (ebk)

DOI: 10.1201/9781003132233

Typeset in Latin Modern font
by KnowledgeWorks Global Ltd.

Publisher's note: This book has been prepared from camera-ready copy provided by the authors.

Dedication

To our families

Contents

Preface

MOTIVATION

This book originated from an effort to produce appropriate course materials for an undergraduate course on Computational Programming and Numerical Methods applicable to upper-level undergraduate students in engineering departments, primarily those majoring in disciplines related to astronautical, aerospace, and mechanical engineering. We have noticed that most scientific programming courses in such programs tend to focus on simple examples in scripted interpreted languages such as MATLAB®. While this mode of instruction simplifies the introduction of the new material and helps students quickly obtain skills needed to complete elementary self-contained homework assignments, it is failing to prepare them for a real world career in scientific computing. Our experiences in attempting to integrate recent university graduates with this type of background into larger scale computational science projects was that this single language mode of instruction left them ill-prepared for the complexities of these efforts. Large-scale projects often result from an amalgamation of code integrated from legacy efforts involving high performance languages such as Fortran or C++. These codes rely on technologies not covered in the introductory courses such as dynamic memory allocation, object-oriented programming, or the use of parallel computational environments. At the same time, the increasing computational power of microcontrollers allows for an in-situ analysis of sensor data. Numerical analysis work is also moving from the traditional desktop realm to a cloud-backed setup, with an interactive interface provided by web browsers. Engineering students outside the computer science departments tend to have limited understanding of these important technologies.

With this challenge in mind, a new course was developed, the content of which provides the basis for this textbook. The goal of the course, and hence this text, is to expose students to the diversity of programming languages, numerical methods, and computational environments one may encounter as a numerical analyst. While we realize that no single book can turn a novice into an expert, our hope is that the reader gains sufficient insight into the field to become a productive member of a computational research group in a short time. At the same time, we are cognizant that a book such as this will suffer from many limitations. Each chapter merely scratches the surface of the covered topic, as an entire volume could be (and have been) written on each of the topics covered here. The focus is limited to the technologies and numerical methods most frequently utilized in our work related to developing simulation codes for the space environment, plasma physics, and rarefied gas dynamics. Where feasible, we include references to cited works. Much of the material discussed here has however over the years become "common knowledge" for which no specific citation is readily available. We thus also include additional reference texts that were found useful in our past studies of the subjects covered.

ORGANIZATION

We attempted to provide the new material in an orderly manner, although this is not an easy task, as some topics depend on each other. We begin by introducing basic concepts of numerical integration in Chapter 1. We demonstrate this approach by developing a Python code for calculating the trajectory of a tennis ball. Here we also introduce fundamental programming concepts such as variables, loops, conditional statements, functions, and random numbers. We close the chapter with a survey of commonly encountered programming languages. Chapter 2 dives deeper into the world of numerical analysis. Here we derive equations for the Finite Difference. We then describe direct and iterative solver algorithms for linear algebra problems. These are used to develop a 2D steady-state heat equation solver in Python. Chapter 3 describes additional numerical analysis topics, such as filtering, sampling from distribution functions, quadrature, root finding, and solutions to non-linear systems. Chapter 4 then serves as a crash course in C++. The chapter begins by introducing the language. We then use concepts from traditional, functional programming to develop a C++ version of the tennis ball integration. We next introduce more advanced concepts such as object oriented programming, templates, and operator overloading and uses them to demonstrate a C++ version of the heat equation solver. Data from both examples is visualized using Paraview.

Chapters 5 and 6 introduce solution methods relevant to engineering analysis, with a focus on gas dynamic simulations. First, in Chapter 5 we present methods based on the Lagrangian treatment, in which parcels of information (such as simulation particles) move about the computational domain. We demonstrate this approach by first developing code simulating non-collisional free-molecular flow past a disc. We then learn how to incorporate molecular collisions using the Direct Simulation Monte Carlo (DSMC) method, and plasma effects using the Particle in Cell (PIC) method. Chapter 6 then introduces the Eulerian approach, in which partial differential equations are solved on a stationary grid. We mainly focus on the advection-diffusion equation, but also introduce solution methods for the vorticity-stream function method as well as other model equations. We also introduce Vlasov solvers. These two chapters are meant to provide just the sufficient introductory material to the field of kinetic and fluid gas dynamics. Students interested in these fields are directed to review external references, including the texts listed in the bibliography section, for solution methods specific to computational fluid dynamics, structure mechanics, and electromagnetics.

Chapter 7 changes gears by introducing interactive simulations running in web browsers. The aim of this chapter is to make students aware that any computer, having just a text editor and a web browser, can be used to develop simulation programs even if such a system does not contain a compiler or an interpreter. Programs running in web browsers are also inherently interactive by the nature of web pages and as such can be used for prototyping algorithms that require mouse-driven feedback. Chapter 8 discusses various scientific computing concepts such as debugging, use of software libraries, unit testing, documentation, version control, and LATEX. In Chapter 9 we introduce high performance computing using multithreading, domain decomposition using MPI, and graphics card processing using CUDA. Here we also discuss code profiling and the use of Linux based clusters. Then, in Chapter 10, we introduce methods for parameter space optimization, which naturally lead to a discussion on machine learning. We develop a simple neural network for classifying input numbers. Finally, Chapter 11 discusses methods for programming embedded systems such as microcontrollers and Field Programmable Gate Arrays (FPGAs). We develop an Arduino program that activates an LED when the ambient light is turned off, and also discuss integration with external sensors via libraries. We also cover the use of FPGA IP blocks, desktop-based testing, and interfacing of Arduino microcontrollers with the FPGA.

COMPANION CODE AND SAMPLE ASSIGNMENTS

This book includes an extensive companion sample code. Only the relevant snippets are included in print due to space constraints. The complete source code can be downloaded from the book website **https://www.scientificprogrammingbook.com/**. On this page, you will also find sample homework problems derived from the work assigned to students at the Department of Astronautical Engineering at USC. Solutions to the assignments can be obtained by emailing Dr. Brieda.

Author Bios

Lubos Brieda holds a Ph.D. in Aerospace and Mechanical Aerospace Engineering from the George Washington University, and a M.Sc. in Aerospace and Ocean Engineering from Virginia Tech. He is the president of Particle in Cell Consulting LLC. He also serves as a part-time lecturer in the Department of Astronautical Engineering at the University of Southern California (USC) where he teaches courses on computational plasma physics and scientific computing, with notes from the latter class (ASTE-404) forming the basis for this text. His other experience includes the role of a Research Engineer at the Air Force Research Laboratory, and a contamination control engineer supporting NASA Goddard Space Flight Center. Dr. Brieda is the author of numerous simulation codes utilized by the aerospace and plasma modeling communities. These include the 3D plasma code Draco, the 3D free-molecular transport code CTSP, and the 2D code Starfish. He also maintains a scientific computing blog found at `particleincell.com/blog` and has previously published a text introducing computational plasma physics, *Plasma Simulations by Example*, CRC Press, 2019. Dr. Brieda can be contacted at `lubos.brieda@particleincell.com`.

Joseph Wang is Professor of Astronautics and Aerospace and Mechanical Engineering at University of Southern California (USC). Prof. Wang received his Ph.D in Aeronautics and Astronautics from Massachusetts Institute of Technology in 1991. Prior to joining the faculty at USC in 2008, he had been a principal member of the engineering staff at NASA's Jet Propulsion Laboratory, and on the aerospace engineering faculty at Virginia Tech. He has also been a visiting professor at Sapienza University of Rome, Kyoto University, and Kyushu Institute of Technology, and a senior visiting scientist at French National Aerospace Research Laboratory. Prof. Wang conducts research in space technology, space physics, and planetary science. He and his students have developed many first-principle based, large-scale computer simulation models utilizing particle-in-cell (PIC), Vlasov simulation, and molecular dynamics simulation in these areas. Prof. Wang can be contacted at `josephjw@usc.edu`.

Robert Martin received his B.S. in Aerospace Engineering from Iowa State University and his M.S. and Ph.D in Computational Science, Mathematics, and Engineering focused on modeling non-equilibrium gas and plasma at the University of California, San Diego. He joined the Air Force Research Laboratory (AFRL) In-Space Propulsion branch in 2011 leading development of a new multiscale/multiphysics modeling framework for spacecraft plasma. At AFRL, he also led several basic research modeling initiatives in compressed kinetic theory, electromagnetic propulsion, very low earth orbit propulsion, and adapting dynamical systems theory to verification and validation. He joined the Army Research Office to run the Modeling of Complex Systems Program in 2021.

1 Scientific Computing Basics

This chapter introduces basic concepts relevant to computer simulations. Specifically, we learn about numerical integration by developing a program simulating the trajectory of a tennis ball. We also review popular programming languages.

1.1 INTRODUCTION

During your studies of physics or engineering, you have surely come across a multitude of example problems designed to illustrate analytical solution methods applied to fundamental physical problems. One of these may have read something like this:

> A tennis player serves a ball such that it leaves the racket at a height of 2.1 m moving completely parallel to the ground at 45 m/s (this is about 100 mph). Ignoring aerodynamic effects, how far will the ball travel before hitting the ground?

Before attempting to solve this problem, we first need to establish some coordinate system. We can let the origin $(x, y) = (0, 0)$ be the spot on the ground directly below where the ball leaves the racket. This is visualized in Figure 1.1. We also let time $t = 0$ be the instant this happens. Therefore, the *initial conditions* can be summarized as

$$x(0) = 0 \text{ m} \quad ; \quad y(0) = 2.1 \text{ m} \tag{1.1}$$
$$v_x(0) = 45 \text{ m/s} \quad ; \quad v_y(0) = 0 \text{ m/s}$$

The syntax $x(0)$ implies the position at time $t = 0$. This notation is often further simplified as x_0.

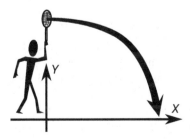

Figure 1.1 Coordinate system for a numerical integration example.

DOI: 10.1201/9781003132233-1

Next, from basic mechanics, we know that velocity v is the time derivative of position x,

$$\frac{dx}{dt} = v \tag{1.2}$$

Similarly, the time derivative of velocity is acceleration, a,

$$\frac{dv}{dt} = a \tag{1.3}$$

We also know from Newton's Second Law that acceleration is related to the total force acting on the object,

$$a = F/m \tag{1.4}$$

where m is the object's mass. Here we are assuming that the mass of the object does not change.

Per the problem statement, we ignore drag (and also lift created by a spinning ball) and only concern ourselves with gravity. Gravity acts only in the vertical y direction. The components of the net force then become

$$F_x = 0 \tag{1.5}$$
$$F_y = -mg \tag{1.6}$$

where $g = 9.81$ m/s^2 is the acceleration due to gravity.

Utilizing Equation 1.4 and substituting the above expression into Equation 1.3, we obtain the following relationship for the horizontal and vertical velocity components:

$$\frac{dv_x}{dt} = 0 \tag{1.7}$$

$$\frac{dv_y}{dt} = -g \tag{1.8}$$

They are here labeled as v_x and v_y, but it is customary to use u, v, and w for the x, y, and z three-dimensional velocity components. These expressions can next be integrated. Starting with an indefinite integral, we obtain

$$v_x = 0t + A \tag{1.9}$$
$$v_y = -gt + B \tag{1.10}$$

The integration constants A and B are evaluated using the initial condition. From Equation 1.1 we determine that $A = 45$ m/s and $B = 0$ m/s. In general, units will be omitted through-out this book. However, in your work, it is important to ensure that units properly balance.

These two equations are subsequently integrated one more time to obtain an expression for the position,

$$x = 45t + C \tag{1.11}$$

$$y = -\frac{1}{2}gt^2 + D \tag{1.12}$$

Again, using the initial condition, we find that $C = 0$ m and $D = 2.1$ m. The horizontal and vertical position of the ball, given as a function of time, becomes

$$x = 45t \tag{1.13}$$

$$y = -\frac{1}{2}gt^2 + 2.1 \tag{1.14}$$

Our first goal is to determine at what time t the ball hits the ground, $y(t) = 0$. From 1.14, we have

$$t = \sqrt{\frac{2 * 2.1}{g}}$$

$$t \approx 0.654 \; [\text{s}] \tag{1.15}$$

This t can then be used with Equation 1.13 to compute the corresponding horizontal displacement. We find the ball impacts the ground at $x \approx 29.4$ m.

1.2 NUMERICAL INTEGRATION

Nothing in the above mathematical treatment should be surprising. This is just upper-level high school mathematics. But, let's pretend for a moment that you have completely forgotten all your integration fundamentals. Wouldn't it be nice if there was an alternative scheme that could be used to *estimate* the impact location without actually needing to derive the analytical expression given by Equation 1.14?

Turns out there is, and it serves as the basis for much of modern scientific computing. Equation 1.2 gives an expression for the *rate of change* of velocity. The dx and dt differential elements correspond to infinitesimal increments of position and time. By taking the dt to the equation's right side, we can write

$$dx = v \cdot dt \tag{1.16}$$

Given an infinitesimal change in time dt, position changes by $v \cdot dt$. But what if instead of using the mathematical infinitely small increment, we use one that is small, but still finite and hence useful in practical calculations? We denote this finite-sized chunk with Δ,

$$\Delta x = v\Delta t \tag{1.17}$$

assure Equation 1.17 is a direct counterpart of Equation 1.16 for the finite-sized increment. This can be easily confirmed by integrating the first equation over finite limits,

$$\int_{x_1}^{x_2} dx = \int_{t_1}^{t_2} vdt \tag{1.18}$$

which, for constant velocity, becomes

$$x_2 - x_1 = v(t_2 - t_1) \tag{1.19}$$

Since $\Delta x \equiv x_2 - x_1$ and $\Delta t \equiv t_2 - t_1$, these equations are identical. This formulation also makes it obvious that given the current position $x_1 = x(t)$, the new position Δt seconds in the future, $x_2 = x(t + \Delta t)$, is obtained from

$$x_2 = x_1 + v\Delta t \qquad (1.20)$$

Similarly, utilizing Equation 1.3, we can write

$$v_2 = v_1 + a\Delta t \qquad (1.21)$$

There is a caveat. As noted above, the integration in Equation 1.19 holds only if the velocity is constant. In utilizing Equations 1.20 and 1.21, we are implicitly assuming that velocity or acceleration does not change during the Δt time step. Now, in our tennis ball example, acceleration and v_x are indeed constant, but v_y is not. The gravitational force accelerates the ball, and as such, the vertical component of velocity is changing continuously. However, if we assume that Δt is "sufficiently" small, it may not be too incorrect to assume that velocity also remains mostly constant during the time interval. Specifically, we require

$$\lim_{\Delta t \to 0} \Delta v = 0 \qquad (1.22)$$

As the size of the time step decreases toward zero, the change in velocity should also vanish. Ideally, this scaling will be at least linear, so that if Δt is decreased by a factor of two, Δv becomes a half or less of the prior value.

Combined together, these two equations allow us to advance velocity and position in a step-by-step manner. We start by assigning initial values to properties of interest. We then march the solution forward through multiple time steps. At each step, we use the rate of change equations to update the value of the property fields. This approach can be summarized by the following pseudo-code:

```
1)  let  t = 0                    # initialize time counter
2)  let  x = 0 and y = 2.1        # assign initial position
3)  let  vx = 45 and vy = 0       # assign initial velocities
4)  let  gx = 0 and gy = -9.81    # assign gravitational acceleration
5)  let  dt = 0.1                 # assign time step
6)  let  t = t + dt               # increment time counter
7)  let  x = x + vx*dt            # update x-component of position
8)  let  y = y + vy*dt            # update y-component of position
9)  let  vx = vx + gx*dt          # update x-component of velocity
10) let  vy = vy + gy*dt          # update y-component of velocity
11) return to 6) if y>0           # continue until ground impact
```

The scheme can be tested with a pen and paper. Table 1.1 lists the values of x and y after a specified number of $\Delta t = 0.1$ s *iterations*, with 0 indicating the *initial conditions*.

We can note that the x component of velocity remains constant, as expected since $a_x = 0$. The y component magnitude increases due to the gravitational acceleration acting in the vertical direction. The algorithm terminates after 7 steps with $t = 0.7$ and $x = 31.50$. Given that the analytical solution

Table 1.1

Positions and Velocity Estimates

it	t	x	y	v_x	v_y
0	0.00	0.00	2.10	45.00	0.00
1	0.10	4.50	2.00	45.00	-0.98
2	0.20	9.00	1.81	45.00	-1.96
3	0.30	13.50	1.51	45.00	-2.94
4	0.40	18.00	1.12	45.00	-3.92
5	0.50	22.50	0.63	45.00	-4.91
6	0.60	27.00	0.04	45.00	-5.89
7	0.70	31.50	-0.65	45.00	-6.87

predicted $t = 0.654$ and $x = 29.4$, this estimate is not too bad, especially given that at no point did we utilize the analytical integrals of the rate equations!

But a discrepancy exists. Our Δt time step limits the precision of the numerical scheme. We simply cannot resolve events happening at intervals smaller than Δt. The workaround is obvious: we re-calculate the results using a smaller value of Δt. For example, we may want to utilize $\Delta t = 0.01$. Decreasing the time step by a factor of 10 also implies that we need to perform $10\times$ as many calculations to cover the same time period. Perhaps you may feel comfortable making the 7 sets of calculations by hand but repeating these steps 70, or 70 million, times may not sound too appealing!

This is where computers come in. After all, despite the fact that nowadays office workstations tend to be used mainly for preparing presentation slides or sending emails, computers were initially utilized to perform computations. A simple integration scheme as given above can be implemented within a spreadsheet program such as Microsoft Excel or LibreOffice Calc. Figure 1.2 shows an example implementation. We specify values for Δt, g_x, and g_y in cells C1, C2, and C3. Then, row 6 is populated with the initial conditions corresponding to iteration 0. Below it, in row 7, we write the following expressions:

cell	expression	purpose
A7	=A6+1	$it = it + 1$
B7	=B6+$C$1	$t = t + \Delta t$
C7	=C6+E6*$C$1	$x = x + v_x \Delta t$
D7	=D6+F6*$C$1	$y = y + v_y \Delta t$
E7	=E6+$C$2*$C$1	$v_x = v_x + a_x \Delta t$
F7	=F6+$C$3*$C$1	$v_y = v_y + a_y \Delta t$

We then highlighted cells A7 through F7 and "drag" them down for the desired number of iterations. The spreadsheet program automatically updates the equations so that the formula in A8 becomes '= A7+Implementation1'

	A	B	C	D	E	F
1	dt		0.04			
2	gx		0			
3	gy		-9.81			
4						
5	it	t	x	y	vx	vy
6	0	0.00	0.00	2.10	45.00	0.00
7	1	0.04	1.80	2.10	45.00	-0.39
8	2	0.08	3.60	2.08	45.00	-0.78
9	3	0.12	5.40	2.05	45.00	-1.18
10	4	0.16	7.20	2.01	45.00	-1.57
11	5	0.20	9.00	1.94	45.00	-1.96
12	6	0.24	10.80	1.86	45.00	-2.35
13	7	0.28	12.60	1.77	45.00	-2.75
14	8	0.32	14.40	1.66	45.00	-3.14
15	9	0.36	16.20	1.53	45.00	-3.53
16	10	0.40	18.00	1.39	45.00	-3.92
17	11	0.44	19.80	1.24	45.00	-4.32
18	12	0.48	21.60	1.06	45.00	-4.71
19	13	0.52	23.40	0.88	45.00	-5.10
20	14	0.56	25.20	0.67	45.00	-5.49
21	15	0.60	27.00	0.45	45.00	-5.89
22	16	0.64	28.80	0.22	45.00	-6.28
23	17	0.68	30.60	-0.03	45.00	-6.67

Figure 1.2 Numerical integration implemented in a spreadsheet program.

The dollar signs anchor the column and/or the row index. In Figure 1.2, we also used conditional formatting to flag y cells reaching negative values. This helps us to visually identify the appropriate iteration where the ground impact happens. Using $\Delta t = 0.04$, the prediction improves to $t = 0.68$ and $x = 30.60$. It appears that our numerical scheme is consistent. As Δt decreases, the solution converges to the true solution.

1.3 SCIENTIFIC COMPUTING

The set of instructions in the prior algorithm may not seem all that impressive, but it captures the essence of the majority of scientific codes. Computers can help us with a variety of mathematical tasks, including solving systems of equations, fitting trends to a set of data, or performing automatic classification of input images. However, when people typically think of scientific computing, they tend to envision numerical simulations such as those used to compute the airflow around a race car, or the structural deformation of a bridge in the case of an earthquake.

First-principle laws studied in physics and engineering lead to partial differential equations (PDEs) governing how physical quantities *change* in response to other parameters. For example, from energy conservation (the first law of thermodynamics), we may arrive at the following

$$\frac{\partial U}{\partial t} = -\nabla \cdot \vec{q} \tag{1.23}$$

where U is the internal energy and $\vec{q}$ is the heat flux. This equation simply states that the time rate of change of energy is due to heat flowing into (or out of) a system. Fairly obvious! With some additional definitions for the energy and heat terms ($U = \rho C_p T$ and $\vec{q} = -k\nabla T$), the equation can be written in terms of temperature T,

$$\frac{\partial T}{\partial t} = \alpha \nabla^2 T \qquad (1.24)$$

where $\alpha = k/(\rho C_p)$. The $\nabla^2(\cdot) = \partial^2(\cdot)/\partial x^2 + \partial^2(\cdot)/\partial y^2 + \partial^2(\cdot)/\partial z^2$ term on the right-hand side is called the *Laplacian*, and is a generalization of the second derivative to three dimensions. The above equation is equivalent to

$$\frac{\partial T}{\partial t} = \alpha \left(\frac{\partial^2 T}{\partial x^2} + \frac{\partial^2 T}{\partial y^2} + \frac{\partial^2 T}{\partial z^2} \right) \qquad (1.25)$$

For simple configurations, such as two infinitely large parallel plates, this equation can be solved *analytically* using approaches such as separation of variables. But for real-world problems consisting of complex boundaries and internal heat sources, an analytical solution is simply impossible. This is when we turn to computers. A heat transport simulation program is not much different than our algorithm for integrating the tennis ball location. In fact, we can see this from the structure of Equation 1.25. Just as before, we have a relationship in the form

$$\frac{\partial X}{\partial t} = f(X, t) \qquad (1.26)$$

where X is some quantity of interest. Instead of numerically advancing the position of a single tennis ball, we numerically advance temperatures at multiple spatial locations.

The heat equation example demonstrates the so-called *Eulerian* approach, in which one or more differential equations are solved at stationary spatial points. In the aerospace field, this approach leads to Computational Fluid Dynamics (CFD) codes, which solve governing equations for conservation of mass, momentum, and energy to calculate the spatial variation of gas density, flow velocity, temperature, and pressure by considering information flowing into and out of stationary cells. Another popular class of solution methods is based on *Lagrangian* dynamics in which we let components of the solution propagate through the computational domain. These components may correspond to gas molecules. Instead of integrating PDEs for conservation of mass, momentum, and energy, we simply let the gas molecules bounce around the computational domain. The molecular speeds and positions are integrated through small Δt time steps in the same way the tennis ball moved across the court. The primary difference arises from the need to consider additional forces (such as electrostatics for ionized gases) and to account for inter-molecular collisions and surface impacts. The reaction dynamics generally involves sampling random numbers. These approaches are thus also known as Monte Carlo methods, after the famous casino. Chapters 5 and 6 introduce numerical techniques relevant to these two formulations.

1.4   PROGRAMMING LANGUAGES

While the spreadsheet approach may suffice for a single tennis ball, imagine that we wanted to model hundreds, or perhaps millions, of objects bouncing around. Such a calculation would be extremely time consuming even using the spreadsheet. This is where computer programs comes in. With a bit of custom code, we can have the computer do all the nitty-gritty calculations for us.

You may already be familiar with programming languages such as MATLAB®, Python, C++, Java, or Fortran. They represent just a tiny subset of the vast ecosystem of programming languages in existence. The differences between programming languages are in many ways analogous to the variation in spoken languages around the world. The syntax may be different, but the languages generally describe the same set of ideas. While spoken languages may be grouped into linguistic families, programming languages can be organized according to whether they are compiled or interpreted, or whether they are procedural, functional, or logical. These categories control how the source code gets translated into the instruction set understood by the computer processor, and how the code organizes the specified instructions. And just as many world languages derive from the same lineage, computer programming languages are also derivatives of others. For example, C++, Java, C#, and D all build on the foundation of the C language.

New students of scientific computing often want to know which language to learn. There is no simple answer here. Let's imagine that you want to bake a cake. First, you will need a recipe. A recipe is a just a set of instructions for converting some initial state (the ingredients) into a desired product (the cake). Numerical methods and pseudocodes we have been discussing so far form this recipe. You will also need some way to convert this abstract idea into the actual product. This is where programming languages come in. They allow you to implement the algorithm into a form that can be executed on the hardware system in order to produce the desired result. But just as there is no single oven (or a mixing bowl) ideal for every chef, there is no ideal programming language. Each language has its pros and cons. It is crucial to become a programming polyglot so that you can utilize the best tool available for each particular task. For example, Python, created and released in 1991 by a Dutch programmer Guido van Rossum, is popular with the scientific computing community. One reason is its huge library of linear algebra, signal processing, and plotting functions. With Python, you can solve a linear matrix system $\mathbf{A}\vec{x} = \vec{b}$ with just a single line of code. Results can also be visualized easily using a variety of built-in plotting functions.

But Python is also slow. This is partly due to Python being an interpreted language in which the source code is nominally translated to machine instructions on the fly. Yet, even using a compiler add-on, the performance just does not compare to that of raw C/C++, at least for applications developed using native data structures and operators. The performance difference may not be of concern for a simple script that takes few minutes to run. But it can become a show stopper for a massive simulation that would require months of supercomputer time with Python versus several days with C++. But C++ does not include any built in linear algebra support. Solving a matrix system requires developing your own solver, or at least learning how to interface to external mathematical libraries. C++ also does not contain any graphing support. Results need to be saved to a file and subsequently visualized using a 3rd party program. Therefore, Python is a useful first step for prototyping a new algorithm. Once you have gained the confidence that the code works as intended, you may want to convert it to a higher-performance language. For this reason, you will also find this book focusing on these two languages.

1.4.1 MACHINE LANGUAGE

In the previous section, we mentioned interpreted and compiled languages without going into details of these definitions. The computer, or specifically its "brain", the Central Processing Unit (CPU), does not understand C, Fortran, Python, or any of the languages you may be familiar with. Instead, it implements, in hardware using the internal electrical circuitry, hundreds of elementary instructions. These include operations for moving data from the main Random Access Memory (RAM), performing basic mathematical operations, and conditionally (when some expression is true) jumping the instruction pointer to a new memory address. Each operation is given a numeric code. Let's say that you want to assign to one variable the value of another variable incremented by some scalar,

$$a = 4$$
$$b = a + 2$$

To perform this operation, the CPU receives the following sequence of numbers:

```
199, 69, 248, 4, 0, 0, 0
139, 69, 248
131, 192, 2
137, 69, 252
```

Each line corresponds to a separate instruction, although the line breaks are included only for clarity and are not part of the produced binary code. Each item represent a *byte*, which is a collection of eight 0 or 1 *bits*. The 8 bits "activate" progressively higher powers of 2. The value of the byte is obtained by adding up the contributions. A single byte can thus hold values from 0 to $2^7 + 2^6 + 2^5 + 2^4 + 2^3 + 2^2 + 2^1 + 2^0 = 2^8 - 1 = 255$. A byte is the smallest storage unit for integers. Larger numbers, or even floating-point real values, are represented by collecting bytes into 2, 4, or 8 byte groups. Each instruction starts with the operation identifier code. The remaining values are operation specific and provide the needed data. The first instruction (with code 199) stores the value 4 into a section of RAM called the stack. The CPU arithmetic functions do not directly operate on RAM, instead they utilize small memory banks located directly on the chip called registers. The second operation (139) moves data from the stack position where the value 4 got placed into the register called EAX. Then, operation 131 adds 2 to whatever the value is stored in this register. Finally, the fourth operation (137) moves data from EAX to another position on the stack that our program assigned to the variable b.

This set of numeric op codes is called machine language and is literally all that the CPU understands. As you can imagine, developing a complex program using a scheme like this would be tedious at best, if not outright impossible. This is however how the earliest computers were programmed. We can make our life slightly easier by defining names for the operations, stack pointers, and registers. We could then write

```
1   movl    $0x4,-0x8(%rbp)
2   mov     -0x8(%rbp),%eax
3   add     $0x2,%eax
4   mov     %eax,-0x4(%rbp)
```

Here `rbp` is another register that stores the initial memory address of the stack. The numeric values are given in the hexadecimal notation (1 = 0x1, 10 = 0xa, 15 = 0xf, 17 = 0x11, 255 = 0xff), which is the convention for listing memory addresses. The first instruction places the value 4 at the memory position 8 bytes before the stack end. Data from this location is then moved to EAX. We then add 2 to the value of this register. Finally, the value from EAX is copied to memory address 4 bytes from the stack end. This address maps to the "b" variable.

1.4.2   COMPILED AND INTERPRETED LANGUAGES

This prior listing is written in language called the *assembler*. While this is an improvement over writing machine code directly, it is a marginal improvement at best. We would still need to know the underlying CPU instruction set, and be able to partition our algorithm into these fine steps consisting of elementary mathematical operations and direct memory manipulation. This is where "higher" programming languages come in. They allow us to express ideas using simpler and intuitive constructs, such "b = a + 2". However, in the end, regardless of the language itself, these constructs need to be translated into the machine language. This translation can happen in one of two ways. The code may be *compiled* or *interpreted*. With the former, the set of textual instructions (the source code) is translated by a program called the *compiler* into the binary machine code. This compilation happens for every source files. The multiple files are then *linked* together to generate the application that can be executed directly by, for example, double-clicking on an .exe file in Microsoft Windows.

The compiler is not needed once the code is compiled. Anyone using a computer with a CPU and operating system compatible with the particular instruction set generated during compilation can run the program without needing any additional tools. The only exception are dynamically loaded libraries (.dll or .so) that the program may depend on. C/C++ and Fortran are two example of compiled languages. The popularity of C and C++ is at least partly due to the fact that, while it is a high-level language, it retains close mapping to the underlying workings of the computer. This allows the compiler to generate efficient code. If needed, we can even directly include assembly inside the C++ source code.

Interpreted languages on the other hand utilize another program, called the *interpreter*, to translate the instructions into machine language on the fly. MATLAB, Python, and Javascript are examples of interpreted languages. You need to have the MATLAB or Python environment installed on your system in order to run programs written in these languages. Your program literally consists just of the textual source code and you may never get to see the binary executable. This translation happens at run time leading to slower performance. Instead of devoting all of its resources to performing the calculations, the CPU needs to dedicate clock cycles to the translation. The advantage of this approach is the ability to run computations interactively. You are given access to a console into which you can type new instructions, press Enter, and see the results. This line-by-line interactive access is particularly useful when prototyping new algorithms since you can dynamically respond to the results obtained on the prior step. It is also useful when introducing scientific programming to new students.

1.4.3 EXAMPLE ALGORITHM

There are hundreds, if not thousands, of programming languages in existence, although only a small subset is used regularly. We now provide a short summary of popular languages that you may encounter during your numerical analysis work. To illustrate the difference, let's consider the function

$$f(x) = \max(x^2 - 2x, 0) \tag{1.27}$$

for $x \in [-1, 1]$. We would like to write code that initializes an 11 element-sized array covering this range and also outputs the first and the last entry. In Python, we may write

```
1   # function that evaluates max(x^2-2x,0)
2   def fun(x):
3       z = x*x - 2*x
4       if (z>=0):
5           return z
6       else:
7           return 0
8
9   n = 11
10  y = [0]*n        # array of 11 zeroes
11
12  # evaluate fun(x) for x=[-1,1]
```

```
13  for i in range(n):
14      x = -1.0 + 2*i/(n-1)      # float division by default in Python
15      y[i] = fun(x)
16
17  # show some results
18  print("f(-1)=%g, f(1)=%g"%(y[0],y[n-1]))
```
or using NumPy

```
1  x = np.linspace(-1,1,11) # 11 uniformly spaced values in [-1,1]
2  y = np.maximum(x*x - 2*x,0) # NumPy vector operations
```

1.4.4   FORTRAN

Fortran (standing for Formula Translations, and written as FORTRAN until version 77) was one of the first high-level languages to be developed. It dates back to the 1950s, when engineers at IBM were looking for an alternative to coding the early mainframe computers in assembly. Fortran quickly became *the* language of scientific computing. Some 70 years later, the language persists. There is still a sizeable population of numerical scientists who prefer to code in Fortran, although usually in one of the more modern variants. Even if you do not plan to write new code in Fortran, it is useful to have at least a basic working familiarity with its syntax. We provide a short overview of Fortran syntax in the Appendix. Due to its popularity in the early days of computing, the vast majority of early numerical algorithms were written in Fortran. Many of these legacy codes are still in regular use at national laboratories and research institution. It is not unusual to find out that some popular desktop analysis application is really just a graphical front end built on top of a legacy Fortran solver.

Fortran syntax has undergone several major changes throughout its lifetime. The most dramatic change took place between FORTRAN 77 and the follow-on Fortran 90. The early computers had very little resemblance to modern laptops, tablets, or cell phones. These early *mainframe* systems occupied an entire room, and did not contain a monitor or a keyboard. Instructions were fed in on perforated *punch cards*, with a single line of code per card. The most common card type contained 80 columns, but only the first 72 were actually processed by the popular computer at that time, the IBM 704. Furthermore, columns 1 through 5 were used to denote numeric labels for jump instructions, and column 6 indicated a line continuation. Only columns 7 through 72 were actually available to the program.

This fixed-form syntax carried over from early FORTRAN to the latter versions that utilized tape drives and later hard drives. Even now, when writing a FORTRAN 77 code on a computer with a keyboard, it is imperative to pay attention to character placement. Any text beyond the 72nd column is ignored, and placing a character in column 6 will lead to a syntax error or incorrect logic (since it treats the line as a continuation of the prior one). Comments are denoted by placing the letter c in column 1.

Since punch cards did not distinguish between letter cases, the early code was written by convention in capital letters. Variable names were limited to just six characters, and the name affected the storage type, unless an IMPLICIT NONE directive was used. Variable names starting with I, J, K, L, M, or N were automatically assumed to be of the integer types. Remaining letters produced variables storing real numbers. Here is how our example script would be implemented in FORTRAN 77:

```
1    c23456- program start ─────────────────────────────
2          PROGRAM MAIN
3
4          IMPLICIT NONE
5          INTEGER N, I
6          PARAMETER (N=11)
7          REAL*8 Y(N), X, FMAX
8
9          DO I = 1, N
10             X = -1.0 + 2*(I-1)/(N-1)
11             Y(I) = FMAX(X)
12          END DO
13          WRITE(*,100) Y(1), Y(N)
14    100   FORMAT('f(-1)=',f6.3,',', 'f(1)=',f6.3)
15          STOP
16          END
17
18    c- - -subroutine that evaluates max(x^2-2x,0)
19          REAL*8 FUNCTION FMAX(X)
20          IMPLICIT REAL*8 (A-H, O-Z)
21          Z = X*X - 2.*X
22          IF(Z.GT.0) THEN
23             FMAX = Z
24          ELSE
25             FMAX = 0.
26          END IF
27          RETURN
28          ·END
```

While most of the code should be fairly self-explanatory, there are few areas of interest. First, it was typical to use a comment on the first line to identify the initial 6 columns and to indicate the available columns. Next, PROGRAM MAIN indicates the starting point, which is primarily of use in programs spanning multiple input files. We use IMPLICIT NONE to disable the named-based assignment of variable types. We specify their types on the next 3 lines. The PARAMETER command assigns a constant value to N. REAL*8 gives us the 8-byte (64-bit) double precision floating point values. Y is declared to be an N-sized array. This is also where we declare the return type of the function FMAX. The code then continues with a for-loop, which in Fortran is called a do-loop. The parameter I iterates from 1 to N. We use the index to set the physical location. The equations is slightly altered, however, since due to the 1-based indexing, we need to subtract 1 from I in the calculation. Array components are accessed using (I) instead of the [i] in C or Python.

Results are then written to the screen. The output formatting is peculiar. The two parameters of WRITE on line 13 indicate the file id and the line number to be used for formatting. In this case, it is the line identified with the label 100. It is on this line where we provide the textual strings to be included with the output. The variable data form the previous line is formatted using syntax similar to legacy C printf. The rest of the listing contains the code for the fmax function. Here we demonstrate the declaration of variables using the IMPLICIT keyword. Without it, we would need to specify the type for X (the input argument) and Z, the temporary local variable. Data is returned by assigning it to an object with the name of the function.

This fixed-form source formatting of FORTRAN 77 was eliminated in Fortran 90, which also introduced a plethora of other changes. Among these was the ability to have longer variable names, and also the ability to perform vectorized operations with arrays. It is this support for efficient manipulation of multi-dimensional arrays that is at least partly responsible for Fortran's persistent popularity. Fortran 2003 introduced object-oriented programming, and subsequent versions improved inter-operability with C and introduced parallel processing. Using vector operations, the example code can be written in modern Fortran as

```
1   program main
2   implicit none
3   integer n
4   parameter (n=11)
5   real(kind=8), dimension(n) :: x
6   real(kind=8), dimension(n) :: y
7   integer i
8
9   !linspace(-1,1,11)
10  do i=1,11
11     x(i) = -1.0 + 2.0*(i-1)/(n-1)
12  end do
13
14  !evaluate using vector operations
15  y = max(x*x-2*x,0.)
16  print "(a7,f6.3,a8,f6.3)"," f(-1) =",y(1),",  f(1) =",y(n)
17  end
```

The code no longer needs to be indented by 6 spaces. Lines can be commented out with an exclamation mark. On line 16 we demonstrate the alternate print command for data output. It uses the same format string as write. Here we specify that the data consists of a 7 character alphanumeric string, followed by a floating point number, followed by another string, and then yet another number.

It is not uncommon to have to convert legacy Fortran code to other languages, such as C or Python. When doing so, it is important to pay to the following two areas. First, Fortran array indexing begins with 1, while C and Python start at 0. An array access such as x(2) becomes x[1] in C++. Forgetting to update the indexes will result in memory corruption as data is written to location beyond the array bounds. Secondly, Fortran multi-dimensional arrays are stored in memory in a *column-major order*. This is the opposite of the typical storage in C. Performance of numerical codes operating on large data sets is limited by memory transfer bandwidth and not the speed of the CPU. To reduce this bottleneck, the CPU contains a small on-board memory called the *cache*. Whenever the CPU requests data from RAM, the system bus first checks to see if this data (based on its address) already happens to be in the cache. Only if not found is it copied from RAM. When this happens, data in the contiguous memory space is also copied with the hope that the data the CPU asks for next will be found in this extraneous chunk. When operating on multidimensional arrays, we need to structure our loops such that the data is being accessed more-or-less in the same order it is stored in memory. Otherwise, nearly every operation will produce a *cache miss* which can have tremendous negative impact on code performance. We look into some of these effects in Chapter 8. Due to the difference in the default storage scheme, operations on multi-dimensional arrays need to have their for-loop ordering reversed when migrating Fortran code to C.

1.4.5 BASIC

The same way that Fortran became *the* language for mainframe programming, BASIC (Beginners' All-purpose Symbolic Instruction Code) was the de facto language for programming early personal *microcomputers*. There was nothing really "micro" about these computers by current standards, but they were significantly smaller than the room-sized mainframes of the prior era. They generally consisted of a typewriter-like box containing a keyboard, which would be plugged into the TV to produce a low-resolution text-only (at least at first) output. Many of these early computers, instead of booting into an operating system as we are accustomed to now, would instead boot into a BASIC interpreter. Essentially all you could do with your computer was to run BASIC programs. Popular magazines would arrive with source code for text-based games. Playing the game required manually transcribing the instructions line by line. Even later on, once the Microsoft MS DOS operating system came out, Microsoft's GW- and Quick BASIC (and other variants such as Borland Turbo Basic) remained extremely popular.

The language was created by John Kemeny and Thomas Kurtz at Darmouth College to simplify teaching of programming. Early version of BASIC borrowed heavily from Fortran. At least in its initial version, BASIC used line numbers and GOTO jump-statements to implement logical blocks. Frequent overuse of this statement led to a *spaghetti code* that was difficult to understand due code execution jumping all over the place. Here is a sample BASIC code:

```
10 N = 11
15 DIM Y(N)
20 FOR I = 0 TO N
30 X = -1 + 2*I/(N-1)
50 GOSUB 200
```

```
60 Y(I) = YY
70 NEXT I
80 PRINT "Y(-1) = " ;
82 PRINT Y(0) ;
84 PRINT " , Y(1) = " ;
86 PRINT Y(N-1)
90 END

200 REM custom function
    Z = X*X - 2*X
    IF Z >= 0 THEN
    YY = Z
    ELSE
    YY = 0
    END IF
220 RETURN
```

As you can see, lines may start with numbers. These numbers are then used as labels for GOTO and GOSUB statements. We start off by creating an 11-element array. On line 50, we call a "subroutine" although in this early formulation, this essentially just jumps the execution to the specified line. There is no concept of local variables - they are all global. The RETURN statement on line 220 jumps back to the line following the GOSUB command. The semicolon on line 80 through 84 prevents adding a new line character. Finally, the END statement ends the program.

1.4.6 PASCAL

Pascal was another general language to gain popularity after Basic. It was commonly used to teach programming in high school and university settings in the 1980s and 1990s. The language was developed by a Swiss computer scientist named Niklaus Wirth in the 1970s as a follow on to another early language called ALGOL. ALGOL (and thus Pascal) was the first language to utilize code blocks, denoted by begin and end. Pascal is strongly typed, and introduced novel new data types, including sets, unions, pointers, and structures (called records). Pascal utilizes := for assignment and = for comparison. Unlike BASIC, it is a compiled language. Below is an implementation of the sample algorithm in Pascal:

```
1   program Fun;
2   const
3     N = 11;
4
5   (* custom function *)
6   function fun(x: real): real;
7   var
8       (* local variable declaration *)
9       z: real;
10  begin
11      z := x*x - 2*x;
12      if (z >= 0) then
13          fun := z
14      else
```

```
15        fun := 0;
16   end;
17   (* variable declarations *)
18   type
19      vec11 = array [0..N] of real;
20   var
21      y : vec11;
22      i : integer;
23      x : real;
24
25   (* main program *)
26   begin
27      for i := 0 to N do
28         begin
29            x := -1.0 + 2*i/(N-1);
30            y[i] := fun(x);
31         end;
32      writeln ('y(-1)=',y[0]:5:2, ', y(1)=',y[N-1]:5:2);
33   end.
```

Pascal was developed to encourage good coding practices, however, by modern standards, the syntax may seem a bit rigid. Variables are declared outside the code block in a special var section. Arrays can be given arbitrary ranges. Furthermore, we can define custom types using the type keyword, as illustrated with the vec11 type. Alternatively, we could have written

```
1   var
2      y: array [0..N] of real;
```

Statements end with a semicolon, including after the block delimiters. The only exception is that a period (.) is used at the end of the main program. The writeln statement illustrates how to format the width and precision of data output.

1.4.7 C AND C++

C++ was developed in the early 1980s by Bjarne Stroustrup as an extension to the C language, which itself was developed in the 1970s by Dennis Ritchie at the Bell Labs. C quickly gained popularity due to its high performance and good memory management, which made it particularly suitable to developers of operating system components. This is due to C code essentially being a high-level representation of the underlying hardware model. Just like Fortran and Pascal, C and C++ are strongly-typed languages. Variables are assigned data type on declaration and can only hold data of that type. The basic types such as bytes, integers, and real value of varying precision map directly to the data types the CPU already operates on. C also supports dynamic allocation of arrays of sizes that are not known until run time. Data can also be accessed using memory *addresses*. This is also exactly how the CPU operates. C++, discussed in more detail in Chapter 4, introduced a new keyword called class that offered support for a new programming paradigm called object-oriented programming (OOP). Using this keyword, we can define custom data types that encapsulate their data as well as the functions that operate on them.

While the programming paradigms of C++ have diverged from the classic C coding style, C++ is mostly backwards compatible with native C. The C++ example given below utilizes a new way of performing screen output using cout instead of the legacy printf. Otherwise, this listing could be considered an example of a C code.

```
1   #include <iostream>
2   double fun(double x) {      // custom function
3     double z = x*x - 2*x;
4     if (z>=0) return z; else return 0;
5   }
6
7   int main() {
8     const int n = 11;
9     double y[n];              // using static array
10    for (int i=0;i<n;i++) {
11      double x = -1.0 + 2.0*i/(n-1.0);
12      y[i] = fun(x);
13    }
14
15    std::cout<<"f(-1)="<<y[0]<<" , f(1)="<<y[n-1]<<std::endl;
16    return 0;
17  }
```

On line 1, we use the #include keyword to import standard library functions for handling screen input and output operations. We next declare the custom function. C and C++ functions are declared by specifying the return type (a double-precision real value in this case), followed by the name, and a list of function arguments. The example fun function expects to receive a single double-precision value which will be accessible to the rest of the function code using the x name. On line 7, we declare another function called main. Every C and C++ code must implement this function once, and only once. This is where the code execution begins. This function can receive optional arguments that contain user-provided command line values. We use a static array to store the results. A static array is one for which the memory space is allocated at compile time. A typical code will instead use dynamic arrays which are allocated at run time. Not only do dynamic arrays offer more flexibility, they also let us allocate much larger blocks than their static variant. This is because statically allocated data is limited to a small program memory space called the *stack*. Dynamically allocated arrays utilize the remaining system memory called the *heap*. Next, we use a for loop to iterate over the array. It is important to note that C indexing begins at 0, just as in Python, which differs from the 1-based indexing in MATLAB or Fortran. For each array pass, we calculate the corresponding x value and then use our custom function to evaluate the expression. Finally, we use the C++ "stream operations" to write to the standard console output stream cout. The 0 return value indicates normal exit.

1.4.8   JAVA

Java was originally developed in the late 1990s by James Gosling at Sun Microsystems. It is a derivative of C++, and according to some, it is what C++ should have been had it been implemented correctly. C++ essentially extended the C language by including *classes* (discussed in Chapter 4). However, the core C language functionality has been retained in entirety. Java on the other hand completely eliminated the legacy non-class based programming paradigm. It also eliminated some non-safe aspects of C++, such as the direct use of memory addresses, and introduced automated garbage collection to free unused memory. The standard library includes numerous features not present in C++, including support for graphical user interfaces, networking, and multi-threaded parallel processing (which is now also part of C++ as of C++11 language standard). On the other hand, Java does not support operator overloading, which is a quite handy feature of C++. Also, while Java does not explicitly include the pointer data type, objects are internally represented by pointer-like references. This can lead to a confusion as to whether some data is actually being cloned (called a *deep copy*) or whether just the reference is duplicated (*shallow copy*). Consider this code, with lines 2 and 3 equally valid in both Java and C++:

```
1   MyObject A = new MyObject();  // create new object, in C++ use A()
2   MyObject B = A;       // intentions is to make a duplicate of A
3   B.setData();          // same as A.setData()
```

On the second line, our intention is to make a duplicate copy of A. However, in Java, only the reference is copied. The last line thus ends up modifying A's data, since A and B are just two handles to the same memory location. In C++, this assignment performs the deep copy (although a custom copy operator is required if `MyObject` contains dynamically allocated arrays) with A and B referring to two distinct data objects. Therefore, we need to be careful when porting C++ code to Java, as an identical code snippet can lead to a different behavior.

Unlike C++, Java is not a fully compiled language. The source code is converted into a Java *bytecode*, which runs within a Java Runtime Environment (JRE). The Java bytecode thus acts as a virtual machine language. The JRE converts it to the appropriate CPU instructions based on the hardware architecture it is running on. This approach ensures that the code runs identically on any hardware system, however, it does introduce a small performance hit. While Java is slower than C++, the difference is smaller than when compared to native code in interpreted languages such as Python or MATLAB. Use of Java also requires that the JRE is installed on the client system. Below is our example code implemented in Java. This file needs to be saved in a file called `Fun.java`, since the file name must be identical to the single public class implemented in each file:

```
1   import java.lang.*;    // support for System.out
2   public class Fun {     // must have same name as file
3     public static void main(String[] args) {  // main function
```

```
4      double y [] = new double [11];        // array of doubles
5
6      for (int i=0;i<11;i++) {
7        double x = -1.0 + 2.0*i/10;
8        y[i] = fun(x);
9      }
10
11     // similar to C printf
12     System.out.printf("f(-1)=%g, f(1)=%g\n",y[0],y[10]);
13     }
14
15     // static since not associated with any object instance
16     static double fun(double x) {
17       double z =x*x - 2*x;
18       if (z>=0) return z; else return 0;
19     }
20   }
```

Java is not the only derivative of C++. Other notable languages that attempt to improve on C++ include C# and D.

1.4.9 MATLAB®

There is a good chance that you are already familiar with MATLAB. MATLAB's history dates back to the 1970s when Cleve Moler at University of New Mexico started the development to make it easier for students to utilize matrix manipulation capabilities of Fortran linear algebra libraries LINPACK and EISPACK. Nowadays, MATLAB tends to be the default language for teaching university engineering and physics numerical analysis courses. There are some obvious reason for this: being specifically designed for mathematical analysis, MATLAB contains built-in support for matrix operations and plotting. MATLAB comes with a huge library of functions for solving matrix problems, fitting data, and performing signal processing. The plotting capabilities make it easy to generate x-y line plots, scatter plots, and 2D contour plots. It is also possible to visualize 3D data, but this is generally more straightforward in dedicated tools such as Tecplot, Paraview, or VisIt.

But MATLAB also suffers from several shortcomings. It is an interpreted language. Use of loops is discouraged the code inside the loop block gets recompiled on each pass. MATLAB supports vector operations that allow manipulation of multiple array indexes at once and generally yield faster code. But not all algorithms can be easily converted to a vector formulation. Further, in order to share your code with colleagues, they need to have MATLAB installed on their system. Given that MATLAB is a commercial product, your organization may have a limited number of available licenses. Having said that, it is actually possible to compile MATLAB code into an executable program for distribution, but this comes with the additional expense for the compiler add-on. One workaround is to use an open-source MATLAB emulator such as Octave. Using loops, the example script can be written in MATLAB as

```
1  # main script file
```

```
2   n = 11;
3   y = zeros(n,1);
4
5   for i = 1:n
6       x = -1 + 2*(i-1)/(n-1);
7       y(i) = fun(x);
8   end
9
10  fprintf("f(-1)=%g, f(1)=%g\n",y(1),y(n))
```

with the custom function defined in `fun.m` as

```
1   # custom function defined in fun.m
2   function [y] = fun(x)
3       z = x^2 - 2*x;  # could also just save as y
4       if (z>=0)
5           y = z;
6       else
7           y = 0;
8       end
9   return
```

Note that just like in Fortran, MATLAB array indexing starts with 1. Similarly, we use parentheses instead of square brackets to refer to array items. Many lines end with a semicolon, which suppresses printing the result of that operation to the screen. Using vector operations, the above calculation simplifies to

```
1   x = linspace(-1,1,11)
2   y = max(x.^2-2*x,0)
```

Note the use of the dot in `x.^2`. This forces MATLAB to perform an element-wise exponentiation instead of attempting to perform vector-vector dot product.

1.4.10 JULIA

Another language that has been rapidly gaining followers since its 2012 debut is Julia. Its syntax is roughly as high-level as Python, but runs at speeds comparable to compiled C code without having to rely on vectorized benefits of NumPy. Julia was built for data science. Its standard library includes much of the functionality one is accustomed to from Python. The built-in visualization allows producing publication quality plots with a variety of backends. The language includes native support to run code in other languages and (along with R and Python) is one of the three languages Jupyter notebooks were designed to support. On the other hand, Julia is a new language. Its user base, while growing, is still quite small which can make finding answers to questions challenging. Also, since the syntax has evolved considerably from earlier version, some online code examples will no longer run. The "just in time" compiler that Julia uses, while allowing the code to run very fast once started, does have a start up penalty. For example, the "time to plot" on a first load of Julia can be in excess of 30 seconds, whereas in Python it may be almost instantaneous. Subsequent plot calls execute faster once this initial compilation penalty has been paid. This can be mitigated somewhat by integrating compiled system images with the code. The code below illustrates the use Julia with our example function.

```
1   #function that evaluates max(x^2-2x,0)
2   function fun(x)        # functional way equivalent to what you had
        in Python code
3       z = x^2 - 2*x
4       if z>=0
5           return z
6       else
7           return 0
8       end
9   end
10
11  # functional for
12  f(x) = (x^2-2*x) >=0 ? (x^2-2*x) : 0 # short functional
        definition using a ternary operator
13
14  n = 11
15  y = zeros(n) #array of 11 zeros
16
17  for i = 0:n-1
18      x = -1.0 + 2*i/(n-1)
19      y[i+1] = fun(x) # Julia indexing starts at 1
20  end
21
22  # show some results
23  println("f(-1) = $(y[1]),  f(1) = $(y[n])") # $ allows insertion
        of code into print statements, precision could be specified
        using $(round(y[1], sigdigits=3))
24
25  # vectorized approach
26  x = range(-1,stop=1,length=n) # equivalent to numpy linspace
27  y = max.(x.^2 .- 2 .*x,0) # Julia uses . to clarify a vectorized
        operation
28
29  # verify results are the same as before
30  println("f(-1) = $(y[1]),  f(1) = $(y[n])")
```

1.4.11 R

The statistical and data analysis communities tend to use the R programming language. This interpreted language dates back to 1993 and is a derivative of the S language which itself was developed to aid in statistical analysis. The R language provides native support for a wide variety of data structures, including vectors, arrays, and lists as well as the associated mathematical operations. Linux provides support for a command line interpreter with capability for easy generation of publication-ready charts.

The listing below illustrates the definition of a custom function using R and the use of a **for** loop for data assignment:

```
1   N <- 11
2   y <- NULL
```

```
 3   # custom function
 4   fun <- function(x) {
 5      z <- x*x - 2*x;
 6      if (z >= 0) {
 7         return(z)
 8      }
 9      else {
10         return(0)
11      }
12   }
13
14   # main program
15   for (i in 0:N) {
16      x <- -1.0 + 2*i/(N-1)
17      y[i+1] <- fun(x)
18   }
19
20   cat("y(-1)=",y[1],", y(1)=", y[N])
```

Instead of using the **for** loop, we can use vector operations to set the values of the $\vec{x}$ vector, and then apply each item through the custom function. This variant is shown below:

```
1   i <- seq(0,N)
2   x <- -1.0 + 2*i/(N-1)
3   y <- lapply(x, fun)
4
5   cat("y(-1)=",y[[1]],", y(1)=", y[[N]])
```

1.4.12 HASKELL

Another relatively popular language is Haskell. Unlike the procedural languages discusses so far, Haskell is a functional language, in which program logic is defined by application of functions. Among its benefits (at least according to its users), is the rigid typing and functional style which makes it harder to make runtime errors. It utilizes a math-like notation that simplifies the comprehension of the implemented algorithms. The reliance on functors (maps of generic data to functions) and monads (function return value wrappers) means that functional dependence is manifested, making operations like asynchronous execution and parallelism much easier (e.g., for computations on a cluster.) On the other hand, the very rigid typing and functional style leads to a relatively steep learning curve, especially for those not already acquainted with functional-oriented programming. Math-like notation and style can lead to difficulties implementing some algorithms that would be straightforward with procedural programming. The listing below illustrates the use of Haskell:

```
1   module Main where
2
3   -- Definition with guards. The Num and Ord tell Haskell that
4   -- the type variable a can be any type, so long as it's numerical
5   -- and ordered, such as Double, Rational, Int, etc.
6   fun :: (Num a, Ord a) => a -> a
```

```
 7   fun  x
 8    |  z >= 0 = z
 9    |  otherwise = 0
10      where z = x*x − 2*x
11
12   −− / Alternative definition using a built−in function
13   fun '  ::  (Num a, Ord a) => a −> a
14   fun '  x = max (f x) 0
15      where f x' = x'*x' − 2*x'
16
17   main  ::  IO ()
18   main = do
19      let
20         y  :: [Double]
21         −− apply fun for [−1, −0.8, −0.6, .. 1]
22         y = fmap fun [−1,−0.8..1]
23
24      print $ "f(−1) = " ++ show (head y) ++ ", f(1) = " ++ show (
                last y)
```

1.4.13   OTHERS

Another languages that you may encounter in your scientific computing career is Lua. This is a scripting language used in desktop applications to provide custom data access, although nowadays it is becoming more commonplace to provide this support via Python. Go is another clone of C, this time designed at Google. It improves C's memory safety and adds support for concurrent processing, along with various syntax simplifications. Bash and Perl are scripting languages popular on the Linux operating system to create short installation or application launching scripts. PHP is used on web servers to dynamically generate web pages. We discuss web programming in Chapter 7, but limit ourselves to the client side. LaTeX is a typesetting "language" that produces formatted .pdf documents from a textual description. It is ubiquitous in scientific writing as it simplifies including equations and references. Figures and tables are placed automatically with text floated around them to minimize white space. In fact, this book is written using LaTeX. We introduce its syntax in Chapter 8.

1.5   PYTHON IMPLEMENTATION

We will now dive into Python by demonstrating how to implement the tennis ball integrator using this popular language. Before starting with Python, you need to download and install the interpreter. As was noted previously, the popularity of Python stems from the huge ecosystem of available libraries (called packages). Packages not included with the basic distribution, and need to be installed manually. Luckily, Python makes this process simple thanks to the included `pip` tool. For example, the popular linear algebra, scientific computing, and plotting packages NumPy, SciPy, and MatplotLib can be installed from the command line using

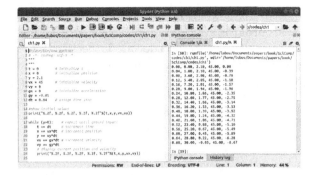

Figure 1.3 Spyder Python integrated development environment.

```
$ pip install numpy scipy matplotlib
```

This example, along with essentially all command-line examples through out the book, is written for the Linux operating system. Linux is heavily utilized by supercomputer clusters and as such it is imperative to become familiar with its use. Similar command-line interface is also available on Microsoft Windows or Apple macOS.

Instead of installing packages one by one, you may prefer to download a Python distribution already bundled with many common libraries. One such a distribution is Anaconda. It additionally includes an integrated development environment (IDE) called Spyder, which is shown in Figure 1.3. Spyder allows you to edit your code and step through it line-by-line to look for bugs.

At this point, you should install Python, Spyder, as well as the above mentioned packages. Next, type in the following code into Spyder:

```
 1   ''' Numerical  integrator  in  Python
 2   '''
 3   t = 0              # initialize  t
 4   x = 0              # initialize  position
 5   y = 2.1
 6   vx = 45            # initialize  velocity
 7   vy = 0
 8   gx = 0             # initialize  acceleration
 9   gy = −9.81
10   dt = 0.04          # assign  time  step
11
12   #show  initial  values
13   print("%.2f,  %.2f,  %.2f,  %.2f,  %.2f"%(t,x,y,vx,vy))
14
15   while (y>0):       # repeat  until  ground  impact
16       t += dt        # increment  time
17       x += vx*dt     # increment  position
18       y += vy*dt
19       vx += gx*dt    # increment  velocity
20       vy += gy*dt
21       # display  current  position  and  velocity
22       print("%.2f,  %.2f,  %.2f,  %.2f,  %.2f"%(t,x,y,vx,vy))
```

You can alternatively download this source code from the companion website at `scientificprogrammingbook.com`. You may have noticed that the listing looks quite similar to the pseudocode in Section 1.2. Next save the code as `ch1.py` and run it by clicking the green triangle in the middle of the top toolbar. If you prefer to work from a command line, navigate to the appropriate folder (such as `my_codes`) using `cd` and execute the script by providing the file name to the `python` interpreter,

```
~$ cd ~/my_codes
~/my_codes$ python ch1.py
0.00, 0.00, 2.10, 45.00, 0.00
0.04, 1.80, 2.10, 45.00, -0.39
0.08, 3.60, 2.08, 45.00, -0.78
...
```

As an aside, there are two main variants of Python, Python 2.7 and Python 3.x. If you happen to have both installed, you may need to use `python3` and `pip3` in lieu of `python` and `pip` in the examples above.

The code output, shown in console on the right side of Figure 1.3, and also partially above, is identical to the output from the spreadsheet, assuming the same value was used for dt. You can next change dt and observe how the output changes. Rerunning the code with smaller time step produces more output lines but also leads to an improved agreement with the expected result.

Let's now go through the code line by line. Lines 1 and 2 illustrate a possible way to include *comments* in Python code. This is an example of a multiline comment in which any text between two sets of matching triple equation marks is ignored. Another option, which is first seen on line 3, is to use the hash (#) mark. It causes the rest of the line to be ignored.

Starting with line 3, we begin assigning initial values to named objects called *variables*. A variable is essentially a memory storage location which is accessed through some given name. A statement such as `gy = -9.81` places the value -9.81 into the storage bucket called `gy`, which itself is located at some "arbitrary" position inside the system main random access memory, or RAM. Variable names are *case sensitive*. `gy` and `Gy` would be two distinct variables. Variable names are no longer limited to a particular length, although for practical purposes, it is the best to name variables with the shortest names that still clearly convey their purpose. Variable names can contain the underscore and numbers, however, a number cannot be the first character. Python automatically creates new variables on first use. Many languages, but not Python, classify variables by the type of the data they can hold. Python does not make such a distinction, and a single variable can hold integer, floating point values, or even text strings. While this may seem like a nice benefit, it can also lead to erroneous results. We may assume that some variable holds a number, but in reality it is currently storing a string. Python's syntax makes it possible to perform algebraic operations with strings and numbers without raising errors, but this may not be what our algorithm had in mind!

The ability to handle multiple data types further requires the language to store additional information about the currently stored item, such as its type. This leads to both additional memory usage as well as additional CPU cycles that need to be devoted to selecting the appropriate operation according to the data type. Finally, many languages, but not Python, support *constants* for storing named values that cannot change once assigned. In compiled languages, constants and the closely related constant expressions can lead to a faster code if the compiler can replace mathematical operations with values that are precomputed at compile time.

On line 13, we use the `print` command to display the initial values. Very early computers did not have a monitor but instead reported their output via an actual printer. The nomenclature has stuck, and in most languages we use some variant of "print" to perform screen or file output. The `print` command uses a peculiar syntax to format the output. This format is based on the C `printf` function (although this function got surpassed by the `cout` console stream in C++). The percent sign `%` denotes a section to be replaced by data from a list of substitution variables or expressions. At the minimum, we need indicate the data type. Some possibilities include `d` (or `i`) for integer, `f`, `e`, `g` for floating point values, and `s` for strings. For the numeric types, we can also specify optional *width* and *precision*. For example, `%10.2f` prints a number with two digits after the decimal point and makes sure that the output takes up at 10 characters. If the value is not wide enough, the output is padded on the left using spaces. The 'f', 'e', and 'g' types display the value in fixed precision, in scientific ("exponential") notation, or the better of these two depending on the data magnitude. The actual data to be substituted is provided in a *tuple* attached to the string via a `%` sign. The code on line 13 simply outputs the current values stored in the t, x, y, v_x, and v_y variables as floating point values with two digits after the decimal point.

Line 15 starts with a `while`. This *keyword* is one of several ways to define a repeating block. A block can either repeat *for* a specified number of iterations (using the `for` keyword), or *while* some expression holds true. Here we literally say: "keep repeating the following block as long as y is positive". Once this expression no longer holds true, implying $y \leq 0$, the loop terminates. When using while-loops, it is important to use an expression that can in fact become false at some point. Otherwise, you end up with what is known as an *infinite loop* and your program will run forever (or at least until you kill it using Ctrl+C, or by closing the program window).

One peculiar feature of Python which is loved by some and disliked by others (including the authors of this book) is Python's use of *white-space* to denote blocks. As you can see, lines 16 through 22 start off indented with an identical number of spaces. It is this indentation that tells Python exactly which lines make up the block that should be repeated. While this syntax naturally leads to a clean-looking code, it can introduce error during refactoring. When copying and pasting code from one function to another, it is imperative to check and adjust the indentation, otherwise, code logic will change.

Other languages use curly braces {...} or keywords such as begin and end to denote blocks.

The first five lines of the while-block increment the values stored in the t, x, y, v_x, v_y variables. Note that we update position before velocity to remain consistent with the algorithm. This ordering has an impact on the solution, as discussed in Section 1.5.9. Here we use the += compound operator. This is a handy shortcut offered by some languages as an alternative to

```
x = x + vx*dt
```

As you can see, we update the value in place. This is acceptable, since the code requires just the latest position and velocity in order to march the solution forward.

1.5.1 FILE OUTPUT

The current code version writes the output to the screen. The values are comma separated to simplify pasting into a spreadsheet program, if so desired. As the value of the time step is decreased, the number of output lines increases. At some point it may no longer be practical to just copy and paste the output. A better way is to have the code write an output file directly. Tabular data can be saved using the *comma-separated values*, or .csv, format, which is supported by all spreadsheet programs. The first line, by convention, specifies the column names. This is followed by an arbitrary number of lines corresponding to the rows. As the name indicates, columns are separated by a comma, although .csv file loaders tend to support arbitrary user-specified delimiters.

In Python we can use the same print command to write to a file by including a file=f argument. Here f is a file object identifier returned by the open command. A file can be opened in one of three ways. We can open it for reading ("r") which limits us to a read-only access. If the file does not exist, the open command fails. The second option is to open the file for writing ("w"). We need to be careful here, since the file is overwritten if it already exists. The final mode lets us append data ("a"), by giving us write access to an existing file. The insertion position is initialized to the end of the file.

The typical syntax looks like this:

```
1  f = open("results.csv","w")
2  print("t,x,y,u,v",file=f)
3  f.close()
```

We create a new file called results.csv using the "w" write mode and write out a single line to it. When done, the file is closed with the close function defined within the f object. We learn more about objects in Chapter 4.

The listing above does not check for success in opening the file. This operation can fail if, for example, the path does not exist, or if we do not have write-access privileges to the specified location. One way to make sure the file was indeed opened is to include the command within a with block:

```
1  with open("results.csv","w") as f:
2    print("t,x,y,u,v",file=f)
```

The code inside the block executes only if the open function succeeds. The file is also closed automatically once the block ends. With this syntax, we modify the code to redirect the output to the file:

```
1   t = 0              # initialize t
2   x = 0              # initialize position
3   y = 2.1
4   vx = 45            # initialize velocity
5   vy = 0
6   gx = 0             # initialize acceleration
7   gy = −9.81
8   dt = 0.04          # assign time step
9
10  with open("results.csv","w") as f:
11    print("%.2f, %.2f, %.2f, %.2f, %.2f"%(t,x,y,vx,vy),
12            file=f)
13
14    while (y>0):      # repeat until ground impact
15      t += dt         # increment time
16      x += vx*dt      # increment position
17      y += vy*dt
18      vx += gx*dt # increment velocity
19      vy += gy*dt
20      print("%.2f, %.2f, %.2f, %.2f, %.2f"%(t,x,y,vx,vy),
21              file=f)
22
23  print("The tennis ball hit the ground after about "
24        "%.3f s at distance %.3f m"%(t,x))
```

Now only the final

```
The tennis ball hit the ground after about 0.680 s at distance 30.600 m
```

message is printed to the screen. The time history of position and velocities is found in results.csv.

1.5.2   ARRAYS AND PLOTTING

Instead of writing the data to a file to be visualized using an external tool, we can generate graphs directly. Plotting support is provided by a package called matplotlib. It implements many of the plotting commands found in another popular scientific computing language, MATLAB. Specifically, this plotting is provided by an internal module called pyplot, and in order to use it, we first need to import it into the program.

Here is how we can create a simple line plot:

```
1   import matplotlib.pyplot as plt
2   x = [0,1,2,3,4,5]          # x is a list
3   y = [0,1,4,9,16,25]
4   plt.plot(x,y,color='purple')
```

The plot command requires as inputs data vectors containing the x and y values. It can also receive additional optional arguments to change the line style or to supply plot label to be shown in the legend. x and y are special types of variables called *list* in Python, but they are just an abstraction of another data type called an *array*. We use the "array" terminology in this section for generality. Arrays offer a way to store multiple data values in a single named variable. The difference between Python lists and "true" arrays is that the list can store non-homogeneous data, such as numbers and strings, while arrays store data of the same type. The individual entries are accessed using an index provided within square brackets. As already noted (but often forgotten by new students), Python, C, and Java array-indexing begins with zero. In other languages, such as MATLAB, Fortran, and Julia, indexing starts with 1. To demonstrate, we can replace individual entries in y as

```
1  y[0] = 2  # replace the first element
2  y[3] = 6  # replace 4th element, y is now [2,1,4,6,16,25]
```

Array sizes are limited only by the available memory (although older 32-bit architectures also imposed limits arising from addressable memory space). The array length is queried using the **len** command, as in

```
ni = len(y)   # ni = 6
```

Often we also need to access the last item in an array. Due to zero-based indexing, the valid index entries range from 0 to "length - 1",

```
y[ni-1] = -42    # replace the last item (25)
```

In Python, the two prior expressions can be combined using negative indexing,

```
y[-1] = -42    # replace the last item (25)
```

When accessing arrays, Python makes sure that the provided index is valid. Otherwise, we get an error,

```
y[7]=2      IndexError: list assignment index out of range
```

Other languages, namely C/C++, do not perform such bounds checking. This increases their performance, since CPU clock cycles are not used up on this check. On the other hand, the lack of such "guardrails" makes it possible to inadvertently overwrite memory leading to program crashes (the infamous *segmentation fault*) or spurious results.

Python allows us to start with an empty zero-element list, and resize it dynamically. This is accomplished with the help of **append** function defined for the list object. As an example, we can write

```
1  x = []           # create a empty list (array)
2  x.append(-1.2)   # x = [-1.2]
```

In order to add plotting to our simulation program, we need to perform the following tasks:

1. Initialize empty arrays to store $\vec{t}$, $\vec{x}$, $\vec{y}$, $\vec{v}_x$, and $\vec{v}_y$
2. Add data to the arrays in the loop
3. Generate plots of interest such as $\vec{x}$ vs. $\vec{y}$ or $\vec{v}_y$ vs. $\vec{t}$

Previously, we just updated the values stored in variables such as x,

```
x = x + vx*dt
```

It is important to note that here we are essentially taking the existing value of x from the prior loop iteration and using it compute the value at the new iteration. Previously we did not care about storing the "old" values, and hence we could get by with this simple replacement. However, we now need to store the entire time history in order to generate plots. We accomplish this by appending the latest value to the end of the initially empty x array,

```
1  # add new position array entry using the latest x and vx
2  x.append(x[-1] + vx[-1]*dt)
```

We thus replace the initial integration code with

```
1  import matplotlib.pylab as pl
2
3  t = [0]           # initialize t
4  x = [0]           # initialize position
5  y = [2.1]
6  vx = [45]         # initialize velocity
7  vy = [0]
8  gx = 0            # initialize acceleration
9  gy = -9.81
10 dt = 0.04         # assign time step
11
12 while (y[-1]>0):     # repeat until ground impact
13     t.append(t[-1]+dt)           # increment time
14     x.append(x[-1] + vx[-1]*dt)  # increment position
15     y.append(y[-1] + vy[-1]*dt)
16     vx.append(vx[-1] + gx*dt)    # increment velocity
17     vy.append(vy[-1] + gy*dt)
18
19 print("The tennis ball hit the ground after about "
20       "%.3f s at distance %.3f m"%(t[-1],x[-1]))
21
22 # plot x vs. y using the specified styling
23 pl.plot(x,y,color='purple',linewidth=3,dashes=[6,2],
24         marker='o',markersize=12, fillstyle='full',
25         markerfacecolor='yellow',label='tennis ball')
26 pl.xlabel('x (m)')    # add x-axis label
27 pl.ylabel('y (m)')    # add y-axis label
28 pl.grid()             # show x-y grid
29 pl.legend(loc=3)      # show legend in bottom left corner
30 plt.savefig('trace.eps', format='eps')
```

The **plot** command contains multiple optional arguments demonstrating how to style the graph. We also include commands to add labels to the x and y axes, and to display a legend in the bottom left corner (at location slot "3"). The final command saves the figure to a file using the vector .eps format. It is this plot that is displayed in Figure 1.4. Vector-formats save data using analytical shapes, such as lines or circles, and thus do not suffer from degradation when zoomed-in and are thus preferred in publishing. Pdf, another supported format, also uses vector graphics. Raster, or bitmap, formats such as .png or .jpg on the other hand save the individual pixels, resulting in pixelated representation when the image is zoomed into.

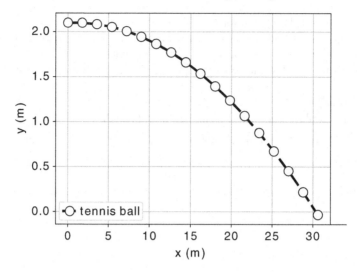

Figure 1.4 Tennis ball trace produced by the code in Listing 1.5.2.

1.5.3 LIST INITIALIZATION

General arrays store the data by allocating a block of contiguous memory to store the items. Python lists are a bit different since they allow storing mixed data types. Specifically, Python lists are arrays of references to the actual data objects which may be scattered throughout the RAM. In the previous example, the array grows dynamically since we append a new item on each pass. Simply expanding the space whenever more storage is needed is typically not feasible, since the memory space beyond the current extents of the array may already be assigned to other variables. Therefore, resizing an array requires allocating a completely new larger array into which the data from the original array is copied to. As you can imagine, doing all this copying is quite wasteful. Therefore, the best practice is to resize arrays as little as possible. In a real production code, we definitely would not want to be extending an array element by element on each loop iteration. Instead, if you know how large an array needs to be, you should always begin by preallocating the sufficient space.

 Let's say that we know that we the simulation will run for at most 10 iterations. We could create an x vector containing initially 10 zeros with the following declaration:

```
a = [0,0,0,0,0,0,0,0,0,0]
```
This same allocation can be accomplished using Python's *repetition* operator,

```
a = [0]*10    # this also generates a list of 10 zeros
```
This expression gives us a list with valid entries ranging from a[0] through a[9].

1.5.4 NUMPY ARRAYS

Yet, there is another way. Python also contains support for proper arrays that store homogeneous data in a contiguous memory block. Such a storage results in faster performance due to memory caching. In the scientific computing community, we create these arrays using functions from the *NumPy* library. This library provides support for general vector and matrix mathematics. Take a look at this example:

```
1  import numpy as np
2  a = np.ones(10)        # array of 10 ones
3  b = np.zeros(10)       # array of 10 zeros
4  b[5:] = 2
5  c = 2*a + b            # c = [2, 2, 2, 2, 2, 4, 4, 4, 4, 4]
```

We start off by importing the NumPy package. We then create two floating point value arrays (this is the default NumPy array type), containing 10 elements each. The elements of $\vec{a}$ are all set to 1.0, while all elements of $\vec{b}$ are set to zero. Next, on line 4, we utilize syntax you may be familiar with from MATLAB to modify a subset of the elements. Specifically, we assign the value of 2 to elements `b[5]` through `b[9]`. The syntax of this operator is

```
[start:end:step]
```

with every step-th element in range `[start]:[end-1]` being affected. Using the mathematical range syntax, this can also be written as [start, end). The start or the end index may be omitted, in which case they default to 0 and the array length. The step defaults to 1. Therefore, for 10-element array, `b[5:]` is identical to `b[5:10]`. The benefit of excluding the end value is that if we decide to use a large array, we don't need to go through our code to modify the end index. We confirm that the values are assigned as expected using Spyder's console:

```
a
Out[1]:  array([1., 1., 1., 1., 1., 1., 1., 1., 1., 1.])

b
Out[2]:  array([0., 0., 0., 0., 0., 2., 2., 2., 2., 2.])
```

Then finally on line 5, we utilize NumPy's vector support to assign $2\vec{a} + \vec{b}$ to a new variable $\vec{c}$. The resulting array contains a sequence of 5 twos, followed by 5 fours. This code is not only cleaner, but also noticeably faster. Expanding this idea calculation to a larger, 10 million element array, on an Intel i7 workstation the computation takes over 5 seconds when implemented using a for-loop. It takes only 0.025 seconds when implemented using NumPy vector operations as shown above. This is an increase by a factor of 200! This timing is also identical to that of a for-loop implementation in C++. This is because NumPy is actually written in highly optimized C, and utilizing the library vector operations we skip the Python interpreter.

Using the NumPy arrays, we modify our code to read

```
1  import matplotlib.pylab as plt
2  import numpy as np
```

```
3
4    # allocate arrays
5    ni = 100              # array size
6    t = np.zeros(ni)
7    x = np.zeros(ni)
8    y = np.zeros(ni)
9    vx = np.zeros(ni)
10   vy = np.zeros(ni)
11
12   #set initial values
13   t[0] = 0
14   x[0] = 0
15   y[0] = 2.1
16   vx[0] = 45
17   vy[0] = 0
18
19   #additional parameters
20   gx = 0
21   gy = -9.81
22   dt = 0.04
23
24   #initialize array indexthis
25   i = 0
26   while (y[i]>0):        # repeat until ground impact
27       t[i+1] = t[i] + dt
28       x[i+1] = x[i] + vx[i]*dt
29       y[i+1] = y[i] + vy[i]*dt
30       vx[i+1] = vx[i] + gx*dt
31       vy[i+1] = vy[i] + gy*dt
32       i = i+1              # increment array index
33
34   print("The tennis ball hit the ground after about "
35         "%.3f s at distance %.3f m"%(t[i],x[i]))
36
37   # plot x vs. y
38   plt.plot(x[:i+1],y[:i+1])
```

There are few important points to pay attention to. First, on line 5, we assign the value of 100 to a variable called `ni`. For historical reasons, it is common to use i, j, and k to represent indexes in the first, second, and third spatial dimension. `ni` simply means "number of i" entries. On lines 6 through 10, we use the NumPy `zeros` command to allocate ni-sized zero-valued arrays. Then on lines 13 through 17, we assign the initial values to the first item in each respective array. The acceleration terms are kept as *scalars* since they remain constant.

On line 25, we initialize the variable `i` to zero. As noted above, we are using this variable to keep track of the current position in the array. Since the array is fixed sized, we no longer use `y[-1]` to check for the ground impact. Instead, the latest assigned value is located at the `[i]` position. Therefore, on line 26, we modified the while-loop condition to use `y[i]`.

Inside the loop, we use the current [i] values of time and velocity to compute the new, future values. Since `x[i]` corresponds to the position $x(t)$, `x[i+1]` corresponds to $x(t + \Delta t)$. After we are done assigning all values, the

i index is incremented on line 32. This increment makes the next iteration of the while-loop use the just-set values as the "old" data.

Line 34 prints the solution obtained at the loop termination. We use the value of i to determine which values to print. Specifically, x[i] is the x position at which y became negative. Plotting is limited to the indexes that were actually modified, in other words items [0] through [i]. Since the range operator returns items exclusive of the second index, we use x[:i+1] to select items x[0] to, and including, x[i].

1.5.5 FOR-LOOPS

Iterating over a range of data and having to keep track of the index is so commonplace that there is a loop operator specifically designed for this task. These types of loops are called *for*-loops since they execute for a specified number of iterations. Let's say that we want to display numbers 0 through 9. One way to do so using our familiar while-loop is

```
1   i = 0
2   while ( i <10):
3       print ( i )
4       i += 1
```

More concise alternative is to use the for-loop:

```
1   for i in range(10):
2       print ( i )
```

Note that we no longer need to initialize the i variable and also do not need to manually increment it. In the Python implementation, the range(10) function returns an *iterator* for the 10 integers in $[0, 10)$ (zero inclusive, 10 exclusive). The for-loop steps through them consecutively until all values are used up. The currently used value is assigned to the loop control variable i. If needed, the range command also exists in a 3-argument form

```
range(start, end, step)
```

where the last value indicates the skip between consecutive values,

```
1   for i in range(3, 13, 2):
2       print i
```

outputs 3, 5, 7, 9, and 11. The default increment of 1 is used if the step is not specified.

1.5.6 CONDITIONAL STATEMENTS

One difficulty with the prior algorithm is that the for-loop continues to run even after the ball impacts the ground. We would like to do instead is to terminate the loop as soon as the exit condition $y < 0$ is met. The majority of languages use the if keyword for *conditional statements*. In Python, we have

```
1   if (condition):
2       # code to evaluate when condition true
3   elif (condition2):
```

```
4        # code to evaluate when condition 2 is true
5   else :
6        # code to evaluate otherwise
```

The `elif` (else if) and `else` clauses are optional and there can be multiple `elif` sections. As an example

```
1   if a>0:
2        print ("Positive!")
3   elif a<0:
4        print ("Negative!")
5   elif a==0:
6        print ("Zero!")
7   else :
8        print ("Not a number")
```

Note that equality is compared using two equal signs, `==`. Multiple logical conditions can be combined with `and` and `or` statements. The former is true when both conditions are true, while the second one is true if at least one condition is true. Another operand, `xor` is true when only one of the two conditions is true. `not` is used to negate expressions. For example, we can write

```
if (x>=0 and x<b) or (b>10):
```

Inside any `for` or `while` block, we can use the `break` keyword to terminate the loop prematurely. The listing from page 33 can then be simplified as:

```
1   for i in range(ni−1):      # i = [0,1,2,3,..,ni−2]
2        if (y[i]<0):
3             break        # exit loop if ground impact
4        t[i+1] = t[i] + dt
5        x[i+1] = x[i] + vx[i]*dt
6        y[i+1] = y[i] + vy[i]*dt
7        vx[i+1] = vx[i] + gx*dt
8        vy[i+1] = vy[i] + gy*dt
```

Note that we are no longer initializing nor incrementing i. Each loop iteration assigns data to the i+1 item, and hence the loop is limited to $i = ni - 2$. This final iteration, assuming the code gets there, would assign to the final allocated array slot, [ni-1].

Conditional statements allow us to add additional physics. For instance, instead of terminating the simulation on ground impact, we can let the ball bounce off. Without getting too much into the details, collisions can be classified as *elastic* or *inelastic* and as *specular* and *diffuse* depending on whether energy is conserved and whether the tangential direction is retained post-collision. A simple way to approximate elastic specular collision with wall at $y = 0$ is to flip the sign on the vertical coordinate y to bring the ball back above ground. The sign on v_y is also flipped assign the new post impact velocity direction upward from the surface. The v_y and v_x components can be scaled by some $\alpha < 1$ factor to approximate energy loss in an inelastic collision. The for-loop can then be modified as:

```
1   for i in range(ni−1):      # i = [0,1,2,3,..,ni−2]
2        if (y[i]<0):           # check for ground impact
3             y[i] = −y[i]       # flip back to above ground
```

```
4          alpha = 0.5        # specify bounciness
5          vy[i] *= -alpha # scale and flip y-velocity
6          vx[i] *= alpha  # scale x-velocity
7
8       t[i+1] = t[i] + dt
9       x[i+1] = x[i] + vx[i]*dt
10      y[i+1] = y[i] + vy[i]*dt
11      vx[i+1] = vx[i] + gx*dt
12      vy[i+1] = vy[i] + gy*dt
```

Note that in this version we have removed the loop termination on ground impact. The loop now continues for the specified number of iterations.

1.5.7   FUNCTIONS

The for-loop listed in the above example is the actual algorithm that computes the trajectory of a single object given some initial conditions provided in the [0] index of the utilized arrays. We can envision that in a simulation with multiple tennis balls, we will need to execute this exact set of instructions repeatedly for each object. Programming languages allow us to create custom named blocks of codes that can receive input arguments, and return back other set of return values. These objects are called *functions*, in the same way

$$y = f(x) \tag{1.28}$$

$f(x)$ is a mathematical function of x. In Python, we define functions using the **def** keyword. Python function can return back an arbitrary number of values using the **return** keyword. As an example, this is how we would create, and use, a custom function to evaluate $x^2 + 3y$:

```
1  def my_fun(x,y):
2     return x*x + 3*y
3
4  z = my_fun(-2.2,1.2)
5  print ("my_fun(-2.2,1.2) = %g"%z)
```

which outputs

```
my_fun(-2.2,1.2) = 8.44
```

With this in mind, we can isolate the code specific to computing the trajectory of a single ball into a single function. This function receives as inputs the initial conditions and returns the resulting $\vec{x}$, $\vec{y}$, $\vec{v}_x$, and $\vec{v}_y$ arrays. The following code listing implements this function-based integration algorithm:

```
1  import matplotlib.pylab as plt
2  import numpy as np
3
4  # global inputs
5  ni = 200
6  dt = 0.01
7  gx = 0
8  gy = -9.81
9
```

```
10    # returns [x,y,vx,vy] given initial conditions
11    def integrate (x0,y0,vx0,vy0):
12        t = np.zeros(ni)   # allocate arrays
13        x = np.zeros(ni)
14        y = np.zeros(ni)
15        vx = np.zeros(ni)
16        vy = np.zeros(ni)
17
18        t [0]  = 0      # set initial values
19        x [0]  = x0
20        y [0]  = y0
21        vx [0] = vx0
22        vy [0] = vy0
23
24        for i in range(ni-1):
25            if (y[i]<0):   # check for ground impact
26                y[i] = -y[i]
27                alpha = 0.5
28                vy[i] *= -alpha
29                vx[i] *= alpha
30            t [i+1] = t [i] + dt
31            x [i+1] = x [i] + vx[i]*dt
32            y [i+1] = y [i] + vy[i]*dt
33            vx [i+1] = vx [i] + gx*dt
34            vy [i+1] = vy [i] + gy*dt
35
36        return x,y,vx,vy
37    # (end of def integrate)
38
39    x,y,vx,vy = integrate (0,2.1,45,0)
40    plt.plot (x,y, '-',LineWidth=3)
41
42    x,y,vx,vy = integrate (0.01,2.3,55,0)
43    plt.plot (x,y, '-.',LineWidth=3)
44
45    x,y,vx,vy = integrate (0,2.0,40,0)
46    plt.plot (x,y, '--',LineWidth=3)
47
48    plt.xlabel ('x (m)')
49    plt.ylabel ('y (m)')
```

We start off by setting several common variables like the time step and the gravitational vector. They are examples of *global* variables as they are defined in the global scope outside any particular function. Global variables are visible for read-access to all functions. The function definition starts on line 11 and continues until line 36. The dashed-line was added for clarity. The function is named **integrate** and needs to be called with four arguments that correspond to the $x_0, y_0, v_{x,0},$ and $v_{y,0}$ initial conditions. The internal code should be quite familiar. The function starts by allocating data arrays and setting the initial values. It then jumps into the for-loop. Upon completion, instead of plotting the data, we return the computed vectors using the **return** command.

Line 39 illustrates how to call and use this function. Here we use the same initial conditions we have been using all along. The resulting trace is then plotted on Line 40, with the code illustrating another way to specify the line

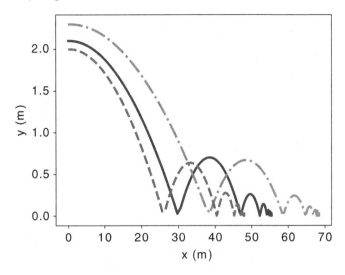

Figure 1.5 Traces of multiple tennis balls.

style. '`-`' produces a solid line. In order to generate another trace, all we need to do is to replicate these two lines. On line 42, we call the `integrate` function again but this time with slightly different initial conditions. This trace uses '`-.`' indicator to select a dash-dot line style. Finally, yet another tennis ball trace is included on lines 45 and 46, this one plotted with a dashed line specified with the '`--`' selector. The resulting plot is shown in Figure 1.5.

1.5.8 RANDOM NUMBERS

In many engineering situations, we may only have a rough estimate for the input conditions. In the case of the tennis player, we may have only approximate values for the height of the serve and the speed with which the ball leaves the racket. But there will be some variation between serves. The initial conditions may be better approximated by including "error bars":

$$x_0 = 0 \pm 0.05 \text{ [m]} \qquad (1.29)$$
$$y_0 = 2.1 \pm 0.2 \text{ [m]}$$
$$v_{x,0} = 45 \pm 15 \text{ [m/s]}$$
$$v_{y,0} = 0 \pm 2 \text{ [m/s]}$$

Given this variation, we may be interested to find out where is the ball's most probable resting position. For this particular example, the answer can derived analytically. Real-world processes tend to include non-linear response that makes such an analytical answer non-trivial. In the above example, we plotted traces for three distinct serves. With just a small modification, namely by adding a for-loop, we can extend the code to perform a stochastic, or Monte-Carlo, simulation. At each iteration, we initialize the serve with random values sampled from the input set given by expressions in Equation 1.29. Given a large number of serves, the code results can be expected to convergence on the most probable settling location.

We demonstrate the use of of random numbers in gas kinetic simulations in Chapter 5. For now, it suffices to know that all programming languages implement some functionality to obtain random numbers. These numbers are not truly random but instead they represent consecutive values sampled from a particularly large sequence. In Python, we obtain random values using

```
1  from random import random
2  R = random ()
```

Here we use a slightly different form of the **import** statement to directly import the **random()** function. This function returns a real value in [0,1). A zero is a possibility, but we never get 1.0. The values are sampled from the *uniform distribution*. Every value in the $[0, 1.0)$ range has equal probability of being picked. Alternately, the input space could be described by the *normal* (or Gaussian) distribution that captures the "bell-curve" in which values near the mean have the highest probability of being selected. We cover this distribution in Chapter 3, and for now just stick with the uniform model. A value in

$$z = a \pm b \tag{1.30}$$

can be obtained using

$$z = a + (-1 + 2\mathcal{R})b \tag{1.31}$$

where $\mathcal{R}$ is a unique random number in $[0, 1)$. The expression in parentheses takes on a value in $[-1, 1)$. Scaling by b, we have $[-b, b)$. With this in mind, we replace lines 37 through 45 with

```
1  for s in range(25): # display 25 random serves
2      x0 = 0 + (-1+2*random())*0.05      # 0 +/- 0.05
3      y0 = 2.1 + (-1+2*random())*0.2     # 2.1 +/- 0.2
4      vx0 = 45 + (-1+2*random())*15      # 45 +/- 15
5      vy0 = 0 + (-1+2*random())*2        # 0 +/- 2
6      x,y,vx,vy = integrate(x0,y0,vx0,vy0)
7      plt.plot(x,y,LineWidth=1)
```

to obtain a plot similar to Figure 1.6. Your output will not be exactly identical, and will in fact change from run to run. This is expected due to the stochastic nature of the simulation. By visually "eye-balling" the plot, it seems that most serves terminate between the 40 and 50 m markers; however, there are a few occasional serves that make it out to 80 m.

1.5.9 CODE VALIDATION

Now that we have a basic simulation, it is important to make sure it is working correctly. To truly validate the code, we would need to find a tennis player with the appropriate height, and after a large number of serves, record where

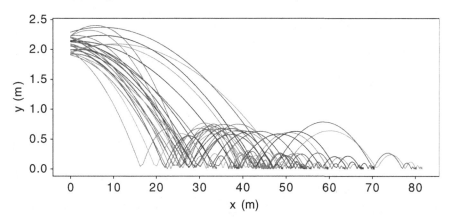

Figure 1.6 Traces of tennis balls with random initial conditions.

the tennis balls have landed. We may find the predictions to be quite off due to factors such as incorrect initial conditions (the serve speed may be incorrect) or the lack of required physical models (drag and lift are not included, and the ground impact model is quite simplistic). But even if these shortcomings were addressed, we may still notice a discrepancy arising from bugs in the implemented code. The check that the equations are implemented as intended (whether they are the correct equations for the particular setup) is known as verification. We discuss these topics in more detail in Chapter 8. But, one popular approach is to reduce the problem sufficiently so that an analytical solution can be found. We then compare the numerical result to theory. In our case, we already have the analytical solution to the entire algorithm given by Equations 1.13 and 1.14. We can thus add the following code to plot the analytical solution for $t \in [0, t_{impact}]$, where $t_{impact} \approx 0.654$ s per Equation 1.15:

```
1  t_th = np.linspace(0,0.654,25)
2  x_th = 45*t_th
3  y_th = 0.5*gy*t_th**2 + 2.1
4  plt.plot(x_th,y_th,':o', color='0.4',markersize=5,label='
       analytical')
5  plt.legend(loc=3)
```

Here we illustrate another benefit of the NumPy library. NumPy vectors can be manipulated using familiar mathematical syntax without needing to iterate through them element-wise. This approach is often faster than a for-loop, since the calculation is done internally within the NumPy library functions, which are implemented in C++. On the first line, we use NumPy's **linspace** function to generate a vector of 25 evenly distributed values between 0 and 0.654. Then on the second line, each element of t_th is multiplied by 45 and the resulting vector, having identical dimensions as t_th, is saved as x_th. We then compute the analytical solution for y. We use t_th**2 to calculate t_{th}^2. The x_{th} vs. y_{th} graph is plotted by a dotted line with circle markers, as can be seen in Figure 1.7.

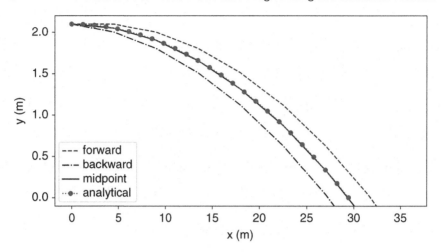

Figure 1.7 Comparison of integration methods for $\Delta t = 0.1$ s.

1.5.10 LEAPFROG METHOD

The trace from the algorithm we have been using so far is shown by the dashed line. While the trajectory shape is similar to that given by the analytical solution, there is a clear discrepancy in the horizontal position. We find that rerunning the code with a smaller integration time step Δt leads to a reduction in the discrepancy. The scheme appears to be *consistent*.

Reviewing the integration scheme, we can observe that the "new" position is computed using the "old" value of velocity. In other words,

$$x(t + \Delta t) = x(t) + v(t)\Delta t \tag{1.32}$$

It is customary to write values at different time points using a superscript notation where x^n implies $x(t)$ and x^{n+1} is $x(t+\Delta t)$.[1] This integration scheme is known as the *Forward Euler Method*. The above equation can thus be written as

$$x^{n+1} = x^n + v^n \Delta t \tag{1.33}$$

The "forward" term implies that we use only "current" time t data to compute future values at $t + \Delta t$.

But, we could alternatively use the future velocity $v(t + \Delta t)$ to advance the position, giving us

$$x^{n+1} = x^n + v^{n+1}\Delta t \tag{1.34}$$

This is known as the *Backward Euler Method,* and it produces the trace shown by the dash-dot line. Now the numerical horizontal displacement seems to lag behind the analytical one. It appears that the correct solution lies between

---

[1] k is also commonly utilized for the time index. We utilize n here to avoid confusion with the z-direction index found in 3D algorithms.

the forward and the backward method, implying that we should utilize the average of the two velocities. This *midpoint* method,

$$x^{n+1} = x^n + 0.5(v^{n+1} + v^n)\Delta t \tag{1.35}$$

shown by the solid line, indeed produces the desired solution. This algorithm is also known as the *Leapfrog method*. The velocity term is really just an approximation for $v(t + 0.5\Delta t)$. In order to cut down on the number of calculations, which could be critical in large simulations, we let the velocity to be known at half-time step intervals. We then have

$$v(t + 0.5) = v(t - 0.5) + a(t)\Delta t$$
$$x(t + 1) = x(t) + v(t + 0.5)\Delta t$$

which can be indexed as

$$v^{n+1} = v^n + a^n\Delta t$$
$$x^{n+1} = x^n + v^{n+1}\Delta t$$

This algorithm requires that on the first iteration we rewind the velocity backward by a half time-step,

$$v^{-0.5} = v^0 - a^0\Delta t/2 \tag{1.36}$$

Alternatively, we could apply only a half acceleration on the first integration. The comparison of these three methods is shown in the listing below:

```
1   for i in range(ni-1):
2       t[i+1] = t[i] + dt
3       vx[i+1] = vx[i] + gx*dt
4       vy[i+1] = vy[i] + gy*dt
5
6       #forward method
7       xf[i+1] = xf[i] + vx[i]*dt
8       yf[i+1] = yf[i] + vy[i]*dt
9
10      #backward method
11      xb[i+1] = xb[i] + vx[i+1]*dt
12      yb[i+1] = yb[i] + vy[i+1]*dt
13
14      #midpoint method
15      xm[i+1] = xm[i] + 0.5*(vx[i+1]+vx[i])*dt
16      ym[i+1] = ym[i] + 0.5*(vy[i+1]+vy[i])*dt
```

The important takeaway here is realizing that integration of a general equation of type

$$\frac{\partial \psi}{\partial t} = f(\psi, t) \tag{1.37}$$

should be performed with f evaluated at the midpoint time, $t + 0.5\Delta t$. This is in fact the foundation of the popular Crank-Nicolson method which will be discussed in Chapter 6.

2 Finite Difference and Linear Algebra

In this chapter, we introduce the Finite Difference Method for solving differential equations. We discuss matrix representation for systems of equations, and learn how to solve linear systems using direct and iterative methods. We implement these algorithms in Python.

2.1 TAYLOR SERIES

The example discussed in Chapter 1 utilized the following approximation of the first derivative:

$$\frac{df}{dx} \approx \frac{\Delta f}{\Delta x} = \frac{f(x_2) - f(x_1)}{x_2 - x_1} = \frac{f(x + \Delta x) - f(x)}{\Delta x} \tag{2.1}$$

We arrived at this expression by replacing the infinitesimal differential elements df and dx with finite sized increments Δf and Δx. This is the first-order approximation arising from a method known as the Finite Difference Method (FDM). It can be derived formally by considering the Taylor Series. This infinite series, named after a 17th-century English mathematician Brook Taylor, tells us that we can calculate the value of a smooth, infinitely differentiable function $f(x)$ at some other point $x + \Delta x$ if we know the value of the function, and all its derivatives, at x,

$$f(x + \Delta x) = f(x) + \sum_{n=1}^{\infty} \frac{(\Delta x)^n}{n!} \frac{d^{(n)} f(x)}{dx^{(n)}} \tag{2.2}$$

For $\Delta x << 1$, the $(\Delta x)^n$ terms quickly become tiny. Combined with the rapidly increasing factorial $n!$ term in the denominator, the higher-order terms (HOT) can be seen to vanish rapidly. Therefore, it is customary to group them and write

$$f(x + \Delta x) = f(x) + \frac{\Delta x}{1!} \frac{df}{dx}(x) + \frac{\Delta^2 x}{2!} \frac{d^2 f}{dx^2}(x) + \frac{\Delta^3 x}{3!} \frac{d^3 f}{dx^3}(x) + \text{HOT} \tag{2.3}$$

2.1.1 FIRST DERIVATIVE

By further ignoring the $d^2 f/dx^2$ and the higher-order terms, we can write

$$f(x + \Delta x) = f(x) + \frac{\Delta x}{1!} \frac{df}{dx}(x) + \text{HOT} \tag{2.4}$$

DOI: 10.1201/9781003132233-2

which can be rearrange to yield the now familiar

$$\frac{df}{dx}(x) = \frac{f(x + \Delta x) - f(x)}{\Delta x} + O(1) \tag{2.5}$$

This representation is known as Euler Forward Difference. It is only first-order accurate, implying that the estimation error scales linearly with the change in Δx. We indicate this scaling using the "Big O" notation $O(n)$, where n is the error order.

We can also use Taylor Series to evaluate the function at $x - \Delta x$. Equation 2.3 then turns into

$$f(x - \Delta x) = f(x) - \frac{\Delta x}{1!}\frac{df}{dx}(x) + \frac{\Delta^2 x}{2!}\frac{d^2 f}{dx^2}(x) - \frac{\Delta^3 x}{3!}\frac{d^3 f}{dx^3}(x) + O(1) \tag{2.6}$$

Similarly, by ignoring the second-order terms, we obtain another expression for the first derivative,

$$\frac{df}{dx}(x) = \frac{f(x) - f(x - \Delta x)}{\Delta x} + O(1) \tag{2.7}$$

This expression is known as the Euler Backward Difference and is also first-order accurate.

It is important to notice that in Equation 2.6 all odd terms retain the negative sign, while the even terms remain positive. By subtracting this equation from 2.3, we obtain

$$f(x + \Delta x) - f(x - \Delta x) = 2\frac{\Delta x}{1!}\frac{df}{dx}(x) + 2\frac{\Delta x}{3!}\frac{d^3 f}{dx^3}(x) + O(2) \tag{2.8}$$

The even terms cancel out and vanish. Rearranging we obtain

$$\frac{df}{dx}(x) = \frac{f(x + \Delta x) - f(x - \Delta x)}{2\Delta x} + O(2) \tag{2.9}$$

Equation 2.9 is known as the *central difference*. This representation of the first derivative is *second-order accurate*. Halving Δx leads to a 4× reduction in the estimation error. Notice that the derivative is calculated using values "to the left" and "to the right" of the location x, hence the "central" naming convention.

Alternatively, the same way we used Taylor series to define the value at point $x + \Delta x$, we can estimate the value at $x + 2\Delta x$,

$$f(x + 2\Delta x) = f(x) + 2\Delta x \frac{\partial f}{\partial x}(x) + 2\Delta^2 x \frac{\partial^2 f}{\partial x^2}(x) + \text{HOT} \tag{2.10}$$

We next proceed to subtract Equation 2.4 multiplied by 4. The second derivative terms then cancel naturally, and we obtain the second-order forward difference scheme,

$$\frac{\partial f}{\partial x}(x) = \frac{-3f(x) + 4f(x + \Delta x) - f(x + 2\Delta x)}{2\Delta x} + O(2) \tag{2.11}$$

We can similarly derive a second order backward difference scheme,

$$\frac{\partial f}{\partial x}(x) = \frac{f(x - 2\Delta x) - 4f(x - \Delta x) + 3f(x)}{2\Delta x} + O(2) \tag{2.12}$$

These *one-sided* forms allow us to estimate the derivatives using two points either "to the left" or two points "to the right" of the location of interest x, while retaining the second-order accuracy of the central difference scheme.

2.1.2   SECOND DERIVATIVE

Instead of subtracting, Equation 2.6 can be added to 2.3 to obtain

$$f(x + \Delta x) + f(x - \Delta x) = f(x) + \frac{\Delta^2 x}{2!}\frac{d^2 f}{dx^2}(x) + \frac{\Delta^4 x}{4!}\frac{d^4 f}{dx^4}(x) + \text{HOT} \tag{2.13}$$

Now the odd terms vanish implicitly. The above equation can then be rearranged to isolate the second derivative,

$$\frac{d^2 f}{dx^2}(x) = \frac{f(x - \Delta x) - 2f(x) + f(x + \Delta x)}{\Delta^2 x} + O(2) \tag{2.14}$$

This is the standard central difference for the second derivative and is second-order accurate.

2.2   FINITE DIFFERENCE DISCRETIZATION

Let's now consider a simple model differential equation. At steady state, $\partial()/\partial t = 0$, and with no sources, the form of the energy equation presented in 1.24 reduces to

$$\nabla^2 T = 0 \tag{2.15}$$

This type of equation is called the Laplace equation. It is named after the 18th-century French mathematician Pierre-Simon Laplace. It is a reduced form of the general Poisson's equation $\nabla^2 \psi = \vec{b}$ for the case with the forcing vector $\vec{b} = 0$. That equation is named after yet another 18th-century French mathematician, Siméon Denis Poisson.

Expanding the Laplace operator (or Laplacian) ∇^2, we have

$$\frac{\partial^2 T}{\partial x^2} + \frac{\partial^2 T}{\partial y^2} + \frac{\partial^2 T}{\partial z^2} = \vec{b} \tag{2.16}$$

Here we include for generality some arbitrary source term $\vec{b}$ corresponding to volumetric heating. Next, let's consider just the one-dimensional form,

$$\frac{d^2 T}{dx^2} = \vec{b} \tag{2.17}$$

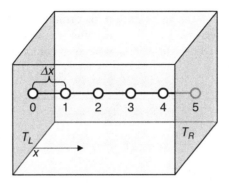

Figure 2.1 One dimensional domain used to illustrate a heat equation solver.

Utilizing Equation 2.14, the above relationship can be approximated as

$$\frac{T(x - \Delta x) - 2T(x) + T(x + \Delta x)}{\Delta^2 x} \approx b(x) \qquad (2.18)$$

Let's say that we are interested in solving Equation 2.17 for a brick-shaped conductor as shown in Figure 2.1. The left and right faces are held at some fixed temperature. We also assume that the distance between them is significantly smaller than the width and height in the y and z directions. Moving along a line connecting the left and right faces, the conductor appears infinitely large in y and z and any variations in these directions vanish (since they take place over an infinitely large distance). We thus have $\partial T / \partial y = \partial T / \partial z = 0$, and the solution reduces to a one-dimensional form.

We next subdivide the internal region into small spatial increments. Using uniform size Δx, the position of the first *node* (shown by the white circles) is x_0. The second node is found at $x_0 + \Delta x$, the third node is at $x_0 + 2\Delta x$ and so on. The region between nodes is called a *cell*. We have constructed a one-dimensional uniform Cartesian mesh. These meshes utilize uniform cell size, and the cell edges (lines connecting adjacent nodes) are aligned with the coordinate directions. Uniform Cartesian meshes are ubiquitous in numerical simulations due to their simplicity. Their disadvantage is that they introduce difficulties in representing conical geometrical boundaries. Utilizing zero-based indexing, the position of the grid node i is given by

$$x_i = x_0 + i\Delta x \qquad ; i \in [0, n_i - 1] \qquad (2.19)$$

where n_i is the number of nodes in the x direction. The cell spacing needed to span the distance from x_0 to x_m using $n_i - 1$ cells (the number of cells is one less than the number of nodes for every dimension) is calculated from

$$\Delta x = \frac{x_m - x_0}{n_i - 1} \qquad (2.20)$$

Now, let's consider the node at $i = 3$. From Equation 2.18 we see that approximating the derivative involves utilizing temperature values $T(x - \Delta x)$, $T(x)$, and $T(x + \Delta x)$. These positions happen to coincide with our grid nodes. For $x = x_3$ we have

$$\frac{T(x_3 - \Delta x) - 2T(x_3) + T(x_3 + \Delta x)}{\Delta^2 x} = b_3 \qquad (2.21)$$

or

$$\frac{T(x_2) - 2T(x_3) + T(x_4)}{\Delta^2 x} = b_3 \qquad (2.22)$$

Or, in general,

$$\frac{T(x_{i-1}) - 2T(x_i) + T(x_{i+1})}{\Delta^2 x} = b_i \qquad (2.23)$$

Estimating the second derivative at node index i involves utilizing values of the function at the point in question, along with the grid node to the left and to the right. This can be visualized with the *stencil* shown in Figure 2.2. In order to simplify the notation, we let $T(x_i) = T_i$. We then write

$$\frac{T_{i-1} - 2T_i + T_{i+1}}{\Delta^2 x} = b_i \qquad (2.24)$$

Figure 2.2 Computational stencil for the second derivative d^2/dx^2.

2.2.1 BOUNDARY CONDITIONS

Similar equation can be written for every point $i \in [1, ni - 2]$. However, attempting to use it on the boundaries $i = 0$ and $i = n_i - 1$ would require values of T outside the computational domain. Therefore, a different approach is needed here.

There are two primary types of boundary conditions: Dirichlet and Neumann. The *Dirichlet* boundary condition specifies the value of the property we are solving for. In this case, it implies a fixed temperature. A *Neumann* boundary instead specifies the rate of change of the property in the direction normal to the bounding surface, $\partial T/\partial \hat{n}$. Heat flux $\vec{q}$ is given by $\vec{q} = k\nabla T$. Dotting both sides of the equation with the normal vector yields $q_\perp/k = \partial T/\partial \hat{n}$. Therefore, in the case of the heat equation, the Neumann boundary prescribes the heat flux across the boundary. A zero Neumann boundary implies that there is no heat flux. It also indicates that there is no change in the property of interest (T) across the boundary, since $\partial T/\partial \hat{n} = 0$ This condition is commonly used to model expansion to free space. It suggests that the boundary is so far away from any sources that moving slightly farther away results in no observable change in the computed value. A third kind of boundary condition, called *Robin*, is a linear combination of Dirichlet and Neumann, $aT(x) + b\partial T/\partial \hat{n} = g$.

2.2.2 SYSTEM OF EQUATIONS

Let's now consider this one-dimensional domain discretized into a grid made up of 6 nodes, i.e. $n_i = 6$. Equation 2.24 holds on the four internal nodes. We also assign a fixed temperature on both ends, $T(x_0) = T_L$ and $T(x_m) = T_R$. We thus end up with the following system of equations:

$$T_0 = T_L \tag{2.25}$$
$$\alpha T_0 - 2\alpha T_1 + \alpha T_2 = b_1 \tag{2.26}$$
$$\alpha T_1 - 2\alpha T_2 + \alpha T_3 = b_2$$
$$\alpha T_2 - 2\alpha T_3 + \alpha T_4 = b_3$$
$$\alpha T_3 - 2\alpha T_4 + \alpha T_5 = b_4$$
$$T_5 = T_R$$

where $\alpha = 1/\Delta^2 x$. We have a total of 6 equations and 6 "unknowns", T_0 through T_5. The unknowns correspond to the 6 grid nodes on which we need to set the value of T_i. Given that these equations are linearly independent (none of the equations can be written by adding scaled versions of any of the other equations), the system has a unique solution. Finding this appropriate $\vec{T}$ is called solving the system.

2.2.3 GAUSSIAN ELIMINATION

Consider the following system

$$
\begin{array}{ccccccccc}
a_{0,0}T_0 & + & a_{0,1}T_1 & + & a_{0,2}T_2 & + & a_{0,3}T_3 & = & b_0 \\
0 \cdot T_0 & + & a_{1,1}T_1 & + & a_{1,2}T_2 & + & a_{1,3}T_3 & = & b_1 \\
0 \cdot T_0 & + & 0 \cdot T_1 & + & a_{2,2}T_2 & + & a_{2,3}T_3 & = & b_2 \\
0 \cdot T_0 & + & 0 \cdot T_1 & + & 0 \cdot T_2 & + & a_{3,3}T_3 & = & b_3
\end{array}
\tag{2.27}
$$

where the a_0 terms are arbitrary non-zero coefficients. Of importance is noting that each row has an increasingly longer list of leading zeros. The last equation contains only a single non-zero term, and as such, we can solve for T_3 directly,

$$T_3 = b_3/a_{3,3} \tag{2.28}$$

Once a value of T_3 is computed, it can be substituted into the second from last equation,

$$a_{2,2}T_2 + a_{2,3}\,T_2 = b_2 \tag{2.29}$$

which can then be solved directly for T_2,

$$T_2 = (b_2 - a_{2,3}T_3)/a_{2,2} \tag{2.30}$$

We next substitute values of T_2 and T_3 into the next higher-up equation, and continue so until the solution for the entire system is obtained. This solution method is called *back substitution*. In order to utilize it, is necessary to convert the linear system into an *upper-triangular*, or *echelon*, form. The name arises from the shape of the region in which non-zero coefficients can be found.

To illustrate the approach, let's now consider the following system of equations:

$$L0: \quad 0x_0 + 2x_1 + 1x_2 = 4 \tag{2.31}$$

$$L1: \quad 1x_0 + 0x_1 + 1x_2 = 2 \tag{2.32}$$

$$L2: \quad 1x_0 + 2x_1 + 0x_2 = 2 \tag{2.33}$$

The actual ordering of equations does not matter, and as such, we can reorder the equations as needed. This is known as Rule 1. The left and right sides of any equation can also be multiplied by an arbitrary non-zero constant without changing the relationship. In other words

$$1x_0 + 2x_1 + 0x_2 = 2 \tag{2.34}$$

is identical to

$$2(1x_0 + 2x_1 + 0x_2) = 2(2) \tag{2.35}$$

which simplifies to

$$2x_0 + 4x_1 + 0x_2 = 4 \tag{2.36}$$

This is known as Rule 2. Finally, we can add any single equation, possibly scaled by a non-zero constant per Rule 2, to any other equation. This is Rule 3. To demonstrate, adding L0 to L2 leads to

$$[1x_0 + 2x_1 + 0x_2] + [0x_0 + 2x_1 + 1x_2] = [2] + [4] \tag{2.37}$$

By subtracting 4 from both sides,

$$[1x_0 + 2x_1 + 0x_2] + \{[0x_0 + 2x_1 + 1x_2] - [4]\} = [2] \tag{2.38}$$

But by definition of L0, $0x_0 + 2x_1 + 1x_2 = 4$ and hence the term in the curly braces reduces to

$$[1x_0 + 2x_1 + 0x_2] + \{[4] - [4]\} = [2] \tag{2.39}$$

or

$$1x_0 + 2x_1 + 0x_2 = 2 \tag{2.40}$$

which is just the original equation L2.

These three rules lead to a solution method called the *Gaussian Elimination*. Its objective is to use them to convert a system of equations into the upper triangular form. The solution is then obtained with back-substitution. Since we require a non-zero coefficient in the upper-left corner, let's swap $L0$ and $L2$ to obtain

$$L0: \quad 1x_0 + 2x_1 + 0x_2 = 2 \tag{2.41}$$

$$L1: \quad 1x_0 + 0x_1 + 1x_2 = 2 \tag{2.42}$$

$$L2: \quad 0x_0 + 2x_1 + 1x_2 = 4 \tag{2.43}$$

This swap is known as *pivoting*.

Next, we require the first coefficient of L1 to be zero. We can eliminate this leading coefficient by utilizing Rule 3 to perform

$$L1 - L0 \to L1 \tag{2.44}$$

or

$$1x_0 + 0x_1 + 1x_2 - 1x_0 - 2x_1 - 0x_2 = 2 - 2 \tag{2.45}$$
$$0x_0 - 2x_1 + 1x_2 = 0 \tag{2.46}$$

We now have

$$L0: \quad 1x_0 + 2x_1 + 0x_2 = 2 \tag{2.47}$$
$$L1: \quad 0x_0 - 2x_1 + 1x_2 = 0 \tag{2.48}$$
$$L2: \quad 0x_0 + 2x_1 + 1x_2 = 4 \tag{2.49}$$

This is now almost the correct form except for the non-zero coefficient 2 multiplying the x_1 term in Equation L2. We can eliminate it using

$$L2 + L1 \to L2 \tag{2.50}$$

or

$$0x_0 + 2x_1 + 1x_2 + 0x_0 - 2x_1 + 1x_2 = 4 + 0 \tag{2.51}$$
$$0x_0 + 0x_1 + 2x_2 = 4 \tag{2.52}$$
$$x_2 = 2 \tag{2.53}$$

The resulting upper-triangular system is

$$\begin{aligned} L0: \quad & 1x_0 + 2x_1 + 0x_2 = 2 \\ L1: \quad & \qquad\;\; -2x_1 + 1x_2 = 0 \\ L2: \quad & \qquad\qquad\quad 1x_2 = 2 \end{aligned} \tag{2.54}$$

From the L2 equation, we see that $x_2 = 2$. Substituting this value into L1 leads to

$$-2x_1 + 1 \cdot 2 = 0 \tag{2.55}$$

or $x_1 = 1$. Finally, substituting x_1 into L0, we obtain

$$1x_0 + 2 \cdot 1 + 0 \cdot 2 = 2 \tag{2.56}$$

or $x_0 = 0$.

2.2.4 MATRIX REPRESENTATION

The above system of equations can be written in a more concise form by utilizing *matrix representation*. A matrix is essentially a two-dimensional array of coefficients,

$$
\mathbf{A} = \begin{bmatrix}
a_{0,0} & a_{0,1} & a_{0,2} & \cdots & a_{0,nc-1} \\
a_{1,0} & a_{1,1} & a_{1,2} & \cdots & a_{1,nc-1} \\
\vdots & \vdots & \vdots & \ddots & \vdots \\
a_{nr-1,0} & a_{nr-1,1} & a_{nr-1,2} & \cdots & a_{nr-1,nc-1}
\end{bmatrix}
\tag{2.57}
$$

The above matrix consists of n_r rows and n_c columns, for the total of $n_r \times n_c$ $a_{r,c}$ coefficients. We again utilize the zero-based indexing and thus the first row and column start at index zero. The maximum row and column index is $n_r - 1$ and $n_c - 1$, respectively.

2.2.5 MATRIX ALGEBRA

Matrices with identical dimensions can be added or subtracted from each other. The operation is applied to each coefficient independently,

$$
\begin{bmatrix}
a_{0,0} & a_{0,1} \\
a_{1,0} & a_{1,1} \\
a_{2,0} & a_{2,1}
\end{bmatrix}
+
\begin{bmatrix}
b_{0,0} & b_{0,1} \\
b_{1,0} & b_{1,1} \\
b_{2,0} & b_{2,1}
\end{bmatrix}
=
\begin{bmatrix}
a_{0,0} + b_{0,0} & a_{0,1} + b_{0,1} \\
a_{1,0} + b_{1,0} & a_{1,1} + b_{1,1} \\
a_{2,0} + b_{2,0} & a_{2,1} + b_{2,1}
\end{bmatrix}
\tag{2.58}
$$

Matrices can also be multiplied together. Multiplying matrix $\mathbf{A}$ by another matrix $\mathbf{B}$ involves assigning to each row r and column c of the target $\mathbf{C}$ the dot product of $\mathbf{A}$'s row r with $\mathbf{B}$'s column c. A dot product of two vectors $\vec{a}$ and $\vec{b}$ is given by

$$
\vec{a} \cdot \vec{b} = a_0 b_0 + a_1 b_1 + \ldots + a_{n-1} b_{n-1}
\tag{2.59}
$$

This operation is feasible only if the two vectors have identical sizes. In the case of matrix multiplication, it implies that the number of columns of matrix $\mathbf{A}$ must equal to the number of rows of matrix $\mathbf{B}$. Letting $\vec{a}_{i,:}$ correspond to the coefficients on the i-th row of matrix $\mathbf{A}$,a nd $\vec{b}_{:,j}$ be the j-th column of $\mathbf{B}$, matrix multiplication can be written as

$$
\mathbf{A} \cdot \mathbf{B} = \begin{bmatrix}
\vec{a}_{0,:} \cdot \vec{b}_{:,0} & \vec{a}_{0,:} \cdot \vec{b}_{:,1} & \cdots & \vec{a}_{0,:} \cdot \vec{b}_{:,nc-1} \\
\vec{a}_{1,:} \cdot \vec{b}_{:,0} & \vec{a}_{1,:} \cdot \vec{b}_{:,1} & \cdots & \vec{a}_{1,:} \cdot \vec{b}_{:,nv-1} \\
\vdots & \vdots & \ddots & \vdots \\
\vec{a}_{nr-1,:} \cdot \vec{b}_{:,0} & \vec{a}_{nr-1,:} \cdot \vec{b}_{:,1} & \cdots & \vec{a}_{nr-1,:} \cdot \vec{b}_{:,nc-1}
\end{bmatrix}
\tag{2.60}
$$

We can illustrate this with a concrete example. Let

$$
\mathbf{A} = \begin{bmatrix}
1 & 2 \\
3 & 4 \\
5 & 6
\end{bmatrix}
\qquad
\mathbf{B} = \begin{bmatrix}
0 & 1 & 2 \\
3 & 4 & 5
\end{bmatrix}
\tag{2.61}
$$

then

$$\mathbf{A} \cdot \mathbf{B} = \begin{bmatrix} (1*0+2*3) & (1*1+2*4) & (1*2+2*5) \\ (3*0+4*3) & (3*1+4*4) & (3*2+4*5) \\ (5*0+6*3) & (5*1+6*4) & (5*2+6*5) \end{bmatrix}$$

$$= \begin{bmatrix} 6 & 9 & 12 \\ 12 & 19 & 26 \\ 18 & 29 & 40 \end{bmatrix}$$

As illustrated, multiplication of a 3×2 and 2×3 matrix results in a 3×3 product. Multiplying a $n \times n$ *square* matrix with a $n \times 1$ vector results in another $n \times 1$ vector,

$$\begin{bmatrix} a_{0,0} & a_{0,1} & \cdots & a_{0,n-1} \\ a_{1,0} & a_{1,1} & \cdots & a_{1,n-1} \\ \vdots & \vdots & \ddots & \vdots \\ a_{n-1,0} & a_{n-1,1} & \cdots & a_{n-1,n-1} \end{bmatrix} \begin{bmatrix} x_0 \\ x_1 \\ \vdots \\ x_{n-1} \end{bmatrix} = \begin{bmatrix} b_0 \\ b_1 \\ \vdots \\ b_{n-1} \end{bmatrix} \quad (2.62)$$

Writing out the terms, we obtain

$$a_{0,0}x_0 + a_{0,1}x_1 + \cdots a_{0,n-1}x_{n-1} = b_0 \quad (2.63)$$
$$a_{1,0}x_0 + a_{1,1}x_1 + \cdots a_{1,n-1}x_{n-1} = b_1 \quad (2.64)$$

$$\vdots$$

$$a_{n-1,0}x_0 + a_{n-1,1}x_1 + \cdots a_{n-1,n-1}x_{n-1} = b_{n-1} \quad (2.65)$$

This is just a linear system as discussed in Equation 2.27. The system is linear since each equation is composed only of a combination of scalar multiples of components of $\vec{x}$. There are no higher-order terms such as x_i^2 or $x_1 x_2$. The system in Equation 2.62 can written in a concise form by utilizing matrix notation as

$$\mathbf{A}\vec{x} = \vec{b} \quad (2.66)$$

2.2.6   MATRIX INVERSE

Typically, the coefficients of $\mathbf{A}$ and $\vec{b}$ are known, but values of $\vec{x}$ need to be solved for. If instead of the matrix $\mathbf{A}$ we had a scalar coefficient a, the above equation could be solved for x through

$$\left(\frac{1}{a}\right) ax = \left(\frac{1}{a}\right) b \quad (2.67)$$

or

$$1x = \left(\frac{1}{a}\right) b \quad (2.68)$$

A similar concept exists for matrices. There may exist an *inverse* matrix $\mathbf{A}^{-1}$ for which

$$\mathbf{A}^{-1}\mathbf{A} = \mathbf{I} \quad (2.69)$$

where $\mathbf{I}$ is the *identity matrix*,

$$\mathbf{I} = \begin{bmatrix} 1 & 0 & 0 & \cdots & 0 \\ 0 & 1 & 0 & \cdots & 0 \\ 0 & 0 & 1 & \cdots & 0 \\ \vdots & \vdots & \vdots & \ddots & \vdots \\ 0 & 0 & 0 & \cdots & 1 \end{bmatrix} \tag{2.70}$$

This matrix serves the same purpose as the number 1 does in scalar math: multiplying by it has no impact on the result:

$$\mathbf{I}\vec{x} = \vec{x} \tag{2.71}$$

Solution to the linear system $\mathbf{A}\vec{x} = \vec{b}$ could thus be obtained by multiplying both sides by the inverse,

$$\mathbf{A}^{-1}\mathbf{A}\vec{x} = \mathbf{A}^{-1}\vec{b} \tag{2.72}$$

or

$$\vec{x} = \mathbf{A}^{-1}\vec{b} \tag{2.73}$$

To illustrate, let's consider the following system

$$\begin{bmatrix} 1 & 2 & 0 \\ 0 & 2 & 1 \\ 1 & 0 & 1 \end{bmatrix} \cdot \begin{bmatrix} x_0 \\ x_1 \\ x_2 \end{bmatrix} = \begin{bmatrix} 2 \\ 4 \\ 2 \end{bmatrix} \tag{2.74}$$

The inverse of the matrix is given by

$$\mathbf{A}^{-1} = \begin{bmatrix} 1/2 & -1/2 & 1/2 \\ 1/4 & 1/4 & -1/4 \\ -1/2 & 1/2 & 1/2 \end{bmatrix} \tag{2.75}$$

Left-multiplying both sides of Equation 2.74 by this matrix yields,

$$\begin{bmatrix} 1/2 & -1/2 & 1/2 \\ 1/4 & 1/4 & -1/4 \\ -1/2 & 1/2 & 1/2 \end{bmatrix} \begin{bmatrix} 1 & 2 & 0 \\ 0 & 2 & 1 \\ 1 & 0 & 1 \end{bmatrix} \cdot \begin{bmatrix} x_0 \\ x_1 \\ x_2 \end{bmatrix} =$$
$$\begin{bmatrix} 1/2 & -1/2 & 1/2 \\ 1/4 & 1/4 & -1/4 \\ -1/2 & 1/2 & 1/2 \end{bmatrix} \begin{bmatrix} 2 \\ 4 \\ 2 \end{bmatrix} \tag{2.76}$$

or

$$\begin{bmatrix} 1 & 0 & 0 \\ 0 & 1 & 0 \\ 0 & 0 & 1 \end{bmatrix} \cdot \begin{bmatrix} x_0 \\ x_1 \\ x_2 \end{bmatrix} \equiv \begin{bmatrix} x_0 \\ x_1 \\ x_2 \end{bmatrix} = \begin{bmatrix} 0 \\ 1 \\ 2 \end{bmatrix} \tag{2.77}$$

which is the same solution as what was obtained by Gaussian Elimination in Section 2.2.3.

2.2.7 MATRIX TYPES

Matrices can be classified based on their dimensions and coefficients. An $n_r \times n_c$ matrix is *square* if the number of rows and columns is equal, $n_r = n_c$. Matrix with zeros everywhere except for ones on the diagonal, $a_{ij} = \delta_{ij}$ (using the *Kronecker delta* notation, $\delta_{ij} = 1$ if $i = j$, zero otherwise) is an *identity matrix*, typically written as **I**. A matrix is *symmetric* if $a_{ij} = a_{ji}$ for every i and j

A square matrix is *invertible* if the inverse $\mathbf{A}^{-1}$ exists. Otherwise, the matrix is *singular*. The system $\mathbf{A}\vec{x} = \vec{b}$ has no solution if $\mathbf{A}$ is singular. This implies that we have more unknowns than equations due to one or more rows being linear combinations of others. Such a matrix does not have full *rank* and is said to be *degenerate*. For example,

$$\mathbf{S} = \begin{bmatrix} 0 & 1 & 2 \\ 2 & 3 & 4 \\ 2 & 5 & 9 \end{bmatrix}$$

is singular due to the last row being a linear combination of rows 0 and 1, $L2 = 2L0 + L1$. Despite having three equations, only two of them provide unique information about the three unknowns.

The identity matrix also illustrates two other matrix types. It is *sparse* since the majority of coefficients are zero. The counterpart of a sparse matrix is a *dense* matrix. It is also *banded*, since non-zero coefficients are limited to one or more diagonal bands. A *tridiagonal* matrix has all non-zeros limited to the main diagonal and the two diagonals immediately to the left and right of it. As will be shown soon, in engineering simulations we often encounter matrices with the non-zero coefficients limited to 5 or 7 bands not immediately adjacent to each other. Such matrixes are often called *pentadiagonal* or *septadiagonal* however this notation is meant to be used for matrixes in which the bands are adjacent, which is not the case here. A matrix is *diagonally-dominant* if the magnitude of the coefficient on the main diagonal is larger (or at least equal) to that of any other coefficient, $|a_{ii}| > |a_{ij}|$ for any $j \neq i$. Finally, a matrix is *positive definite* if for any non-zero vector $\vec{x}$, $\vec{x}^T \mathbf{A}\vec{x} > 0$. These two last definitions are important when selecting a solution method since some algorithms converge only for matrices exhibiting these properties.

Below is an example of a diagonally-dominant tridiagonal matrix

$$\mathbf{A} = \begin{bmatrix} 2 & 1 & 0 & 0 \\ 0.1 & -5 & 0.2 & 0 \\ 0 & 0.2 & 3 & -2 \\ 0 & 0 & -0.3 & 1 \end{bmatrix}$$

And here is an example of a banded matrix with 5 bands. The a through e entries denote coefficients with possibly non-zero values. All other coefficients are zero.

$$\mathbf{A} = \begin{bmatrix} c & d & 0 & 0 & 0 & e & 0 & 0 & 0 \\ b & c & d & 0 & 0 & 0 & e & 0 & 0 \\ 0 & b & c & d & 0 & 0 & 0 & e & 0 \\ 0 & 0 & b & c & d & 0 & 0 & 0 & e \\ 0 & 0 & 0 & b & c & d & 0 & 0 & 0 \\ a & 0 & 0 & 0 & b & c & d & 0 & 0 \\ 0 & a & 0 & 0 & 0 & b & c & d & 0 \\ 0 & 0 & a & 0 & 0 & 0 & b & c & d \\ 0 & 0 & 0 & a & 0 & 0 & 0 & b & c \end{bmatrix}$$

In the above example, the coefficients on each row do not need to be the identical. However, a banded matrix with uniform coefficients is called the *Toeplitz matrix*.

2.3 MATRIX SOLVERS

Unfortunately, computing the matrix inverse is prohibitively slow for large matrices and is not computationally effective for real-world engineering problems. This is where *matrix solvers* come in. They are algorithms that obtain the solution $\vec{x}$ to a linear system $\mathbf{A}\vec{x} = \vec{b}$ without computing the matrix inverse explicitly. You may already be familiar with the MATLAB "backslash":

```
1  A = [1,2,0;0,2,1;1,0,1];
2  x = [0;1;2];
3  b = A*x          % compute b
4  x_solved = A\b   % solve back for x
```

Outside of using MATLAB, we have three options. First, we could use a language such as Python that provides matrix support. The NumPy package introduces many linear algebra routines, with naming conventions often following that of MATLAB. The above example can be written in Python as

```
1  import numpy as np
2  A = np.array([[1,2,0],[0,2,1],[1,0,1]])
3  x = np.array([0,1,2])
4  b = np.dot(A,x)   # matrix-vector multiplication
5  x_solved = np.linalg.solve(A,b) # solve Ax=b for x
6  print(x,x_solved)
```

Note that we are using **numpy.dot** to perform the matrix-vector multiplication and **numpy.linalg.solve** to solve the system.

In languages such as C++ or Java that do not implement matrix math as part of their standard library, we can take advantage of external mathematical libraries such as BLAS/LAPACK or PETSc. These libraries are discussed in Chapter 8. Just to demonstrate the approach, a LAPACK version of the above example implemented in C++ reads

```
1  #include <iostream>
2  #include <lapacke.h>
3  using namespace std;
4
```

```
5   int main(int argc, char *argv[]) {
6      constexpr lapack_int n = 3;
7      lapack_int info, nrhs = 1;
8      double A[n*n] = {1,2,0,0,2,1,1,0,1};
9      double b[n] = {0,1,2};
10     int ipiv[n];
11
12     // (d)ouble precision (ge)neral matrix (s)ol(v)e of A*x=b
13     info = LAPACKE_dgesv(LAPACK_ROW_MAJOR,n,nrhs,A,n,ipiv,b,n);
14
15     cout<<"b = [";
16     for (int i=0;i<n;i++) cout<<" "<<b[i]<<" ";
17     cout<<"]"<<endl;
18
19     return 0;
20  }
```

We can also write our own solver. Implementations in libraries such as LAPACK are highly optimized, and as such, lead to superior performance compared to a quickly built in-house version. However, sometimes such an in-house implementation is necessary if, for instance, the target system does not contain the needed libraries. Matrix solvers can be divided into two types: direct and iterative. *Direct solvers* produce the exact solution (assuming there is one) following an algorithm consisting of a prescribed sequence of steps that is independent of $\vec{b}$. The Gaussian elimination method is one example of a direct solver. *Iterative solvers* obtain an approximate solution $\mathbf{A}\vec{x'} = \vec{b} + \vec{R}$, where $\vec{R}$ is a residue vector after each pass (iteration). Given a large number of iterations, the solution ideally converges to the true solution, $\vec{x'} \to \vec{x}$, and $\vec{R} \to 0$.

2.4 SOLVER ALGORITHMS

2.4.1 GAUSS-JORDAN ELIMINATION

We are now ready to illustrate several solver algorithms. We have already seen how to use the Gauss-Jordan elimination on a simple system in Section 2.2.3. Let's now consider the system described in Equation 2.26, with $T_L = T_R = 0$ and $b = -100$. For $x_m - x_0 = 1$ m and 5 cells, $\Delta x = 0.2$, and hence $1/\Delta^2 x = 25$. Utilizing matrix notation, we have

$$
\begin{bmatrix}
1 & 0 & 0 & 0 & 0 & 0 \\
25 & -50 & 25 & 0 & 0 & 0 \\
0 & 25 & -50 & 25 & 0 & 0 \\
0 & 0 & 25 & -50 & 25 & 0 \\
0 & 0 & 0 & 25 & -50 & 25 \\
0 & 0 & 0 & 0 & 0 & 1
\end{bmatrix}
\begin{bmatrix}
T_0 \\
T_1 \\
T_2 \\
T_3 \\
T_4 \\
T_5
\end{bmatrix}
=
\begin{bmatrix}
0 \\
-100 \\
-100 \\
-100 \\
-100 \\
0
\end{bmatrix}
\tag{2.78}
$$

The Gauss-Jordan elimination operates on both sides of the equation and thus it is customary to combine the $\mathbf{A}$ matrix and the vector $\vec{b}$ into a single *augmented* matrix $[\mathbf{A}|\vec{b}]$,

$$
\begin{bmatrix}
1 & 0 & 0 & 0 & 0 & 0 & 0 \\
1 & -2 & 1 & 0 & 0 & 0 & -4 \\
0 & 1 & -2 & 1 & 0 & 0 & -4 \\
0 & 0 & 1 & -2 & 1 & 0 & -4 \\
0 & 0 & 0 & 1 & -2 & 1 & -4 \\
0 & 0 & 0 & 0 & 0 & 1 & 0
\end{bmatrix}
\tag{2.79}
$$

Rows 1 through $n_r - 2$ were simplified by dividing both sides by 25. This system is almost already in the required echelon form. As such, *pivoting*, or rearranging rows, is not necessary. We simply need to eliminate the $a_{r-1,r}$ coefficient for each row $r > 0$. For $r = 1$ we have

$$
L1 - 1 * L0 \rightarrow L1
\tag{2.80}
$$

leading to

$$
\begin{bmatrix}
1 & 0 & 0 & 0 & 0 & 0 & 0 \\
0 & -2 & 1 & 0 & 0 & 0 & -4 \\
0 & 1 & -2 & 1 & 0 & 0 & -4 \\
0 & 0 & 1 & -2 & 1 & 0 & -4 \\
0 & 0 & 0 & 1 & -2 & 1 & -4 \\
0 & 0 & 0 & 0 & 0 & 1 & 0
\end{bmatrix}
\tag{2.81}
$$

Next, we eliminate $a_{2,1}$ via

$$
L2 + 0.5 * L1 \rightarrow L2
\tag{2.82}
$$

giving us

$$
\begin{bmatrix}
1 & 0 & 0 & 0 & 0 & 0 & 0 \\
0 & -2 & 1 & 0 & 0 & 0 & -4 \\
0 & 0 & -1.5 & 1 & 0 & 0 & -6 \\
0 & 0 & 1 & -2 & 1 & 0 & -4 \\
0 & 0 & 0 & 1 & -2 & 1 & -4 \\
0 & 0 & 0 & 0 & 0 & 1 & 0
\end{bmatrix}
\tag{2.83}
$$

We next use

$$
L_3 + (1/1.5) * L_2 \rightarrow L_3
\tag{2.84}
$$

and

$$
L_4 + (3/4) * L_3 \rightarrow L_4
\tag{2.85}
$$

to obtain the upper triangular augmented matrix

$$
\begin{bmatrix}
1 & 0 & 0 & 0 & 0 & 0 & 0 \\
0 & -2 & 1 & 0 & 0 & 0 & -4 \\
0 & 0 & -1.5 & 1 & 0 & 0 & -6 \\
0 & 0 & 0 & -1.33 & 1 & 0 & -8 \\
0 & 0 & 0 & 0 & -1.25 & 1 & -10 \\
0 & 0 & 0 & 0 & 0 & 1 & 0
\end{bmatrix}
\tag{2.86}
$$

Here we already see that $T_5 = 0$. Next from $-1.25T_4 + 1T_5 = -10$, we have $T_4 = 8$. Continuing moving upward, we arrive at the solution

$$T = [0, 8, 12, 12, 8, 0] \tag{2.87}$$

We can confirm that this is indeed the correct answer by computing the analytical solution of

$$\frac{d^2T}{dx^2} = -100 \tag{2.88}$$

Integrating twice leads to

$$T = -50x^2 + Ax + B \tag{2.89}$$

The integration constants are obtained from the boundary conditions. At $x = 0$, $T = 0$, and hence $B = 0$. Similarly, $T(1) = 0$, and hence $A = 50$. The analytical solution is given by

$$T = -50x^2 + 50x \tag{2.90}$$

At the position of the second node, $x_1 = \Delta x = 0.2$, $T_1 = T(x_1) = 8$. Similarly, for $x_2 = 2\Delta x = 0.4$, $T_2 = 12$, matching our numerical approximation. As a final note, Gauss-Jordan elimination can also be used to compute the matrix inverse by starting from the augmented matrix $[\mathbf{A}|\mathbf{I}]$ to obtain $[\mathbf{I}|\mathbf{A}^{-1}]$.

2.4.2 PYTHON IMPLEMENTATION

While doing this elimination "by hand" may be feasible for the small 6×6 matrix, it becomes quickly intractable as the number of unknowns grows. As can be seen, eliminating the coefficient $a_{r-1,r}$ requires using

$$L_r - \left(\frac{a_{r-1,r}}{a_{r-1,r-1}} \right) L_{r-1} \rightarrow L_r \tag{2.91}$$

We can write a Python algorithm to perform this elimination automatically. Note that we do not include "pivoting" (row and column re-ordering), as it is not needed for the matrix at hand. However, for some matrices, this pivoting procedure can improve accuracy by avoiding floating point operations with operands of vastly different magnitudes can even become necessary to avoid division by zero errors. Chapter 2 of Bewley 2018 provides an example of algorithms that incorporate partial or complete pivoting. Numerical libraries such as LAPACK also already implement pivoting.

```
1  import numpy as np
2  import matplotlib.pyplot as plt
3
4  def solveGJ(A,b):
5      # make augmented matrix
6      G = np.zeros((ni,ni+1))      # one extra column
7      G[:,:-1] = A                 # copy A into [0:ni-1,0:ni-1]
```

```
8    G[: , −1] = b                    # copy b into the last column
9
10   # perform G–J elimination
11   for r in range(1, ni−1):
12       # show the elimination rule
13       #print("L%d + %gL%d −> L%d"%(r,−(G[r,r−1]/G[r−1,r−1]),r−1,r))
14
15       # eliminate coefficients
16       G[r ,:] = G[r ,:] − (G[r,r−1]/G[r−1,r−1])*G[r−1,:]
17
18       # optionally display the matrix
19       if (False):
20           for r in range(ni):
21               for c in range(ni+1):
22                   print("%6.2f"%G[r,c],end=' ')
23               print()
24           print()
25
26   # solution vector
27   x = np.zeros(ni)
28
29   # back substitution, loop from r=ni−1 to r=0
30   for r in range(ni−1,−1,−1):
31       s = 0  # compute sum a[r,r+1:]*T[r+1:]
32       for c in range(r+1,ni):
33           s += G[r,c]*x[c]
34       # get value for T_gj[r]
35       x[r] = (G[r,ni]−s)/G[r,r]
36   return x
37
38   L = 1              # domain size
39   ni = 16            # number of nodes
40   dx = L/(ni−1)      # cell spacing
41   b0 = −100
42
43   A = np.zeros((ni,ni))  # ni*ni empty matrix
44   b = np.zeros(ni)       # RHS vector
45
46   A[0,0] = 1         # left boundary
47   b[0] = 0
48   A[ni−1,ni−1] = 1 # right boundary
49   b[ni−1] = 0
50   for r in range(1,ni−1):
51       A[r,r−1] = 1/(dx*dx)   # coefficient for T[r−1]
52       A[r,r] = −2/(dx*dx)    # coefficient for T[r]
53       A[r,r+1] = 1/(dx*dx)   # coefficient for T[r+1]
54       b[r] = b0              # forcing vector
55
56   T_gj = solveGJ(A,b)
57
58   # plot solution
59   x_gj = np.linspace(0,L,ni)
60   plt.figure(figsize=(8,4))
61
62   x = np.linspace(0,L,100)
63   plt.plot(x,0.5*b0*x*x−0.5*b0*x,color=(0.2,0.2,0.2),
```

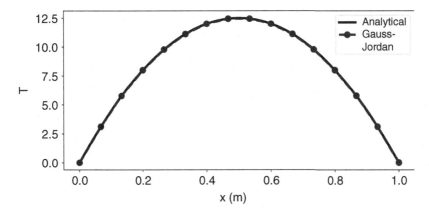

Figure 2.3 Gauss-Jordan solution compared to the analytical relationship.

```
64                linewidth=4, label='Analytical')
65
66   plt.plot(x_gj,T_gj,label='Gauss-Jordan',
67                color="0.3",
68                linewidth=4, linestyle='--',
69                marker='o', markersize=10)
70
71   plt.legend()
72   plt.xlabel('x (m)')
73   plt.ylabel('T')
74   plt.savefig('gj.png',dpi=300)
```

The actual Gauss-Jordan elimination is performed on line 16. We utilize NumPy vector operation to operate on an entire row at a time. The back-substitution begins line 29. For each row starting from the bottom, we compute the dot product of the terms to the right of the diagonal,

$$s_r = \sum_{i=r+1}^{n_r} a_{r,i} \cdot T_i \qquad (2.92)$$

Solution for T_r is then given by

$$T_r = (b_r - s_r)/a_{r,r} \qquad (2.93)$$

For $n_r = 16$, we obtain the plot in Figure 2.3. An excellent agreement can be seen. This file is saved using the .png raster image format. Alternatively, the image can be saved as vector graphics using

```
1    plt.savefig('gj.pdf')
```

2.4.3 TRIDIAGONAL ALGORITHM

While useful to demonstrate a solution technique, the Gauss-Jordan elimination algorithm is not commonly used in engineering codes. The algorithm has a computational complexity of $O(n^3)$, implying that the number of operations scales with n^3 where n is the number of matrix rows. Increasing the number of unknowns $10\times$ increases the operation count $1,000\times$. This quickly becomes prohibitive for large systems. Furthermore, matrices that arise in engineering analyses tend to be highly sparse. The G-J algorithm reduces the sparsity, thus requiring additional memory storage and computational effort.

However, one notable exception is the tridiagonal system as demonstrated in Section 2.4.1. A tridiagonal matrix can be stored efficiently by storing only the coefficients for the three non-zero diagonal bands,

$$
\begin{bmatrix}
d_0 & e_0 & 0 & 0 & 0 & 0 \\
c_1 & d_1 & e_0 & 0 & 0 & 0 \\
0 & c_2 & d_2 & e_2 & 0 & 0 \\
0 & 0 & c_3 & d_3 & e_3 & 0 \\
0 & 0 & 0 & c_4 & d_4 & e_4 \\
0 & 0 & 0 & 0 & c_5 & d_5
\end{bmatrix}
\begin{bmatrix}
x_0 \\ x_1 \\ x_2 \\ x_3 \\ x_4 \\ x_5
\end{bmatrix}
=
\begin{bmatrix}
y_0 \\ y_1 \\ y_2 \\ y_3 \\ y_4 \\ y_5
\end{bmatrix}
\tag{2.94}
$$

As shown, d_i, c_i, and e_i are the coefficients for row i on the main diagonal, immediately to the left of it, and immediately to the right. These are the only possible locations for the non-zero coefficients. Given the coefficient matrix $\mathbf{A}$ described by vectors $\vec{c}$, $\vec{d}$, and $\vec{e}$, and a forcing vector $\vec{y}$, the solution is obtained using the *Thomas* (or tridiagonal) algorithm:

{forward sweep}
$e_0/d_0 \to e_0$
$y_0/d_0 \to y_0$
for $i \in [1, n-1]$ **do**
    $e_i / (d_i - c_i d_{i-1}) \to e_i$
    $(y_i - c_i y_{i-1}) / (d_i - c_i d_{i-1}) \to y_i$
end for
{back substitution}
$y_{n-1} \to x_{n-1}$
for $i \in [n-2, 0]$ **do**
    $y_i - e_i x_{i+1} \to x_i$
end for

This algorithm arises from applying Gauss-Jordan elimination to a tridiagonal system. The listing below illustrates an implementation in Python:

```
1  # solves tridiagonal system A=[c/d/e]*x = y
2  def triDiagSolve(c, d, e, y, x):
3      n = len(e)    # vector length
4      # forward sweep
5      e[0] = e[0]/d[0]
6      y[0] = y[0]/d[0]
7      for i in range(1,n-1):
8          e[i] = e[i]/(d[i] - c[i]*e[i-1])
9          y[i] = (y[i] - c[i]*y[i-1])/(d[i]-c[i]*e[i-1])
```

```
10
11     # back substitution
12     x[n-1] = y[n-1]
13     for i in range(n-2,0,-1):
14         x[i] = y[i] - e[i]*x[i+1]
```

2.5 ITERATIVE SOLVERS

Thomas Algorithm is an example of a *direct solver*. After a single forward sweep and a single back substitution, we have the exact solution. This can be contrasted with *iterative solvers* that produce only an approximate solution after each solver iteration. At iteration n, we have

$$\mathbf{A}\vec{x}^n = \vec{b}^n + \vec{r}^n \tag{2.95}$$

where $\vec{r}$ is the *residue*. It represents the error between the approximate and the true solution, $\vec{r} = \mathbf{A}\vec{x} - \vec{b}$. If $\vec{x}^n$ were the true solution $\vec{x}^n = \vec{x}_{true}$, then $\vec{r}^n = 0$. The magnitude of a vector $|\vec{r}|$ is called the *norm*. We expect that as $n \to \infty$, $|\vec{r}^n| \to 0$. In other words, the norm of the residue decreases with additional solver iterations. The most common l^2 (L2) norm is

$$|\vec{r}| = \sqrt{\sum_{i=0}^{n_r-1} r_i^2} \tag{2.96}$$

However, to make the calculation agnostic of the grid size, in numerical simulations we typically utilize the node-averaged value,

$$\overline{|\vec{r}|} = \sqrt{\frac{\sum_{i=0}^{n_r-1} r_i^2}{n_r}} \tag{2.97}$$

2.5.1 JACOBI METHOD

Consider the system from Equation 2.26. On any internal row, we have

$$x_{i-1} - 2x_i + x_{i+1} = b_i/\alpha \tag{2.98}$$

This matrix is *diagonally-dominant*. The equation on the left-hand side is most strongly affected by the x_i term since $|a_{ii}| > |a_{ij}|$ for any other j. Therefore, perhaps we can obtain a new approximation for the unknown by solving Equation 2.98 for it, $x_i^{n+1} = b_i/\alpha - x_{i-1}^n - x_{i+1}^n$ or in general

$$x_i^{n+1} = \frac{b_i - \sum_j \left[a_{ij} x_j^n (1 - \delta_{ij}) \right]}{a_{ii}} \tag{2.99}$$

$\delta_{ij} = 1$ if $i = j$, 0 otherwise. Alternatively, using matrix notation we have

$$\vec{x}^{n+1} = (\mathbf{IA})^{-1} \left[\vec{b}^n - (\mathbf{A} - \mathbf{I})\vec{x}^n \right] \tag{2.100}$$

This method is known as Jacobi Iteration Method, and is named after a 19th century mathematician Carl Gustav Jacob Jacobi. It is very simple to implement. We repeatedly loop over all matrix rows, and for each, evaluate Equation 2.99 to compute the new estimate for x_i. After completing the pass, the new vector $\vec{x}^{n+1}$ is "copied down" to serve as the "old" data $\vec{x}^n$ in the following iteration. Below is a version in Python operating on a full (dense) matrix. This code can be substituted into the listing on page 61.

```
1   def solveJacobi(A, b):
2       nr = A.shape[0]    # number of rows
3       x = np.zeros(nr)
4       x_new = np.zeros(nr)
5       # repeat 1000 times
6       for it in range(1000):
7           for r in range(nr):
8               x_new[r] = (b[r] - np.dot(A[r,:],x) + A[r,r]*x[r])/A[r,r]
9           x[:] = x_new[:]    # copy down solution
10      return x
11
12  T_j = solveJacobi(A,b)    # call the solver
```

2.5.2   GAUSS-SEIDEL METHOD

In the Jacobi method, we store the new approximations for $\vec{x}$ in a separate $\vec{x}_{new}$ vector, which is then "copied down" at the end of each iteration. But what if we just replace the values in place? This gives us the *Gauss-Seidel Method* (G-S). We iterate through the matrix rows in a consecutive order. As such, for row r, the solution vector already contains the new values for $i \in [0, r-1]$. We can write this as

$$x_i^{n+1} = \frac{b_i - \sum_{r=0}^{i-1} a_{ir} x_r^{n+1} - \sum_{r=i+1}^{n_r} a_{ir} x_r^n}{a_{ii}} \qquad (2.101)$$

The Gauss-Seidel Method is even easier to implement than Jacobi. We simply eliminate any use of the **x_new** vector and use **x** directly

```
1   def solveGS(A, b):
2       nr = A.shape[0]    # number of rows
3       x = np.zeros(nr)
4       # repeate 1000 times
5       for it in range(1000):
6           for r in range(nr):
7               x[r] = (b[r] - np.dot(A[r,:],x) + A[r,r]*x[r])/A[r,r]
8       return x
```

2.5.3   CONVERGENCE CHECK

The solver runs for 1000 iterations. This limit was selected arbitrarily. We can visualize the behavior of the solver by plotting the solution after an increasingly higher number of iterations. This is shown in Figure 2.4.

As the number of solver iterations increases, the produced solution approaches the true answer. This suggests that we should use a large number of

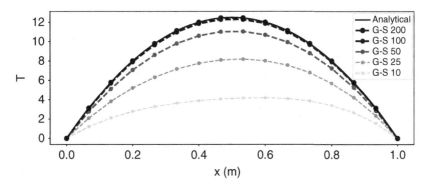

Figure 2.4 Iterative solver convergence toward the analytical solution.

iterations to ensure that the correct answer is produced. While this is certainly possible, it is not computationally efficient, as the solver will continue to run long after the (sufficiently) correct answer is produced. Instead, we implement a *convergence check* by comparing the average norm of the *residue* to some specified tolerance,

---

for n = 1:num_iterations **do**
    compute new solution x
    compute r = b - Ax
    if ave_norm(r)<tolerance **then**
        exit solver loop
    end if
end for

Since computing the norm takes CPU cycles, we generally do not check for convergence at every iteration. An implementation that performs a convergence check once every 25 solver iterations looks as follows:

```
1   def solveGS(A, b):
2       nr = A.shape[0]      # number of rows
3       x = np.zeros(nr)
4       for it in range(1000):
5           for r in range(nr):
6               x[r] = (b[r] - np.dot(A[r,:],x) + A[r,r]*x[r])/A[r,r]
7
8           # convergence check every 25 steps
9           if it%25==0:
10              r = b-np.dot(A,x)     # r = b-A*x
11              s = np.dot(r,r)       # s = sum(r[i]*r[i])
12              norm = np.sqrt(s/nr)
13              print("%d: %.2g"%(it,norm))
14              if (norm<1e-6): break  # exit for loop
15
16      return x
```

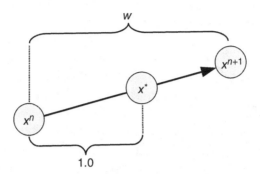

Figure 2.5 Use of the SOR method to predict the future value of x.

This check can also be included in the Jacobi algorithm. Not only is Gauss-Seidel simpler to implement, it also converges faster. For $n = 16$, $|\vec{r}| \leq 10^{-6}$ is reached after 827 iterations with Jacobi, but after only 414 iterations with Gauss-Seidel.

2.5.4 SUCCESSIVE OVER RELAXATION (SOR)

With Jacobi or G-S, we obtain a new estimate for each x_i. Let's temporarily call this new value x_i^*. We also know the "old" value, x_i^n. Therefore, given the change from x_i^n to x_i^* in a single iteration, perhaps we can obtain a better estimate for x_i^{n+1} by extrapolating the rate of change? This is depicted visually in Figure 2.5. From similar triangles,

$$\frac{x^* - x^n}{1.0} = \frac{x^{n+1} - x^n}{w} \tag{2.102}$$

or

$$x^{n+1} = x^n + w(x^* - x^n) \tag{2.103}$$

This correction is known as *Successive Over Relaxation* (or SOR) and can be implemented as follows:

```
1  # in loop over matrix rows
2  x_star = (b[r] - np.dot(A[r,:],x) + A[r,r]*x[r])/A[r,r]
3  x[r] = x[r] + w*(x_star-x[r])
```

It is widely used for accelerating the convergence of Jacobi and Gauss-Seidel algorithms. In fact, it would be unusual to find a GS solver that does not utilize SOR. With the commonly used $w = 1.4$, the GS-SOR solver converges in 170 iterations. SOR is however not a panacea. If the initial vector $\vec{x}$ is not a close approximation to the true solution, the SOR scheme can predict the movement of the solution vector in the incorrect direction. When coupled with a non-linear forcing vector, $\vec{b}(\vec{x})$, this can lead to *divergence*. In that case, we may need to utilize $w < 1$ *under-relaxation* to help stabilize the solver until the solution vector reaches the vicinity of the true value. We can then switch back to over-relaxation to accelerate the final convergence.

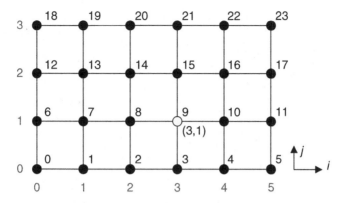

Figure 2.6 Two-dimensional Cartesian grid indexing.

2.6 2D HEAT EQUATION SOLVER

Let's now return to the original heat equation, $\nabla^2 \vec{T} = \vec{b}$. Our previous formulation assumed that there is no variation in the y and z direction, leading to a 1D problem. We can similarly assume that there is no variation in z, which could approximate the solution in a very deep slab. We then have

$$2\text{D:}\quad \frac{\partial^2 T}{\partial x^2} + \frac{\partial^2 T}{\partial y^2} = b \tag{2.104}$$

In three dimensions all three components are present,

$$3\text{D:}\quad \frac{\partial^2 T}{\partial x^2} + \frac{\partial^2 T}{\partial y^2} + \frac{\partial^2 T}{\partial z^2} = b \tag{2.105}$$

Following the same approach used for the 1D problem, we *discretize* the physical domain into a computational grid. Example of such a grid used for a two-dimensional problem is shown in Figure 2.6. This Cartesian mesh consists of a collection of nodes at regular intervals. These nodes represent locations at which the solution will be known. This discretized approach is necessary in order to capture the continuous physical domain using a finite number of values that can be stored in computer memory. Each grid node is referenced by its i and j coordinate, $T_{i,j}$. Consider the highlighted node $T_{3,1}$. It is located at the intersection of $i = 3$ and $j = 1$ grid lines. Further, utilizing Δx and Δy for the node spacing, the position of any node can be calculated from

$$x = x_0 + i\Delta x \tag{2.106}$$
$$y = y_0 + j\Delta y \tag{2.107}$$

with similar equation existing for z. The x_0 and y_0 are the position of the origin, which is the grid node at $i = 0$ and $j = 0$.

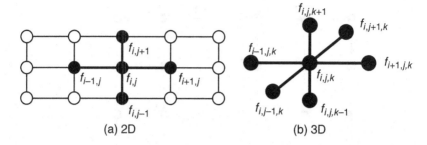

Figure 2.7 Computational stencil for the 2D and 3D Laplace operator.

Next, applying the standard central difference, Equation 2.14, to both terms in Equation 2.104 leads to

$$\frac{T(x - \Delta x, y) - 2T(x,y) + T(x + \Delta x)}{\Delta^2 x} +$$
$$\frac{T(x, y - \Delta y) - 2T(x,y) + T(x, y + \Delta y)}{\Delta^2 y} = b(x,y) \qquad (2.108)$$

Again, we note that all referenced T's coincide with grid nodes. Therefore, letting $T(x,y) = T_{i,j}$, the above equation can be rewritten as

$$\frac{T_{i-1,j} - 2T_{i,j} + T_{i+1,j}}{\Delta^2 x} \frac{T_{i,j-1} - 2T_{i,j} + T_{i,j+1}}{\Delta^2 y} = b_{i,j} \qquad (2.109)$$

In three dimensions, we have a similar expression including one additional term, and values referenced by indexes i, j, and k, i.e. $T_{i,j,k}$.

The resulting computational stencil for the 2D and 3D representation of the Laplacian are visualized in Figure 2.7. We can see that in two dimensions, there are only 5 nodes that contribute to the solution. This can be contrasted with the 3 nodes in 1D, per Figure 2.2. In three dimensions, the number of nodes increases to 7. Rewriting the Laplacian in two dimensions results in a linear system in which each equation has at most 5 terms,

$$\alpha T_{i-1,j} + \alpha T_{i+1,j} + \beta T_{i,j-1} + \beta T_{i,j+1} - 2(\alpha + \beta)T_{i,j} = b_{i,j} \qquad (2.110)$$

where $\alpha = 1/\Delta^2 x$ and $\beta = 1/\Delta^2 y$. On the boundaries, we specify either the Dirichlet or the Neumann condition. For the former, we use

$$T_{0,j} = T_{left} \qquad (2.111)$$

to fix temperature on all nodes with $i = 0$. For the Neumann boundary, $\partial T / \partial \hat{n}$, we can use the first order one-sided expression for the first derivative,

$$T(y + \Delta y) = T(y) + \Delta y \frac{\partial T}{\partial y}(y) + \text{HOT} \qquad (2.112)$$

For the zero-slope condition in the y direction on the upper $j = n_j - 1$ edge, this reduces to

$$T_{i,n_j-1} - T_{i,n_j-2} = 0 \tag{2.113}$$

The above expression is only first-order accurate. We could alternatively use the second-order accurate form per Equation 2.11; however, doing so breaks the banded nature of the resulting coefficient matrix.

Next, we would like to reformat the linear system into a form that can be used with our existing matrix solvers. In order to do so, we "flatten" the multidimensional node index into a one-dimensional "unknown" (or equation) index. We use the term "unknown" loosely as this indexing also includes the known Dirichlet values. Other solution methods, such as the Finite Element Method, which are beyond the scope of this book, do indeed decompose the domain into a set of unknown and known nodes. The flattening is done by assigning consecutive indexes to the nodes first starting with the x direction, and followed by y and z. This is shown in Figure 2.6 and is given by:

$$\text{2D:} \quad U(i,j) = j \cdot n_i + i \tag{2.114}$$

$$\text{3D:} \quad U(i,j,k) = k \cdot n_i n_j + j \cdot n_i + i \tag{2.115}$$

These expressions assume that the indexing starts with 0. In one-indexed languages, such as MATLAB® or Fortran, we would instead write $u = (k - 1)n_i n_j + (j-1)n_i + i$.

With this scheme, the node at the origin has an unknown index of 0. The neighbor in the $+x$ direction has an index of 1. This numbering increases until we reach the final node in the x direction, with index $u = n_i - 1$. We then wrap around to $T_{0,1,0}$, which gets node index n_i. Finally, once the entire $k = 0$ two-dimensional plane is completed, we move onto node $T_{0,0,1}$ with index $u = n_i n_j$. Now let $u_0 = U(i,j,k)$ be the index of node i,j,k. The neighbor in the $+x$ direction, $(i+1,j,k)$ has index

$$U(i+1,j,k) = k \cdot n_i n_j + j \cdot n_i + (i+1)$$
$$= u_0 + 1 \tag{2.116}$$

Similarly, in the y and z directions, we find

$$U(i,j+1,k) = k \cdot n_i n_j + (j+1) \cdot n_i + i)$$
$$= u_0 + n_i \tag{2.117}$$

and

$$U(i,j,k+1) = (k+1) \cdot n_i n_j + j \cdot n_i + i)$$
$$= u_0 + n_i n_j \tag{2.118}$$

The stencil neighbors are separated by 1, ni, and $n_i \cdot n_j$ nodes in the x, y, and z directions, respectively. Using this one-dimensional indexing, the discretized Laplacian equation 2.110 can be written as

$$\beta T_{u-n_i} + \alpha T_{u-1} - 2(\alpha + \beta)T_u + \alpha T_{u+1} + \beta T_{u+n_i} = b_u \tag{2.119}$$

For each matrix row u, the non-zero coefficients are limited to five non-zero bands as illustrated in Equation 2.2.7. The bands are located along the main diagonal, the two bands immediately to the left and right, and another set of bands offset by n_i columns. In 3D, we obtain,

$$\gamma T_{u-n_i n_j} + \beta T_{u-n_i} + \alpha T_{u-1} - 2(\alpha + \beta + \gamma)T_u + \\ \alpha T_{u+1} + \beta T_{u+n_i} + \gamma T_{u+n_i n_j} = b_u \qquad (2.120)$$

resulting in a matrix with seven non-zero bands. The outermost bands are offset by $n_i n_j$ from the diagonal.

2.6.1 SPARSE MATRIX

Consider a 2D simulation using a grid with 10,000 nodes. This is a relatively small size since many engineering simulations use computational meshes with millions of nodes. This number corresponds to the size of the $\vec{x}$ array and hence the full matrix contains $10,000 \times 10,000 = 100 \times 10^6$ coefficients. With 8 bytes per double-precision floating point number, storing the full matrix requires about 762 Mb = 0.75 Gb of RAM. This buffer will be almost completely made up of zeros since there are at most 5 non-zero coefficients per row. On the other hand, storing only the non-zero coefficients requires just $5 \times 10,000$ data elements, or about 0.38 Mb. This is a 2000× reduction in memory use. Such sparse storage also speeds up computations since CPU cycles are not wasted by adding a huge number of zeros.

In Python, support for sparse matrixes is provided by the `scipy` module

```
1  import numpy as np
2  from scipy.sparse import csr_matrix
3  from scipy.sparse.linalg import spsolve
4
5  S = csr_matrix([[0,1,2],[2,3,4],[0,2,3]])
6  x = np.array([1, 2, 3])
7  b = S.dot(x);
8  print('b=',b)
9  print('x_solved = ',spsolve(S,b))
```

Alternatively, for a banded matrix structure, we can just store the coefficients using 3, 5, or 7 unique vectors. This was indeed already demonstrated in our example implementation of the Thomas Algorithm in Section 2.5. In the subsequent example, we label these vectors $\vec{a}$, $\vec{b}$, $\vec{c}$, $\vec{d}$, $\vec{e}$, $\vec{f}$, and $\vec{g}$. They store the coefficients for $x_{i-n_i n_j}$, x_{i-n_i}, x_{i-1}, x_i, x_{i+1}, x_{i+n_i}, and $x_{i+n_i n_j}$, respectively. We will learn how to generate a sparse matrix in C++ in the following chapter.

2.6.2 PUTTING IT ALL TOGETHER

We now demonstrate the above concepts in practice. Let's consider a 2D domain between $x_0 = (0, 0)$ and $x_m = (2.0, 1.0)$ discretized into 80 cells in the

x direction and 60 cells in the y direction. The following equations hold:

$$\nabla^2 T = -100 \quad ; \vec{x} \in \Omega \tag{2.121}$$

$$T = 1 \quad ; \vec{x} \in \Gamma_{y^-} \tag{2.122}$$

$$\partial T/\partial x = 0 \quad ; \vec{x} \in \{\Gamma_{x^-} \text{ or } \Gamma_{x^+}\} \tag{2.123}$$

$$\partial T/\partial y \quad ; \vec{x} \in \Gamma_{y^+} \tag{2.124}$$

$$T = 1 + 9\mathcal{R} \quad ; \vec{x} \in \vec{x}_R \tag{2.125}$$

where Ω denotes the region away from the boundaries, and Γ is the bounding surface on which the boundary conditions are prescribed. Essentially, we are prescribing the zero Neumann boundary condition on all boundaries except for the $j = 0$ edge. We also include several internal Dirichlet points on which we assign a random fixed temperature in order to make the solution more interesting.

Let's first consider a version utilizing SciPy's sparse matrix. We start by importing libraries providing linear algebra, sparse matrix, plotting, and random number generation support. We then define the number of grid nodes and compute Δx and Δy. We also use the `lil_matrix` function to initialize an empty sparse matrix. This is one of several available sparse matrix types. It stores the data as a "list of lists". Its primary benefit is that it supports dynamic assignment (in other words, growth) of data.

```
1   import numpy as np
2   from scipy.sparse.linalg import spsolve
3   import scipy.sparse as sp_mat
4   import matplotlib.pyplot as plt
5   from random import random as rand
6
7   ni = 80          # number of nodes in x
8   nj = 60          # number of nodes in y
9   nu = ni*nj       # total number of nodes
10  Lx = 2           # length in x
11  Ly = 1           # length in y
12  dx = Lx/(ni-1)   # cell spacing in x
13  dy = Ly/(nj-1)   # cell spacing in y
14  A = sp_mat.lil_matrix((nu,nu))
15  b = np.zeros(nu)
```

Next, we assign the matrix coefficient. To do so, we loop through all grid nodes. For each, we first compute the flattened u index per Equation 2.114. We then assign different coefficients based on the node grid coordinates. For any points along the x^- ($i = 0$), x^+ ($i = n_i - 1$), or y^+ ($j = n_j - 1$) edge we assign the Neumann zero boundary condition per Equation 2.113. Lines 6, 7, and 8 in the following listing produce $T_u - T_{u+1} = 0$. On y^- ($j = 0$), we specify $T_i = 1$. Finally, on any other (i.e. non-boundary) node, we specify the expression for the two-dimensional Laplacian per Equation 2.109.

```
1   for j in range(nj):
2       for i in range(ni):
3           u = j*ni+i       # node (row) index
4           # boundary conditions
```

```
5            if (i==0):          # Neumann on xmin
6                A[u,u]   = 1;
7                A[u,u+1] = -1;
8                b[u]     = 0.0;
9            elif (i==ni-1): # Neumann on xmax
10               A[u,u]   = 1;
11               A[u,u-1] = -1;
12               b[u]     = 0.0;
13           elif (j==0):     # Dirichlet on ymin
14               A[u,u]   = 1;
15               b[u]     = 1.0;
16           elif (j==nj-1): # Neumann on ymax
17               A[u,u]   = 1;
18               A[u,u-ni] = -1;
19               b[u]     = 0.0;
20           else:
21               A[u,u-ni] = 1/(dy*dy)
22               A[u,u-1]  = 1/(dx*dx)
23               A[u,u+1]  = 1/(dx*dx)
24               A[u,u+ni] = 1/(dy*dy)
25               A[u,u]    = - 2/(dx*dx) - 2/(dy*dy)
26               b[u]      = -10
```

Next, to make the output more interesting, we randomly sprinkle in some fixed Dirichlet nodes. We use a loop to pick 20 random i and j coordinates such that $i \in [1, n_i - 2]$ and $j \in [1, n_j - 2]$. In other words, we limit the selection to the internal, non-boundary region Ω. We clear any previously set coefficient, and proceed to set just the 1 on the diagonal to mark a Dirichlet node. We assign the corresponding b_u to a random value $\in [1, 10]$.

```
1   # assign random internal points
2   for p in range (20):
3       i = 2+int((ni-2)*rand())
4       j = 2+int((nj-2)*rand())
5       u = j*ni+i
6       A[u,:] = 0      # clear row
7       A[u,u] = 1      # Dirichlet node
8       b[u] = 1+9*rand() # random value
```

Finally we use the `spsolve` command from the `scipy.sparse.linalg` library to solve the $\mathbf{A}\vec{T} = \vec{b}$ system.

```
1   T = spsolve(A,b);      # sparse matrix solver
```

The resulting solution vector $\vec{T}$ is one-dimensional. In order to plot it, we need to reshape it back to the corresponding 2D form. This involves taking an array of consecutive node values numbered according to Figure 2.6 and converting them to the corresponding 2D grid. Python has a `reshape` function to perform this operation. Also, in order to show the spatial coordinates, we need to generate a vector of x and y values corresponding to the i and j grid node positions. Here we use the `linspace` command to get n_i or n_j uniformly spaced values:

```
1   T2d = np.reshape(T,(nj,ni))  # convert to 2D matrix
2   xs = np.linspace(0,Lx,ni)    # x-coordinates for plotting
3   ys = np.linspace(0,Ly,nj)    # y-coordinates for plotting
```

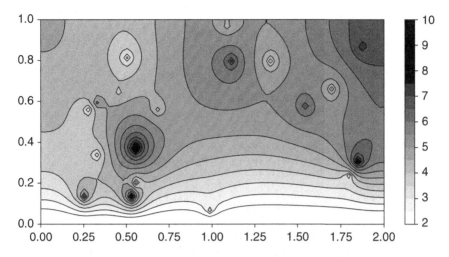

Figure 2.8 Example output from the two-dimensional heat equation solver.

Finally, we use the `contourf` and `contour` commands to plot the contour fill and the contour lines using 12 distinct levels. We also show the colorbar. The result is finally saved to a file. In this case we save the plot as .pdf vector image. Example of a possible output is shown in Figure 2.8. Note that the actual output will be different every time you run the code due to the random location of the internal fixed points.

```
1   #plot countour levels and lines
2   lv = np.linspace(2,10,17)
3   plt.contourf(xs,ys,T2d,levels=lv,cmap='gray_r')
4   plt.colorbar(ticks=np.linspace(2,10,9));
5   plt.contour(xs,ys,T2d,levels=lv,colors='black')
6   plt.tight_layout()
7   plt.savefig('heat2d.pdf')
```

2.6.3 ITERATIVE SOLVER VERSION

Now let's take a look at how the above code could be implemented if we did not have access to the NumPy sparse matrix library and its solvers, and the matrix dimensions were too fine to use a regular dense matrix. We again start by importing the needed libraries (there are fewer of them now) and defining the computational domain. Instead of using a single `A` matrix container, we utilize one-dimensional vectors to store the coefficients for the non-zero bands. We do not have the $\vec{a}$ and $\vec{h}$ bands due to operating on a 2D domain. Also, since $\vec{b}$ is used to store matrix coefficients, we use $\vec{y}$ for the right-hand side vector. Our matrix system is given by $\left[\vec{b}|\vec{c}|\vec{d}|\vec{e}|\vec{f}\right]\vec{T} = \vec{y}$. It is implemented as

```
1   import numpy as np
2   import matplotlib.pyplot as plt
```

```
3   from random import random as rand
4
5   ni = 80                 # number of nodes in x
6   nj = 60                 # number of nodes in y
7   nu = ni*nj              # total number of nodes
8   Lx = 2                  # length in x
9   Ly = 1                  # length in y
10  dx = Lx/(ni-1)          # cell spacing in x
11  dy = Ly/(nj-1)          # cell spacing in y
12
13  b = np.zeros(nu)        # coefficients for T[i,j-1]
14  c = np.zeros(nu)        # coefficients for T[i-1,j]
15  d = np.zeros(nu)        # coefficients for T[i,j]
16  e = np.zeros(nu)        # coefficients for T[i+1,j]
17  f = np.zeros(nu)        # coefficients for T[i,j+1]
18  y = np.zeros(nu)        # right hand side vector
```

Next, we develop a Gauss-Seidel solver. It is identical to the algorithm discussed in Sections 2.5.2 through 2.5.4, except that it operates on a matrix defined by 5 one-dimensional vectors. In order to compute the dot product between the matrix row and the solution vector, we need to multiply the coefficients from the vectors by the corresponding x's, i.e.:

$$A_{u,:} \cdot \vec{x} \equiv b_u T_{u-n_i} + c_u T_{u-1} + d_u T_u + e_u T_{u+1} + f_u T_{u+n_i} \qquad (2.126)$$

For any node on the $j = 0$ boundary, we have $b_u = 0$. However, trying to evaluate $0 \cdot T_{u-n_i}$ would cause a crash. On these nodes $u - n_i < 0$, and as such, we are attempting to access memory outside the allocated space. Therefore, we modify the above dot product expression to perform the multiplication only for the terms with non-zero coefficients. This is shown on line 7 in the following listing. The "nodiag" name indicates that the diagonal term is excluded from the dot product calculation. This diagonal term is retained subsequently when computing the residue.

```
1   def solveGSSOR(b,c,d,e,f,y,ni):
2       nu = b.size    # number of rows / nodes
3       x = np.zeros(nu)
4       for it in range(10000):
5           for u in range(nu):
6               # row-vector dot product minus diagonal term
7               dot_nodiag = ((b[u]*x[u-ni] if b[u]!=0 else 0) +
8                             (c[u]*x[u-1]  if c[u]!=0 else 0) +
9                             (e[u]*x[u+1]  if e[u]!=0 else 0) +
10                            (f[u]*x[u+ni] if f[u]!=0 else 0))
11              x_star = (y[u] - dot_nodiag)/d[u]
12              x[u] = x[u] + 1.4*(x_star-x[u])
13
14          # convergence check every 25 steps
15          if it%25==0:
16              s = 0
17              for u in range(nu):
18                  dot = ((b[u]*x[u-ni] if b[u]!=0 else 0) +
19                         (c[u]*x[u-1]  if c[u]!=0 else 0) +
20                         (d[u]*x[u]) +
21                         (e[u]*x[u+1]  if e[u]!=0 else 0) +
```

```
22                       (f[u]*x[u+ni] if f[u]!=0 else 0))
23              res = y[u]-dot    # r = b-A*x
24              s += res*res  # sum of r^2
25          norm = np.sqrt(s/nu)
26          print("%d: %.2g"%(it ,norm))
27          if (norm<1e-6):
28              print("%d: %.2g"%(it ,norm))
29              break  # exit for loop
30      return x
```

The rest of the code is similar to the sparse matrix version. We next assign coefficients according to the node type. However, all references to A[u,u], A[u,u+ni], and b[u] are replaced with d[u], f[u], and y[u]. For example,

```
1       # boundary conditions
2       if (i==0):         # Neumann on xmin
3           d[u] = 1;
4           e[u] = -1;
5           y[u] = 0.0;
```

The solver is then called as:

```
1   T = solveGSSOR(b,c,d,e,f,y,ni);   # GS-SOR solver
```

The results are identical (within the random variation) to those obtained previously; however, this code takes much longer to run. In fact, we can capture exactly how long the calculation takes by recording the elapsed time,

```
1   import time
2
3   t1 = time.perf_counter()
4   T = solveGSSOR(b,c,d,e,f,y,ni);   # GS-SOR solver
5   t2 = time.perf_counter()
6   print ("solver took %.4g seconds"%(t2-t1))
```

Our SOR-accelerated Gauss-Seidel iterative solver takes about 37.2 seconds, compared to only 0.0151 seconds for the built-in **spsolve** matrix solver. This is about 2500× longer. This example goes to show that whenever possible, we should take advantage of optimized library-provided matrix solver routines.

3 Numerical Analysis

We continue our discussion of numerical analysis techniques by introducing algorithms for data fitting, filtering, numerical integration, and solving non-linear systems, among others.

3.1 DATA FITTING

3.1.1 POLYNOMIAL FITS

In many scientific computing applications, the need arises to interpolate between discretely sampled data points. This can result from attempting to fit an analytical expression to experimental data, or as a building block in deriving more complex numerical methods. A first obvious approach for interpolating $n+1$-discrete points is to fit the parametric n^{th}-order polynomial that exactly coincides with the $n+1$ points.

Given the set of $i = 1...n + 1$ points, $\{x_i, y_i\}$, the coefficients of the nth order polynomial, $P_{a_i}(x) = P(x; \vec{a}) = a_0 + a_1 x + a_2 x^2 + \cdots + a_n x^n$, can be set by solving the linear system for the parametric coefficients, a_i, resulting from requiring that $P_{a_i}(x_i) = y_i$ at each of the i points. The construction of this system can be written in matrix form as

$$
\begin{bmatrix} P_{a_i}(x_0) \\ P_{a_i}(x_1) \\ \vdots \\ P_{a_i}(x_n) \end{bmatrix} = \underbrace{\begin{bmatrix} 1 & x_0 & x_0^2 & \cdots & x_0^n \\ 1 & x_1 & x_1^2 & \cdots & x_1^n \\ \vdots & \vdots & \vdots & \ddots & \vdots \\ 1 & x_n & x_n^2 & \cdots & x_n^n \end{bmatrix}}_{\mathbf{V}} \underbrace{\begin{bmatrix} a_0 \\ a_1 \\ \vdots \\ a_n \end{bmatrix}}_{\vec{a}} = \underbrace{\begin{bmatrix} y_0 \\ y_1 \\ \vdots \\ y_n \end{bmatrix}}_{\vec{y}} \tag{3.1}
$$

Solving $\mathbf{V}\vec{a} = \vec{y}$ involves inverting the Vandermonde matrix, $\mathbf{V}$, using the techniques such as Gaussian Elimination described in Section 2.2.3.

Rather than using $n+1$ points for determining the unique n^{th} order polynomial fit, if derivative information at points is available, it can be used in the polynomial fit as well. The relationship between the polynomial and its derivative is straightforward $P'_{a_i}(x) = a_1 + 2a_2 x + \cdots + n a_n x^{n-1}$ up to the nth derivative. This means that rows of the Vandermonde matrix can be replaced with derivative conditions rather than point values as long as the resulting system remains linearly independent. As an example, consider fitting the 4th order polynomial between the points (x_0, y_0) and (x_1, y_1) with slopes y'_0 and y'_1 at the left and right points respectively. This polynomial can be found by solving the following system:

DOI: 10.1201/9781003132233-3

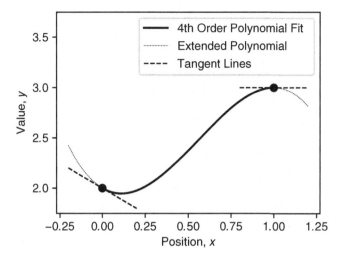

Figure 3.1 Example 4th order polynomial interpolation, $P_{a_i}(x)$, between points $(0,2)$ and $(1,3)$ with slopes -1 and 0 at the end points respectively.

$$\begin{bmatrix} P_{a_i}(x_0) \\ P_{a_i}(x_1) \\ P'_{a_i}(x_0) \\ P'_{a_i}(x_1) \end{bmatrix} = \begin{bmatrix} 1 & x_0 & x_0^2 & x_0^3 \\ 1 & x_1 & x_1^2 & x_1^3 \\ 0 & x_0 & 2x_0 & 3x_0^2 \\ 0 & x_1 & 2x_1 & 3x_1^2 \end{bmatrix} \begin{bmatrix} a_0 \\ a_1 \\ a_2 \\ a_3 \end{bmatrix} = \begin{bmatrix} y_0 \\ y_1 \\ y'_0 \\ y'_1 \end{bmatrix} \tag{3.2}$$

As an example, plugging in points $(0,2)$ and $(1,3)$ with slopes -1 and 0 respectively results coefficients $\vec{a} = [2, -1, 5, -3]$ as shown in Figure 3.1.

While directly solving the linear equations for polynomial fits is mathematically well defined, the Vandermonde matrix is often poorly conditioned making the matrix inversion sensitive to floating point round-off errors and becoming numerically unstable. To avoid this stability issue, an equivalent polynomial can be constructed directly using n^{th} order Lagrange polynomials, $L_k(x)$, designed to have zeros at every point x_i such that $i \neq k$ as:

$$L_k(x) = \alpha_k(x - x_0)(x - x_1) \cdots (x - x_{k-1})(x - x_{k+1}) \cdots (x - x_n) = \alpha_k \prod_{i=0, i \neq k}^{n} (x - x_i)$$

By construction, this polynomial is zero at every $x_{i \neq k}$. Defining α_k as the inverse of the value of the product at point x_k such that $\alpha_k = \left(\prod_{i=0, i \neq k}^{n} (x_k - x_i) \right)^{-1}$, it then can be shown that $L_k(x_i) = \delta_{ik}$. The Lagrange interpolation can then be constructed by summing the $n + 1$ interpolating polynomials, L_k, multiplied by the value of the function, y_k, as $P(x) = \sum_{k=0}^{n} y_k L_k(x)$.

Table 3.1
Interpolation Nodes for Arbitrary Example Function.

x_i	0	1	2	3	4	5	6	7
y_i	2.89	2.99	3.22	3.19	2.48	2.31	1.61	1.76

To demonstrate Lagrange interpolation, let an arbitrary function be

$$y = \log(15 + 10\sin(\pi t/4) + 3\cos(3\pi t/4)) \qquad (3.3)$$

For the interpolation examples, the values $x_i = [0,7]$ are used. The interpolation nodes can be seen in Table 3.1.

In Python, direct Lagrange interpolation can be accomplished as follows:

```
1   import matplotlib.pyplot as plt
2   import numpy as np
3   import math as math
4
5   # setup interpolation nodes
6   xi = np.arange(0,8) # [0,1,2,3,4,5,6,7]
7   yi = np.log(15+10*np.sin(math.pi*xi/4) + 3*np.cos(3*math.pi*xi/4)
        ) # arbitrary function of x
8
9   # setup points where interpolated values will be evaluated
10  x = np.arange(0,80)/10-0.5
11
12  # setup 'true' y values
13  y = np.log(15+10*np.sin(math.pi*x/4) + 3*np.cos(3*math.pi*x/4))
14
15  # directly build P(x)
16  P = 0*x
17  for k in range(0,8):
18      L = 1.0+np.zeros(80)
19      for i in range(0,k):
20          # interpolation and also normalization
21          L = L*(x-xi[i])/(xi[k]-xi[i])
22      # skip i=k for L_k
23      for i in range(k+1,8):
24          L = L*(x-xi[i])/(xi[k]-xi[i])
25      # add scaled L_k contribution to P(x)
26      P = P+yi[k]*L
27
28  # plot results
29  plt.plot(xi,yi,'ko') # plot interpolation points
30  plt.plot(x,y,'k')    # plot reference curve
31  plt.plot(x,P,'k--')  # plot interpolation polynomial
32  plt.ylim([0,5])
33  plt.show()
```

Results for the Lagrange interpolation example are shown in Figure 3.2. Note that this interpolation deviates significantly from the original function in regions between the end nodes and rapidly diverges in the extrapolation

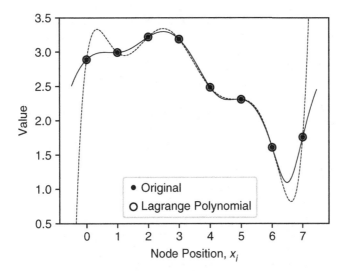

Figure 3.2 Example Lagrange interpolation, $P(x)$, along with the corresponding 8 discrete interpolation nodes and original reference function.

region beyond the end nodes. This is a common issue that emerges as the order of interpolation is increased which motivates the development of piecewise splines and least squares fits discussed in the subsequent sections.

3.1.2 SPLINE FITS

To overcome the tendency of high-order polynomial fits to oscillate, a common alternative interpolation method is to fit lower-order polynomials to subsets of the available data. One example of spline fitting is to utilize Bézier polynomials given by

$$\vec{B}(t) = (1 - t)^3\vec{P}_0 + 3(1 - t)^2 t\vec{P}_1 + 3(1 - t)t^2\vec{P}_2 + t^3\vec{P}_3, \quad t \in [0, 1] \quad (3.4)$$

which can be rewritten as

$$\vec{B}(t) = (1 - t)^3\vec{P}_0 + 3(t - 2t^2 + t^3)\vec{P}_1 + 3(t^2 - t^3)\vec{P}_2 + t^3\vec{P}_3 \quad (3.5)$$

Points 0 and 3 correspond to the end-points, which are also known as *knots*. The other two points are the control points that determine the shape of the curve. The construction of a smooth spline through multiple data points implies that the first and second derivatives should be continuous across the spline boundaries. These conditions lead to two equations (one for each derivative) at each spline interface that can be used to find the control points.

The first derivative is given by

$$\vec{B}'(t) = -3(1 - t)^2\vec{P}_0 + 3(1 - 4t + 3t^2)\vec{P}_1 + 3(2t - 3t^2)\vec{P}_2 + 3t^2\vec{P}_3 \quad (3.6)$$

At the left boundary of segment i we can write

$$\vec{B}_i'(0) = \vec{B}_{i-1}'(1) \tag{3.7}$$

or

$$-3\vec{P}_{0,i} + 3\vec{P}_{1,i} = -3\vec{P}_{2,i-1} + 3\vec{P}_{3,i-1} \tag{3.8}$$

Now, since the curve is continuous, $\vec{P}_{0,i} = \vec{P}_{3,i-1} = \vec{K}_i$, the i-th knot point, and we can simplify this expression as

$$2\vec{K}_i = \vec{P}_{1,i} + \vec{P}_{2,i-1} \tag{3.9}$$

The second derivative is given by

$$\vec{B}''(t) = 6(1-t)\vec{P}_0 + 3(-4+6t)\vec{P}_1 + 3(2-6t)\vec{P}_2 + 6t\vec{P}_3 \tag{3.10}$$

Imposing the condition of continuity at the boundary leads to

$$6\vec{P}_{0,i} - 12\vec{P}_{1,i} + 6\vec{P}_{2,i} = 6\vec{P}_{1,i-1} - 12\vec{P}_{2,i-1} + 6\vec{P}_{3,i-1} \tag{3.11}$$

Simplifying and taking into account the shared knot point, we get equation,

$$-2\vec{P}_{1,i} + \vec{P}_{2,i} = \vec{P}_{1,i-1} - 2\vec{P}_{2,i-1} \tag{3.12}$$

Equations 3.9 and 3.12 are defined only at the internal knots: the places where two segments come together. Mathematically, this means that we have $2(n-1)$ equations for $2n$ unknowns. In order to close the system, we prescribe two more natural boundary conditions, $\vec{B}_0''(0) = 0$ and $\vec{B}_{n-1}''(1) = 0$. In other words, the spline becomes linear at the end points. These two remaining equations are

$$\vec{K}_0 - 2\vec{P}_{1,0} + \vec{P}_{2,0} = 0 \tag{3.13}$$

and

$$\vec{P}_{1,n-1} - 2\vec{P}_{2,n-1} + \vec{K}_n = 0 \tag{3.14}$$

We can further simplify this system by substituting Equation 3.9 into 3.12 to obtain

$$\vec{P}_{1,i-1} + 4\vec{P}_{1,i} + \vec{P}_{1,i+1} = 4\vec{K}_i + 2\vec{K}_{i+1} \quad i \in [1, n-2] \tag{3.15}$$

On the boundary nodes, Equations 3.13 and 3.14 lead to

$$2\vec{P}_{1,0} + \vec{P}_{1,1} = \vec{K}_0 + 2\vec{K}_1 \tag{3.16}$$

and

$$2\vec{P}_{1,n-2} + 7\vec{P}_{1,n-1} = 8\vec{K}_{n-1} + \vec{K}_n \tag{3.17}$$

The resulting tri-diagonal system for P_1 can be solved using the Thomas algorithm. $\vec{P}_2$ is then obtained by utilizing P_1 in Equations 3.9 and 3.14

$$\vec{P}_{2,i} = 2\vec{K}_i - \vec{P}_{1,i} \quad i \in [0, n-2] \tag{3.18}$$

and

$$\vec{P}_{2,n-1} = (1/2)(\vec{K}_n + \vec{P}_{1,n-1}) \tag{3.19}$$

Interactive demo of this algorithm, that utilizes concepts covered in Chapter 7, can be found at particleincell.com/2012/bezier-splines.

3.1.3 LEAST SQUARES FIT

While a nth-order polynomial can be used to fit exactly n points and other combinations of boundary and interface conditions can be used to specify unique splines to interpolate among data points, these methods presume that the reference points are themselves accurate. However, often measurement data is corrupted by significant noise that can obscure underlying lower dimensional trends. While the polynomials can exactly fit the data at all measurement points with unique coordinate values, attempting to interpolate between noisy observations can further exacerbate the oscillation and overshooting problems.

Rather than attempting to exactly fit all of the observation points, assuming that the observed data includes errors allows for fitting lower order models to data by allowing for the addition of small error terms for each observation. A unique specification of this lower order model can be achieved by replacing the exact fitting procedures used for polynomials and splines with a minimization of the total error or model misfit.

Given a set of $N+1$ observation points x_i, y_i, the discrete error vector ϵ_i between the observation and model value, $f(x)$, can be defined as $\epsilon_i = y_i - f(x_i)$. Minimizing the total error between the data and the model can then be accomplished by minimizing some norm of this deviation vector. A common choice is a minimization of the L_2 norm computed as the sum of the squares of the deviation terms, $\sum_{i=0}^{N} \epsilon_i^2$.

While this least squares fit can be sought for any parametric model $f_{a_i}(x)$, it is particularly common to seek a least-squares polynomial fit for polynomial order $m < N$, $P(x; \vec{a}) = a_0 + a_1 x + a_2 x^2 + ... a_m x^m$ as in case of polynomial interpolation. While the resulting linear system consisting of discrete points, x_i, has a form similar to 3.1, note that the $N+1$ observations with only $m+1$ coefficients, $a_0 \cdots a_m$ implies a Vandermonde matrix that is taller than it is wide corresponding to an over-constrained linear system.

Rather than trying to solve this over-constrained system directly, the problem can be converted to that of minimizing the discrete error between the data and the functional fit. In the polynomial case, the total squared error $||\epsilon_i||_2^2$ can be written in terms of a total error, $E_2(\vec{x}, \vec{y}; \vec{a})$, as a function of the constant data points and unknown parameters, a_i as:

$$E_2(\vec{a}) = \sum_i \epsilon_i^2 = \sum_i (y_i - P_{a_i}(x_i))^2 \qquad (3.20)$$

The minimization of E_2 then corresponds to finding the zero of the gradient of the loss with respect to all of the parameters given the constant data points:

$$\frac{\partial E_2(\vec{a})}{\partial \vec{a}} = 2 \sum_i \epsilon_i \frac{\partial \epsilon_i}{\partial \vec{a}} = \vec{0} \qquad (3.21)$$

In the case of a cubic fitting function, $P(x; \vec{a}) = a_3 x^3 + a_2 x^2 + a_1 x + a_0$, we can see that:

$$\frac{\partial \epsilon_i(\vec{a})}{\partial \vec{a}} = \frac{\partial}{\partial \vec{a}}[y_i - a_3 x_i^3 - a_2 x_i^2 - a_1 x_i - a_0] = \begin{bmatrix} -1 \\ -x_i \\ -x_i^2 \\ -x_i^3 \end{bmatrix} \quad (3.22)$$

Rewriting Equation 3.21 for the cubic case then leads to the following four equations:

$$\begin{aligned} a_3 \sum_i x_i^3 + a_2 \sum_i x_i^2 + a_1 \sum_i x_i + a_0 \sum_i(1) &= \sum_i y_i \\ a_3 \sum_i x_i^4 + a_2 \sum_i x_i^3 + a_1 \sum_i x_i^2 + a_0 \sum_i x_i &= \sum_i y_i x_i \\ a_3 \sum_i x_i^5 + a_2 \sum_i x_i^4 + a_1 \sum_i x_i^3 + a_0 \sum_i x_i^2 &= \sum_i y_i x_i^2 \\ a_3 \sum_i x_i^6 + a_2 \sum_i x_i^5 + a_1 \sum_i x_i^4 + a_0 \sum_i x_i^3 &= \sum_i y_i x_i^3 \end{aligned} \quad (3.23)$$

This in turn can be written and solved as a linear system:

$$\begin{bmatrix} \sum_i(1) & \sum_i x_i & \sum_i x_i^2 & \sum_i x_i^3 \\ \sum_i x_i & \sum_i x_i^2 & \sum_i x_i^3 & \sum_i x_i^4 \\ \sum_i x_i^2 & \sum_i x_i^3 & \sum_i x_i^3 & \sum_i x_i^5 \\ \sum_i x_i^3 & \sum_i x_i^4 & \sum_i x_i^5 & \sum_i x_i^6 \end{bmatrix} \begin{bmatrix} a_0 \\ a_1 \\ a_2 \\ a_3 \end{bmatrix} = \begin{bmatrix} \sum_i y_i \\ \sum_i y_i x_i \\ \sum_i y_i x_i^2 \\ \sum_i y_i x_i^3 \end{bmatrix} \quad (3.24)$$

For arbitrary order polynomials, it can be verified that the resulting best-fit equation is equivalent to premultiplying Equation 3.1 by the transposed Vandermonde matrix, $\mathbf{V}^T$, such that the resulting equation is simply $\mathbf{V}^T\mathbf{V}\vec{a} = \mathbf{V}^T\vec{y}$.

This fitting procedure applied to the same eight points used in the Lagrange interpolation test case by replacing the direct Lagrange interpolation with this best-fit solve in the Python code below.

```
1   # build cubic fit P(x)
2   m = 3+1
3   # build Vandermonde matrix
4   V = np.zeros((8,m))
5   for j in np.arange(0,m):
6       V[:,j] = np.power(xi,j)
7
8   # solve Least Squares best fit coefficient vector
9   a = np.matmul(np.linalg.inv(np.matmul(V.T,V)),np.matmul(V.T,yi))
10
11  P = np.zeros(80)
12  for j in np.arange(0,m):
13      P = P+a[j]*np.power(x,j)
```

The least squares cubic fit of the same data points listed in Table 3.1 is shown in Figure 3.3. Note that while the example demonstrated the least squares best fit for polynomial interpolants, the same methodology can be used for finding a least squares fit for any linear combination of non-degenerate functions, $p_j(x_i)$ such that the best fit function can be written as the linear combination $P(x_i; \vec{a}) = \sum_j p_j(x_i)a_j$. While the functions p_j may be nonlinear in x, this fitting procedure is generally referred to as "linear least squares" as the dependence of P is linear with respect to the fitting coefficients, a_j.

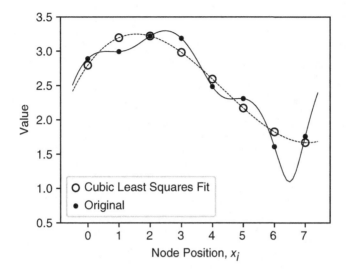

Figure 3.3 Example least-squares cubic polynomial best fit interpolation, $P(x)$, along with the corresponding 8 discrete interpolation nodes and original reference function.

3.1.4 FOURIER SERIES

The Fourier series is one of the most common tools in signal processing. The method exploits the orthogonality of sinusoidal functions to convert an N-point discrete time series signal, A_n, to its discrete frequency domain amplitude compliment, $\hat{A}_k$. The N discrete point approximation of the continuous time series $A(t)$ can then be exactly recovered from the N amplitudes corresponding to each frequency, k. For real data, this conversion is accomplished by projecting the length-N time series data into a coordinate system defined by N orthogonal basis vectors composed of the N sine and cosine waves that fit within a length-N vector, $a_k(t) = \sin(k\omega t)$ and $b_k(t) = \cos(k\omega t)$ where $\omega = 2\pi/N$ and $k = 0 \ldots N/2$. Figure 3.4 shows the mode shapes corresponding to $N=8$.

For each mode basis function, $a_k(t)$ and $b_k(t)$, the corresponding mode amplitudes are calculated by projecting the time series onto the mode basis as $\hat{a}_k = a_k(t) \cdot A(t)$ and $\hat{b}_k = b_k(t) \cdot A(t)$. Due to the symmetry of cosine and the anti-symmetry of sine, it is clear to see that extending the mode amplitudes to negative wavenumbers results in $a_{-k} = a_k$ and $b_{-k} = -b_k$ for the corresponding positive k values. The original time series data can then be recovered from the mode amplitude coefficients as shown in Equation 3.25.

$$A(t) = \frac{1}{N} \sum_{k=-\lfloor \frac{(N-1)}{2} \rfloor}^{\lfloor \frac{(N+1)}{2} \rfloor} \hat{a}_k \cos(k\omega t) + \hat{b}_k \sin(k\omega t) \qquad (3.25)$$

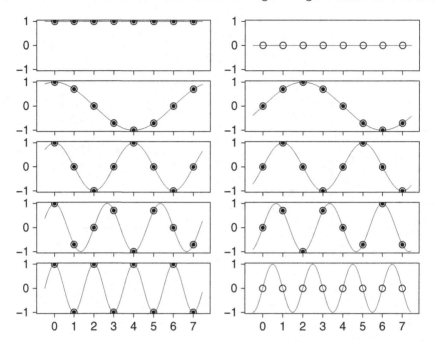

Figure 3.4 Example Fourier series real mode shapes for $N=8$. Left denotes the cosine component, $a_k(t)$ and right denotes the sine component, $b_k(t)$, along with the corresponding 8 discrete points for each mode for $k = [0, \ldots, 4]$ from top to bottom.

Due to the symmetries of the coefficients a_k and b_k, often this is rewritten instead as a sum over positive wavenumbers for real discrete Fourier series in the literature, but in general, these discrepancies are simply artifacts of the index convention adopted. The above notation is adopted to demonstrate more clearly the parallels the complex discrete Fourier series.

The complex Fourier series follows the same form with mode shapes $c_k(t) = e^{ik\omega t}$ for $k = [-\lfloor (N-1)/2 \rfloor, \ldots, \lfloor (N+1)/2 \rfloor]$. The corresponding mode amplitudes can then be calculated with the dot product $\hat{c}_k = c_k(t) \cdot A(t)$. The analogous inverse transform for the complex Fourier series is then obtained by simply multiplying the mode amplitudes by the inverse mode shapes as shown in Equation 3.26.

$$A(t) = \frac{1}{N} \sum_{k=-\lfloor \frac{(N-1)}{2} \rfloor}^{\lfloor \frac{(N+1)}{2} \rfloor} \hat{c}_k e^{-ik\omega t} \tag{3.26}$$

Note that for purely real data, A_n, the complex portions of the complex Fourier series must cancel. This means that, using Euler's formula, the negative wavenumber mode amplitudes are then simply the complex conjugates of the positive mode amplitudes such that $\hat{c}_{(-k)} = \hat{c}_k^*$. This means that the real and imaginary components of the complex Fourier series are related to the real coefficients as $\hat{c}_k = (\hat{a}_k + i\hat{b}_k)$ for $k = [0, \ldots, \lfloor (N+1)/2 \rfloor]$.

The eight-mode Fourier series based on the nodes in Table 3.1 is shown in Figure 3.5. Note that like the Lagrange interpolation, the Fourier series exactly matches the reference node pairs, x_i, y_i. However, rather than extending towards $y = \pm\infty$ beyond the interpolation nodes, the Fourier series is periodic by construction.

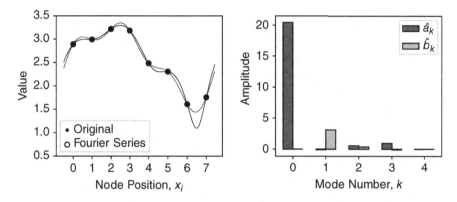

Figure 3.5 Example Fourier series, $A(x)$, along with the corresponding 8 discrete sample nodes and original reference function (left). Right shows the corresponding real Fourier series coefficients.

While this section provides a small sample of how the discrete Fourier transform is constructed, it is much more common to use "Fast-Fourier Transform" (FFT) implementations provided by numerical packages. The various FFT algorithms leverage the special sparse structures to implement the equivalent computations in $O(N \log N)$ rather than $O(N^2)$ resulting from the simplified description provided here. For large N, this difference in computational cost is quite dramatic. Further, because the Fourier transform is an invertible linear operation, multi-dimensional FFT algorithms can be constructed by sequentially transforming the Fourier coefficients of the prior direction with the same algorithms.

Spectral numerical methods for solving differential in the frequency domain are also possible once the system has been converted to Fourier space. Evaluating the derivative of a Fourier series is equivalent to multiplying the coefficients by the complex wavenumbers $-ik\omega$. For linear differential equations, these spectral methods can result in much faster results because the cost of converting to and from frequency space can be much lower than solving the problem directly. For non-linear problems, this is somewhat complicated potentially requiring multiple conversions of terms to and from frequency space resulting in a class of methods called *pseudo-spectral*. This is particularly complicated by nonlinear products causing scattering of energy to high frequencies such that de-aliasing must be incorporated in the conversions to avoid

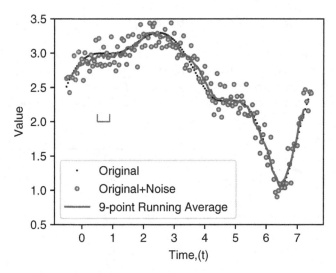

Figure 3.6 Simple running average of noisy discrete sample for arbitrary smooth function. Bracket represents a 9-point constant filter width.

the build up of energy in high frequencies. For more details of these issues, the reader is referred to Bewley 2018 for examples of applications of spectral and pseudo-spectral methods.

3.2 FILTERING

While mathematical models often deal with smooth, continuous, single-valued functions, when interacting with real-world data, the challenge of distinguishing signal from noise is often fundamental. One of the most basic methods for controlling noise is statistical averaging. With regards to discretely sampled time series data, a running average is a simple approach to smoothing noisy data. In the running average, the value of the function is replaced by an (2n+1)-point averaged value, $\tilde{f}$, computed via some weighted average of nearby sample points as depicted in Equation 3.27.

$$\tilde{f}(t_i) = \frac{\sum_{j=-n}^{n} w_j f(t_{(i+j)})}{\sum_{j=-n}^{n} w_j} \tag{3.27}$$

The simplest running average consists of equally weighted samples, $w_i = 1$. Figure 3.6 shows an example of this running average applied to a noisy version of Equation 3.3 with Gaussian random noise, $\mathcal{N}(0, 0.15)$, in time sampled at 20 Hz. This example uses an 9-point average to smooth the function. Note that in 3.6, the running average curve starts at the 5th point and ends at the 5th to last to have the 4 sample sample points before and after. The running average needs to be modified to account for boundary conditions.

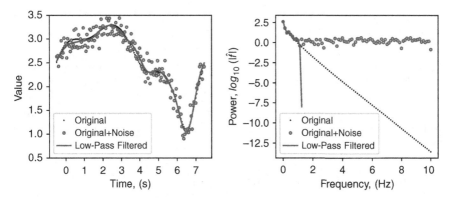

Figure 3.7 Simple running low-pass filter with 1Hz cutoff frequency for arbitrary smooth function. Original, noisy, and filtered time series signals (left) and the corresponding power spectra (right) are depicted.

Filtering in the frequency domain rather than the time domain is also quite common resulting in the prevalence of frequency domain analysis in signal processing. For frequency domain filtering, the discrete time series is first converted to mode amplitudes via the discrete Fourier transform described in Section 3.1.4.

Among the simplest forms of frequency domain filtering is the low pass filter. This filter receives its name from allowing low frequencies to pass through the filter while damping or removing higher frequency components. Figure 3.7 shows the same noisy discrete samples filtered with a low-pass filter. Note that the frequency domain filter assumes a periodic signal and so the filter wraps around the time domain.

In this example, the 1 Hz filter cutoff was selected based on the divergence of the original power spectra from the noise floor. The flat noise floor, referred to as *white-noise* corresponds to random Gaussian noise and so selecting the cut-off at this split provides a good filtered estimate of the original signal. However, note that in the general setting, the true signal is unknown. Further, in some real-world circumstances, the noise spectra may not be perfectly white.

Frequency domain filters are also particularly widely used because the convolution procedure described in Equation 3.27 is potentially quite computationally efficient in the frequency domain. The reason for this efficiency is that convolution in the time domain corresponds to frequency-wise multiplication spectrally. This is particularly true for filters that do not have finite support in the time domain like a Gaussian filter when compared to the running average described has a finite bandwidth, $2n + 1$, non-zero weights.

Inherent to the low-pass filter and running average is an assumption of some intrinsic smoothness. In general, these assumptions are referred to as *regularization* where additional assumptions are made with regards to the form of the solution to make solution of ill-posed problems possible. The least-squares fit of Section 3.1.3 would constitute another type of regularization obtained by assuming some functional form of the solution and minimizing the deviation of the data from this model. Beyond the simple assumptions with regards to function smoothness used here, a wide variety of other assumptions such as sparsity constraints such as methods like TV or L_1-regularization made popular in image processing. For an array of more advanced examples in the context of image processing, the reader is referred to Chan and Shen 2005.

3.3 PROBABILITY DISTRIBUTIONS

Probabilities of random events are familiar phenomena from common activities like rolling dice. For a fair six-sided die, the probability that the die lands on any particular number is $1/6$. For discrete events and discrete outcomes, this notion of probability can be extended to more complex scenarios by enumerating all the possible outcomes and then determining the likelihood of each. The classic example of this is in determining the probability of each possible outcome from $2 - 12$ from rolling a pair of independent fair dice. The most probable outcome, 7, emerges not because any one particular outcome is more likely than others, but because more of the possible outcomes have the same combined total value of the two dice.

As the number of dice is increased, the number of events with the average expectation becomes more and more peaked around the average expected value. The simplest version of this process, for coin flips, can be seen in the *binomial* distribution, $B(n, p)$, for n flips of a coin with probability p chance for success. The chance of $k \in [0, n]$ successes is then the *probability mass function* (PMF):

$$f^{\text{Binom}}(k; n, p) = \frac{n!}{k!(n - k)!} p^k (1 - p)^{(n-k)}$$

While factorials and powers are relatively simple to code, the integer mathematics can overflow for large values of n requiring some care in evaluation. These issues are generally resolved with careful implementations in math and statistics libraries. In the python `scipy.stats` package, the probability of k successes in n attempts can be evaluated using the `binom.pmf` function. The following script evaluates the binomial PMF for 8 trials with a 66% chance of success.

```
1  from scipy.stats import binom
2  n=8
3  for k in range(0,n+1):
4      f = binom.pmf(k=k,n=n,p=0.66)
5      print(k,f)
```

Figure 3.8 shows the resulting binomial PMF. The individual k probabilities sum to a total of 1.0 as every potential outcome of n trials is included. The

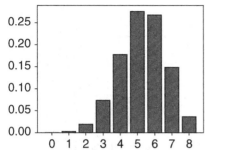 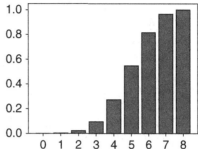

Figure 3.8 Binomial PMF (left) and CDF (right) for 8 trials with 66% chance of success.

cumulative distribution function (CDF) which is a running sum of the probability of k or fewer successes in the n trials can then be seen to accumulate up to 1.0 as depicted in the right half of the figure.

Representative data points can be sampled from this CDF by running a uniform random sample through the inverse of the CDF function. In this discrete case, the inverse function is simply finding the number of successes k that corresponds to the uniform random sample's value in the range $[0, 1]$. Since the CDF is already arranged to be monotonically increasing, this inversion can be accomplished by using `numpy.argmax` to find the number of successes, xi, corresponding to the PMF given a uniform random sample points, yi as shown in the following python code. Figure 3.9 shows the results of this sampling for 12 points from a random uniform distribution, yi, and converting those points into a random sample from the binomial CDF described above.

```
1  import numpy as np
2  def sampleFromPMF(pmf,yi):
3      n = pmf.size
4      cdf = pmf
5      for i in np.arange(1,n):
6          cdf[i] = cdf[i] + cdf[i-1]
7
8      ns = yi.size
9      xi = np.zeros(ns)
10
11     for i in range(ns):
12         xi[i] = np.argmax(cdf>yi[i])
13
14     return xi
```

While outcomes of enumerated discrete events are conceptually simple, often events can occur randomly in a continuous variable such as time. A classic example of this type of random event is the number of independent events that occur in an interval of time for a fixed average rate of occurrence, λ. Unlike the case for a discrete number of trials, the number of possible events has no hard upper bound though the probability becomes vanishingly small for numbers of events that vastly exceed the average expected rate. A

yi	xi
0.07040532	3
0.75043943	6
0.45927765	5
0.85591395	7
0.94483828	7
0.02198352	3
0.77702647	6
0.57127844	6
0.83580346	7
0.51616982	5
0.79996385	6
0.19902407	4

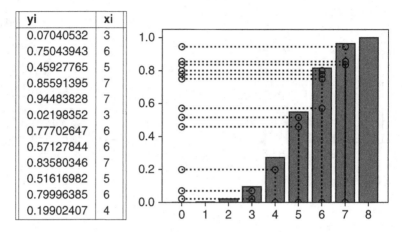

Figure 3.9 Randomly sampled number of successes from binomial PMF for 8 trials with 66% chance of success resulting from inverting the binomial CDF. The uniform random sample points yi are connected to the corresponding random sample of the PMF, xi, via dashed lines depicting the CDF inversion.

Poisson distribution, $f^{\text{Pois}}(k; \lambda) = \lambda^k e^{-k}/k!$, describes the probability of each potential outcome number of events, k. Like the binomial distribution, the `poisson` distribution is available in the Python `scipy.stats` package.

When the random events are also sampled from a continuous rather than discrete set of outcomes, a probability *density* function (PDF) must be used instead of a PMF. The PDF can be conceptualized as the limit of a PMF where the number of discrete classes in finite width x is subdivided into the limit of infinitesimal width, $dx \to 0$. Rather than a probability, the function has units of probability per unit length such that the probability, P, of the random variable, x, taking a value in class $x_{(i)}$ of any value in the range $[x_0 - x_1]$ is:

$$P(x_{(i)}) = \int_{x_0}^{x_1} f(x)dx$$

The most famous continuous probability distibution, the Gaussian or *normal distribution*, has the PDF:

$$f^{\text{Gauss}}(x; \mu, \sigma) = \frac{1}{\sigma\sqrt{2\pi}} e^{\frac{-(x-\mu)^2}{2\sigma^2}}$$

Here, x is the continuous outcome while μ and σ are the mean and standard deviation of the distribution respectively. This distribution emerges from a wide class of random independent events as the number of events becomes large due to the famous central limit theorem. For this reason, it is encountered widely in statistics and physics and commonly used in scientific computing.

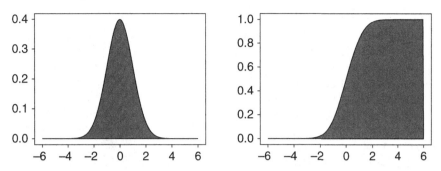

Figure 3.10 Gaussian PDF (left) and CDF (right) for $\mu = 0$ and $\sigma = 1$.

Converting samples from a uniform distribution to a Gaussian distribution can be accomplished in a manner similar to the discrete distributions. The uniform random samples are converted using the inverse cumulative distribution function. For the case of a Gaussian distribution, a special function called the *error function* (erf) is defined to be the CDF of a Gaussian distribution centered at zero such that the CDF of the Gaussian distribution has the form:

$$F^{\text{Gauss}}(x; \mu, \sigma) = \frac{1}{2}\left[1 + \text{erf}\left(\frac{x - \mu}{\sqrt{2}\sigma}\right)\right]$$

The error function, however, is simply the integral of the Gaussian distribution, $\text{erf}(x) = \int_0^x e^{-t^2}\, dt$ such that the CDF is normalized to the range $(0, 1)$ for $x \in (-\infty, \infty)$. Due to the prevalence of Gaussian distributions, many programming languages include numerical approximations to evaluate both the *erf* and its inverse as built-in functions or available in standard libraries. In the `scipy.special` package, these are implemented as the `erf` and `erfinv` functions respectively. Figure 3.10 shows a Gaussian PDF and CDF for the $\mu = 0$ and $\sigma = 1$ case. Using the same random sample as the binomial case, the conversion of uniform samples to Gaussian samples is shown in Figure 3.11.

While drawing samples from known probability densities is useful for problems like initializing stochastic particle based methods like those of Chapter 5 that rely on random sampling, the inverse problem of density estimation is also a critical tool of scientific computing. The most basic tool of density estimation is the histogram. In building a histogram, a region in which samples are found is somehow partitioned and the count of samples in each partition is divided by the volume of the region in the partition. Note that the term volume just refers to the size of the region in whatever dimension/coordinates intended. For a one-dimensional distribution, the volume may just be the length. The point is that if the probability density is integrated over the volume, a mass or expected sample count within the volume is the result.

yi	xi
0.07040532	−1.47277899
0.75043943	0.67587322
0.45927765	−0.10225371
0.85591395	1.06214007
0.94483828	1.59674106
0.02198352	−2.0144049
0.77702647	0.76218925
0.57127844	0.17962991
0.83580346	0.97735571
0.51616982	0.04054283
0.79996385	0.84149212
0.19902407	−0.8451123

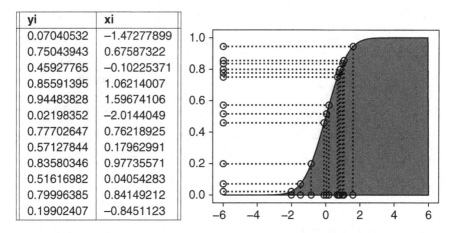

Figure 3.11 Uniform random samples converted to normally distributed samples via the Gaussian CDF. The uniform random sample points yi are connected to the corresponding random sample of the PDF, xi, via dashed lines depicting the CDF inversion.

The simplest histogram is a uniform binning where the region $[x_0, x_1]$ is divided into n equal partitions. The number of samples in each bin of the histogram can then just be counted by determining which cell each sample lies within taking care that samples that land on edges are not double-counted. The estimated density $\widetilde{\rho}_i$ of the ith bin is then found by dividing the number of samples in the bin N_i by the volume of the bin, V_i:

$$\widetilde{\rho}_i = \frac{N_i}{\Delta x}$$

An additional complication to this procedure is the trade off between bias and variance errors in the resulting density estimate. In the limit of many bins as n becomes much larger than the number of samples, the majority of bin will have zero samples and therefore an estimated density of 0. The remaining bins will have at least one sample and therefore an estimated density of $1/\Delta x$. As Δx becomes smaller with increased bin count n, the estimated density in these small regions becomes larger and larger and are an inaccurate estimate of the true density $\rho(x)$. At the other extreme, a single bin at $n = 1$ estimates a single density ρ for the entire region. While the sampling fluctuations around the average expected number of samples for the true density decreases, $\widetilde{\rho}(x)$ is a poor approximation of $\rho(x)$ assuming the true distribution is not uniform in the region of interest. Figure 3.12 depicts this bias-variance trade off for an array of samples and bin widths for data sampled from a Gaussian distribution.

While any of these density estimates will converge to the best n-piecewise approximation of the true density as the number of samples increases, the convergence rate is slow with errors generally decreasing with the square of

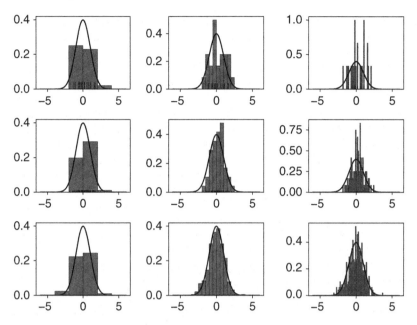

Figure 3.12 Histogram density estimates for random $N = [24, 96, 384]$ (top-bottom) random samples of weight $1/N$ each from a Gaussian PDF with $\mu = 0$ and $\sigma = 1$. Different number of bin counts $n_b = [6, 24, 96]$ for $x \in [-6, 6)$ (left-right) depict the bias-variance trade off with Δx.

the expected number of samples per bin. For true convergence of $\widetilde{\rho}(x) \to \rho(x)$ additionally requires the convergence of both $\Delta x \to 0$ and $N_i \to \infty$ simultaneously. However, the optimal balance of bin width and samples per bin is problem dependent. For additional information this convergence of nonparametric density estimates as well as alternative estimates like the "k Nearest Neighbor" density estimate, the reader is referred Chapter 6 of Fukunaga 1990.

3.4   QUADRATURE

In the prior chapters, we introduced the Finite Difference method to numerically solve differential equations. In scientific computing, there also often arises the need to numerically evaluate integrals. Basic numerical integration parallels the relationship between integration and summation. While the basis of integration results from passing a finite sum $\sum_i f(x_i)\Delta x$ to $\int f(x)dx$ in the limit of $\Delta x \to dx \to 0$, with finite computational effort, nonzero Δx must be used and the various techniques are referred to as Riemann sums.

The most basic integration rule is the *rectangle rule*. Here, the area of each segment of the integration $\int_x^{x+\Delta x} f(x)dx$ is approximated by a rectangle of width Δx and height of the curve at the midpoint of the segment, $f(x+\Delta x/2)$.

For integration between x_0 and x_1, n equal intervals of width $(x_1 - x_0)/n$ are summed such that:

$$\int_{x_0}^{x_1} \approx \sum_{i=1}^{n} f\left(x_0 + (i - 1/2)\Delta x\right)\Delta x) \qquad (3.28)$$

Note that the rectangle rule uses n function evaluations all of which are internal to the domain of integration. The next higher level of approximation, the *trapezoidal rule* builds the numerical approximation from $n-1$ trapezoidal segments that include function evaluations at the endpoints. For the trapezoidal rule, the numerical integral based on the area formula of a trapezoid with $\Delta x = (x_1 - x_0)/(n - 1)$ is:

$$\int_{x_0}^{x_1} \approx \sum_{i=1}^{n-1} \frac{f\left(x_0 + (i - 1)\Delta x\right) + f\left(x_0 + i\Delta x\right)}{2}\Delta x) \qquad (3.29)$$

Note that each $f(x_i)$ contributes to the sum twice except for the two end points. An equivalent approximation can be obtained by:

$$\left[\sum_{i=1}^{n} f\left(x_0 + (i - 1)\Delta x\right) - \frac{f(x_0) + f(x_1)}{2}\right]\Delta x$$

This expression is equivalent to a rectangle rule shifted to center evaluations on the end points where the two end evaluations are width $\Delta x/2$ and evaluated at the end point. Figure 3.13 shows the numerical integration of the standard normal Gaussian distribution using 7 and 14 function evaluations for each of the rectangle and trapezoidal rules.

3.5   ROOT FINDING

The term root refers to the value x for which some function $f(x) = 0$. A function can have multiple roots. Consider the Taylor series approximation for the function value,

$$f(x^{k+1}) = f(x^k) + (x^{k+1} - x^k)f'(x^k) + O(2) \qquad (3.30)$$

If x^{k+1} is the root, then $f(x^{k+1}) = 0$ and

$$x^{k+1} \approx x^k - \frac{f(x^k)}{f'(x^k)} \qquad (3.31)$$

Equation 3.31 is used iteratively to arrive at value of x until $f(x^k) < \epsilon$. This approach is known as the Newton's method.

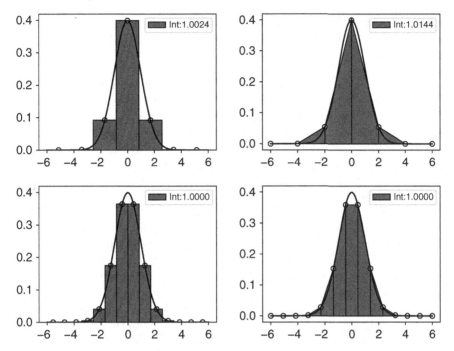

Figure 3.13 Numerical integration of Gaussian PDF for $\mu = 0$ and $\sigma = 1$ using rectangle (left) and trapezoidal (right) integration schemes. The top row shows the integration with 7 function evaluations and the bottom row shows the equivalent with 14 function evaluations.

3.6 ADDITIONAL MATRIX SOLVER ALGORITHMS

3.6.1 MULTIGRID

Consider the 1D Poisson's equation

$$\frac{\partial^2 \phi}{\partial x^2} = b \tag{3.32}$$

with b given by some doubly-integrable analytical function. This formulation allows us to compute the analytical solution ϕ_{true}. Next, consider some iterative scheme to solve the linear system

$$\mathbf{A}\vec{\phi} = \vec{b} \tag{3.33}$$

After k solver iteration, the residue vector satisfies

$$\vec{R}^k = \mathbf{A}\vec{\epsilon}^k \tag{3.34}$$

where $\epsilon^k = \left(\phi^k - \phi_{true}\right)$ is the difference between the approximate solution at iteration k and the true solution.

Having the analytical solution allows one to plot the error vector at each solver iteration. Doing so we notice that the error consists of a high-frequency term that dissipates rather quickly and a low-frequency term that takes much longer to vanish. The high frequency term captures the node-to-node variation. The $\partial^2 \phi / \partial x^2$ operator controls the "acceleration" with which values change in response to their neighbor. This is the high-frequency term. But obtaining good node-to-node local variation does not imply that the solution also satisfies the boundary conditions. This is the low frequency term. It dissipates slowly since, due to the 3-point stencil used in the one-dimensional Finite Difference approximation of the second derivative, it takes $n-1$ iterations for the n-th node to even become "aware" of the boundary.

It may seem natural to solve the system on a coarser grid. A coarser grid effectively "stretches" the arms of the computational stencil. While this approach may be useful for obtaining the initial guess, it does not lead to a practical method for obtaining the solution on the desired fine grid. However, if the ϵ error is smoothly varying (as it generally is), we can use the coarse grid to evaluate the error term. We then use the interpolation of ϵ to the fine grid to correct the solution. This is the basic principle behind multigrid algorithms. There are multiple variants available. The popular "W" uses three grid levels 1, 2, and 3, listed in order from the coarsest to the finest. The grids are visited in a $3-2-1-2-1-2-3$ order, which, when plotted, produces a W shape. Here we limit the description to a simpler two-grid scheme based on Ferziger and Perić 2002:

1. Start by performing one or more iterations on the fine grid,

$$A\left(\phi_f^n\right)^* = b_f \tag{3.35}$$

2. Next compute the residue on the fine grid,

$$R_f^n = A\left(\phi_f^n\right)^* - b_f \tag{3.36}$$

This residue is also used to check for solver termination.

3. Next, interpolate the residue to the coarse grid,

$$R_f \to R_c \tag{3.37}$$

This operation is called restriction. In a two-to-one coarsening, every other fine node overlaps a coarse node. We could simply assign values according to $R_{c,i} = R_{f,2i}$. The downside of this approach is that it completely skips over values on the odd-index fine-grid nodes. Therefore, it is recommended to use a node average such as

$$R_{c,i} = \frac{R_{f,2i-1} + 2R_{f,i} + R_{f,2i+1}}{4} \tag{3.38}$$

This restriction is visualized in Figure 3.14(a).

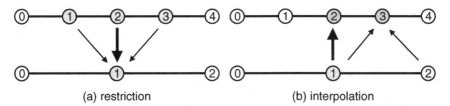

(a) restriction   (b) interpolation

Figure 3.14 Restriction and interpolation operations used to map between coarse and fine grids.

4. Now that we have the residue on the coarse grid, we perform several iterations on the coarse grid to estimate

$$A\epsilon_c = R_c \tag{3.39}$$

In the solver, we need to remember that $(\Delta x)_c = 2(\Delta x)_f$. We obtain the boundary conditions from the desired behavior of the solution vector. On Neumann nodes, the residue is $(\phi_1^* - \phi_0^*)/\Delta x = 0$ which after substitution gives $[(\phi_1 - \epsilon_1) - (\phi_0 - \epsilon_0)]\Delta x = 0 + R_0$. After subtracting $\phi_1 - \phi_0)/\Delta x = 0$, we obtain $\epsilon_0 = \epsilon_1 + R_0\Delta x$. On Dirichlet nodes we have $\epsilon = 0$, since the expected value is provided by the boundary condition.

5. The next step is to interpolate the correction vector to the fine grid,

$$\epsilon_c \to \epsilon_f \tag{3.40}$$

Every other node on the fine mesh overlaps a course grid node, so here we just copy values. On the intermediate nodes (nodes with odd indexes), we set the value to the average of the surrounding course grid nodes,

$$\epsilon_{f,2i+1} = \frac{\epsilon_{c,i} + \epsilon_{c,i+1}}{2} \tag{3.41}$$

This interpolation is visualized in Figure 3.14.

6. Next we correct the solution from

$$\phi_f^n = (\phi_f^n)^* - \epsilon_f \tag{3.42}$$

7. Finally, we return to step 1 and continue until the norm of the residue vector R^n reaches some tolerance value.

This algorithm can be found in the companion samples under `multigrid.py`. The code contains both the GS and MG solvers. The G-S solver requires 17,500 iterations to converge, while the MG solver reaches the same tolerance in only 93 loops of the above algorithm. By adding all mathematical operations, we can estimate that the multigrid algorithm requires about $6.9\times$ fewer operations than Gauss-Seidel. This observation is further confirmed by timing the two algorithms. The MG algorithm produces the solution in 0.127 seconds, compared to 0.863 seconds for G-S. This is a $6.8\times$ reduction in solver run time.

3.6.2 CONJUGATE GRADIENT

Instead of using different grid dimensions, another class of solvers, known as Krylov space solvers, utilize the observation that the gradient of the error points in the direction of the largest variation. As such the solution should move in the $-\nabla\epsilon$ direction in order to minimize the error. This same principle is also utilized in an optimization algorithm called the gradient descent, discussed in Chapter 10. Application of gradient descent to a linear system leads to the conjugate gradient method. Its convergence is further accelerated by utilizing a *preconditioner* matrix, which may be just the diagonal of $\mathbf{A}$. The resulting method is known as the preconditioned conjugate gradient, or PCG. PCG is an attractive algorithm as it can produce solution in just a few iterations. Its downside is that it is applicable only to symmetric positive definite matrices. While many coefficient matrices found in engineering applications satisfy this requirement (especially once the preconditioner is included), not all do. Details of its derivation can be found in numerical analysis books, such as Burden and Faires 2001. Here we only describe the steps.

Given the system $\mathbf{A}\vec{x} = \vec{b}$, one starts with

$$\vec{r}_0 = \vec{b} - \mathbf{A}\vec{x}_0 \tag{3.43}$$

$$\vec{z}_0 = \mathbf{M}^{-1}\vec{r}_0 \tag{3.44}$$

$$\vec{p}_0 = \vec{z}_0 \tag{3.45}$$

During the iteration, one calculates

$$\alpha_k = \frac{\vec{r}_k^T \vec{z}_k}{\vec{p}_k^T \mathbf{A}\vec{p}_k} \tag{3.46}$$

$$\vec{x}_{k+1} = \vec{x}_k + \alpha_k \vec{p}_k \tag{3.47}$$

$$\vec{r}_{k+1} = \vec{r}_k - \alpha_k \mathbf{A}\vec{p}_k \tag{3.48}$$

$$\vec{z}_{k+1} = \mathbf{M}^{-1}\vec{r}_{k+1} \tag{3.49}$$

$$\beta_k = \frac{\vec{r}_{k+1}^T \vec{z}_{k+1}}{\vec{r}_k^T \vec{z}_k} \tag{3.50}$$

$$\vec{p}_{k+1} = \vec{z}_{k+1} + \beta_k \vec{p}_k \tag{3.51}$$

until r_{k+1} is sufficiently small. Then, the solution is found in $\mathbf{x}_{k+1}$. This algorithm is implemented in `heat2D_PCG.py`.

3.6.3 LU AND CHOLESKY DECOMPOSITION

While we have been mostly focusing on iterative solvers, there are instances in which direct solvers may be more suitable. The benefit of direct solvers is that once the appropriate data structures are constructed, the actual matrix solution can proceed rapidly. Direct solvers may be appealing to simulations that consist of a very large number of time steps. The initialization penalty may be overcome by the subsequent rapid calculation during the simulation main loop.

We have already discussed the Gauss-Jordan elimination algorithm in Section 2.4.1. The challenge with this algorithm is that it encodes the forcing vector $\vec{b}$ into the augmented matrix. Since $\vec{b}$ can be expected to change at each time step, the use of G-J in simulation code would require either a complete rebuild of the system (which would defeat the just described benefit of direct solvers) or at least encoding the elimination steps such that the last column of the augmented matrix can be updated without recomputing the rest. But luckily, there is a simpler method available.

Consider the matrix system

$$\mathbf{A}\vec{x} = \vec{b} \tag{3.52}$$

Instead of working with a single matrix, we let matrix $\mathbf{A}$ be given by a product of two matrices

$$\mathbf{A} = \mathbf{L} \cdot \mathbf{U} \tag{3.53}$$

with the important characteristic that $\mathbf{L}$ is a lower triangular and $\mathbf{U}$ is an upper triangular matrix. Thus,

$$\mathbf{A}\vec{x} = \mathbf{L}(\mathbf{U} \cdot \vec{x}) = \vec{b} \tag{3.54}$$

Assuming that this L-U decomposition is available, one first solves

$$\mathbf{L}\vec{y} = \vec{b} \tag{3.55}$$

using forward substitution. The resulting $\vec{y}$ vector is then used to solve

$$\mathbf{U}\vec{x} = \vec{y} \tag{3.56}$$

for $\vec{x}$ using back substitution. If permutation is needed, we solve the modified system

$$\mathbf{PLU}\vec{x} = \mathbf{P}\vec{b} \tag{3.57}$$

The $\mathbf{LU}$ system as given is under-constrained. A unique solution is obtained by requiring that the main diagonal of $\mathbf{L}$ are all 1s. This is known as Doolittle factorization. Alternatively, we could require all ones on $\mathbf{U}$'s diagonal to get the Crout factorization.

The steps given by Equations 3.55 and 3.56 are trivial. The challenge arises in the process of decomposing $\mathbf{A}$ into $\mathbf{LU}$. Optimized algorithms for performing this decomposition are found in all common linear algebra packages. In practice, one never needs to write a LU subroutine from scratch. However, for illustration, we now describe the Doolittle algorithm. The indices of the $\mathbf{A}$ matrix are set per

$$U_{ij} = A_{ij} \text{ for } i, j = 0; \tag{3.58}$$

$$U_{ij} = A_{ij} - \sum_{k=0}^{i-1} L_{ik}U_{kj} \text{ for } i, j > 0 \tag{3.59}$$

Next, the **L** matrix coefficients are computed per

$$L_{ij} = A_{ij}/U_{jj} \ \text{ for } \ i, j = 0; \tag{3.60}$$

$$L_{ij} = (A_{ij} - \sum_{k=0}^{j-1} L_{ik}U_{kj})/U_{jj} \ \text{ for } \ i, j > 0 \tag{3.61}$$

This algorithm can be implemented as shown below:

```
1   def LUdecompose(A):
2       n,nc = np.shape(A)
3       L = np.zeros_like(A)
4       U = np.zeros_like(A)
5
6       print("Performing L-U decomposition")
7       if (A[0,0]==0):
8           print("Can't continue")
9           return
10      L[0,0] = 1.0
11      U[0,0] = A[0,0]/L[0,0]
12
13      for j in range(1,n):    # j = [1,2,3,....,n-1]
14          U[0,j] = A[0,j]/L[0,0]
15          L[j,0] = A[j,0]/U[0,0]
16
17      for i in range(1,n-1):
18          s = 0
19          for k in range(i): s += L[i,k]*U[k,i]
20          L[i,i] = 1.0
21          U[i,i] = A[i,i] - s
22          if (U[i,i]==0):
23              print("Can't continue")
24              return
25          for j in range(i+1,n):
26              s = 0
27              for k in range(i): s+=L[i,k]*U[k,j]
28              U[i,j] = (1/L[i,i])*(A[i,j]-s)
29              s = 0
30              for k in range(i): s+=L[j,k]*U[k,i]
31              L[j,i] = 1/(U[i,i])*(A[j,i]-s)
32
33      s = 0
34      for k in range(n): s += L[n-1,k]*U[k,n-1]
35      L[n-1,n-1] = 1
36      U[n-1,n-1] = A[n-1,n-1]-s
37
38      return L, U
```

The full code is found in **heat2d_LU.py**. This code solves the steady-state heat equation problem with random internal Dirichlet points on a user-specified domain size. It also uses the **spy** function to visualize the sparsity of the system matrix as shown in in Figure 3.15. The black dots correspond to the original matrix **A**, while the lighter and darker shades of gray map to **L** and **U**, respectively. We notice that one drawback of direct solvers is that they increase the density of the coefficient matrix. The primary implication is that

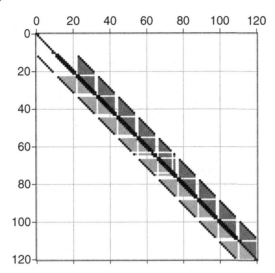

Figure 3.15 Visualization of matrix sparsity for the original matrix (black) and the LU decomposition (grays).

the utilized sparse matrix storage container needs to be flexible to support a variable number of non-zero coefficients per row.

This code also allows us to perform timing studies of the time needed for the initial decomposition, as well as the average time per solution. The solution time is measured by averaging 20 iterations, with the Dirichlet nodes reassigned new random values at each iteration. This timing is then compared to the iterative Gauss-Seidel algorithm in `heat2d_GS.py`, with data summarized in Table 3.2. We can make two observations. First, the solution time using the LU method, LU_s, is faster than for G-S, although the difference is not particularly large. The reason is that our implementation of the LU algorithm utilizes dense matrices in which significant effort is wasted in matrix-vector multiplication with the primarily zero coefficients. A sparse matrix container can perform these calculations more efficiently by considering only the non-zero columns. The second observation we make is that the factorization time LU_f grows rapidly as the number of mesh nodes nn increases. In fact, the calculation time for the Doolittle algorithm scales with $O(n^3)$. We can use the timing data to compute the number of solver iterations for break even, where $LU_s + it_b LU_f = it_b GS$, with the understanding that a more favorable outcome would be expected with optimized codes. For the 50×50 matrix with 2500 mesh nodes, we see that we need to run the solver for almost 3500 iterations before the LU decomposition method becomes superior.

One possibility for speeding up this direct solver involves utilizing alternate decomposition. For Hermetian (symmetric but allowing for complex numbers) positive-definite ($\vec{z}^T A \vec{z} > 0$) matrices, the matrix decomposition can be

Table 3.2

Timing Comparison for G-S and LU Solvers Utilizing Dense Matrix Storage

nn	GS	LU_f	LU_s	it_b
100	0.0023	0.056	0.0025	N/A
400	0.040	3.49	0.024	217
900	0.21	40.12	0.13	506
1600	0.77	307.3	0.55	1404
2500	1.83	1162.0	1.49	3449

written in form

$$\mathbf{A} = \mathbf{LL}^* \tag{3.62}$$

or for strictly real (non-imaginary) coefficients,

$$\mathbf{A} = \mathbf{LL}^T \tag{3.63}$$

This is known as the Cholesky decomposition. The benefit of this method is that it requires only approximately half the number of operations of LU decomposition. It also requires just half the memory, since only a single matrix needs to be stored. The coefficients of the $\mathbf{L}$ matrix are set using the following relationship

$$l_{kk} = \sqrt{a_{kk} - \sum_{j=1}^{k-1} l_{kj}^2} \tag{3.64}$$

$$l_{ik} = (a_{ik} - \sum_{j=1}^{k-1} l_{ij}l_{kj})/l_{kk} \tag{3.65}$$

The Cholesky decomposition method is also found linear algebra libraries.

3.7 EIGENVALUES

Given a linear system $\mathbf{A}\vec{x} = \vec{b}$, the square matrix $\mathbf{A}$ can be thought of as a generic operator transforming the vector $\vec{x}$ into the vector $\vec{b}$. It turns out that there are vectors $\vec{v}$ for which this transformation scales the vector uniformly without changing the relative ratios between components,

$$\mathbf{A}\vec{v} = \lambda\vec{v} \tag{3.66}$$

These vectors are known as *eigenvectors* and the corresponding scalar multipliers λ are called *eigenvalues*. Each matrix will have as many eigenvalues as there are linearly-independent rows, however, some values may be repeated. Eigenvectors show up frequently in image processing, modal analysis, and solutions to special types of matrices. External resources provide various algorithms for their computation. These are omitted here for brevity. However, using Python's Numpy package, eigenvalues can be computed using the `eig` command as shown below:

```
1  import numpy as np
2  A = np.array([[0,2,0],[2,0,0],[1,0,1]])
3  [lam,vec] = np.linalg.eig(A)
```

3.8 NON-LINEAR SYSTEMS

3.8.1 POOR MAN'S LINEARIZATION

So far we have only considered linear systems of form $\mathbf{A}\vec{x} = \vec{b}$ in which the right-hand side forcing vector $\vec{b}$ is a scalar field that is independent of $\vec{x}$. However, governing equations encountered in physics and engineering tend to be of type

$$\mathbf{A}\vec{x} = \vec{b}(\vec{x}) \tag{3.67}$$

The right side is a function of $\vec{x}$, which is the unknown variable we are solving for. One example of such a system is the non-linear Poisson's equation encountered in the field of computational plasma physics, see Chapter 5. It is given by

$$\nabla^2\phi = -\frac{e}{\epsilon_0}\left[n_i - n_0\exp\left(\frac{\phi - \phi_0}{ek_BT_{e,0}}\right)\right] \tag{3.68}$$

The terms in the brackets correspond to the ion and electron number densities. The electron density n_e, given by the second term, is a function of ϕ. Plasmas are generally quasi-neutral such that $n_i \approx n_e$. The forcing term can thus be expected to be weak. The system can then be solved using an iterative solver with a relatively slow convergence rate by utilizing the "old" value x^k to evaluate $\vec{b}$ to advance the solution to iteration $k+1$. This is known as "poor man's linearization". For the Jacobi scheme, we have

$$x_i^{k+1} = \left(b^k - \sum_j \delta_{ij}a_{ij}x_j^k\right)/a_{ii} \tag{3.69}$$

Assuming the system does not diverge, this appraoch will yield the correct solution since at convergence $x^k \approx x^{k+1}$. This method is demonstrated in `heat2d_GS_NL.py`, with the relevant snippet shown below:

```
1  for it in range(10000):
2      for u in range(nu):
3          ...
4          y_nl = y[u] + nl_coeff[u]*np.exp(-x[u]/10)
5          x_star = (y_nl - dot_nodiag)/d[u]
6          x[u] = x[u] + 1.4*(x_star-x[u])
```

3.8.2 NEWTON-RAPHSON LINEARIZATION

The prior method is not easily ported to algorithms such as PCG in which the approximate solution changes significantly between iterations, which can lead to instability and eventual divergence. We then need to utilize Newton-Raphson linearization. It is the generalization of Newton's method for root finding, see Section 3.5, applied to a matrix problem. The non-linear system can be written as

$$\mathbf{A}\vec{x} - b(\vec{x}) \equiv F(\vec{x}) = 0 \tag{3.70}$$

Newton's method for finding roots

$$x^{k+1} = x^k - \frac{f(\vec{x^k})}{f'(x^k)} \tag{3.71}$$

is applied to Equation 3.70 as

$$x^{k+1} = x^k - \mathbf{J}^{-1}F(x^k) \tag{3.72}$$

where $\mathbf{J}$ is called the *Jacobian*. Its components are the derivatives of every function encoded by $F(\vec{x})$ against all x terms,

$$J_{i,j}(x) = \left(\frac{\partial F_i}{\partial x_j}\right)(x) \tag{3.73}$$

As an example, for the non-linear Poisson system simplified as $b = b_0 + C\exp[\alpha(x - x_0)]$, the Jacobian becomes

$$\mathbf{J} = \mathbf{A} - \mathbf{I}\vec{p} \tag{3.74}$$

where

$$\vec{p} = C\alpha\exp[\alpha(\vec{x} - \vec{x}_0)] \tag{3.75}$$

The $\mathbf{I}\vec{p}$ operation converts a vector to a square matrix with its values placed on the diagonal.

Equation 3.72 contains an inverse of the Jacobian, $\mathbf{J}^{-1}$. Instead of computing the inverse directly, which we by now know is inefficient, we solve

$$\mathbf{J}\vec{y} = F(\vec{x}) \tag{3.76}$$

using any linear solver of choice. The solution is subsequently updated per

$$\vec{x}^{k+1} = \vec{x}^k - \vec{y} \tag{3.77}$$

since $y = \mathbf{J}^{-1}F(x)$. This algorithm is demonstrated by the following Python listing:

```
1   def NR(A,b)
2       for k_nr in range(20):
3           p = ...      # evaluate terms
4           I = np.identity(ni)
5           J = A - I*p;
6           F = A*x-b(x);
7
8           # use a linear solver to solve Jy=F
9           y = PCG_linear(J,F)
10
11          # get new solution
12          x = x - y
13
14          # convergence check
15          r = A*x-b(x)
16          if (norm(r)<tol): break;
17      return x
```

It is important to realize that the NR algorithm introduces a secondary solver iteration loop. The linear solver, here assumed to be given by `PCG_linear` uses its own internal iteration loop to find the solution to $\mathbf{J}\vec{y} = \vec{F}$ to its specified tolerance. The Newton-Raphson algorithm uses another tolerance value to test for convergence of its *non-linear* problem. It repeatedly calls the linear solver until convergence at the outer loop is achieved.

4 Introduction to C++

The three prior chapters covered the majority of the mathematical background relevant to scientific computing. The examples were presented mostly in Python. While Python is useful for prototyping, large computer simulations are typically written in high-performance languages such as Fortran, C and C++. While Fortran is commonly found in legacy codes, in this book we utilize C++ due its use in popular solver libraries (such as OpenFOAM) and being the foundation of other popular languages used in desktop (Java), edge (Arduino Programming Language), and high-performance computing (CUDA) environments. This chapter is a crash course on the C++ language. We learn about pointers, references, multidimensional arrays, custom objects, polymorphism, operator overloading, templates, and standard library storage containers. We demonstrate these concepts by developing a C++ version of the tennis ball integrator and a heat equation solver from prior chapters. We also learn how to use Paraview for data visualization.

4.1 C AND C++ BASICS

C and C++ tend to get a lot of bad press and new programmers are led to believe that they are a difficult language to learn. While C++ is a complex language, this is not exactly correct. The overall process of programming in the legacy C language is quite similar to what you are already familiar with from MATLAB® or Python. The primary difference between these languages arises from C being "closer to the metal" by utilizing code constructs that directly map to the underlying hardware architecture. This then leads to some confusion, especially for students unfamiliar with the basics of computer architecture, including the use of addresses to access data in the main system memory. C++ added a new programming paradigm called object-oriented programming (OOP). While OOP is now found in many languages (including Python and MATLAB), it is not commonly used in introductory numerical analysis courses and thus may be unfamiliar. OOP allows one to write modular code that is easier to maintain while being safer due to allowing the program to force data access through validation algorithms. Understanding of OOP is necessary for interfacing with external libraries as all modern software libraries depend on it. We devote the second half of this chapter to OOP concepts.

4.1.1 COMPILERS AND DEVELOPMENT ENVIRONMENTS

C and C++ are compiled languages. Instead of running through an interpreter at run time, the source code is first compiled into a binary executable which is then launched by, for example, double clicking on the executable `.exe` file

DOI: 10.1201/9781003132233-4   **106** in the file explorer. C++ code can be fully developed using a text editor and compiled using command line tools. However, it is more common to utilize a GUI-based integrated development environment (IDE). The IDE allows one to write and launch the code, just as was done using Spyder for launching Python scripts. The IDEs also provide support for code debugging and for benchmarking performance, as discussed in Chapter 8.

Developers using Microsoft Windows tend to utilize Microsoft Visual Studio. This program provides both the graphical front end as well as the necessary compilers. It is possible that your university or company computer already has Visual Studio installed. If not, it can be downloaded from visualstudio.microsoft.com. Visual Studio is offered in several variants. There is a paid professional version, as well as the free Community Edition. The free edition is plenty sufficient to follow along with this book. On Linux, you may more likely encounter Eclipse or NetBeans. These IDEs do not come with compilers. Instead, they utilize the command line GNU Compiler Collection gcc, which is the de facto standard for building C/C++ (as well as Fortran, Go, D, and many others) on Linux. There is a good chance your system already includes gcc, but if not, on Ubuntu, we can install it with the following commands:

```
$ sudo apt update
$ sudo apt install g++
```

This will get you the basic set of tools for compiling C++ code. Alternatively, you can install build-essentials to include additional applications such as make. Test that gcc is properly installed by executing

```
$ g++ --version
g++ (Ubuntu 12.3.0-1ubuntu1~23.04) 12.3.0
```

Make sure you have at least version 5.0 as prior versions do not implement all features utilized in this text.

Next, paste the following code

```
1  int main() {
2      return 0;    // normal exit
3  }
```

into your project's source file (you may need to use the menu bar or the project explorer to manually add a source file if one is not added automatically. Then use the IDE's Run icon (typically a triangle) to compile and launch the code.

Alternatively, if using Linux and you prefer to work from the command-line, save the content to a file called cpp_example.cpp. Build and run the code with

```
$ g++ -O2 cpp_example.cpp -o cpp_example
$ ./cpp_example
```

C++ code is typically built in one of two modes: Debug and Release. The compiler is capable of analyzing the source code to re-arrange different algorithms to improve performance (or to decrease the file size). This is called *optimization.* Optimization in gcc is enabled using the -O2 compiler flag. The number, with valid values including 1, 2, and 3, indicates the level of optimization. There are also additional flags that activate optimizations specific to the hardware in use. Such options provide performance at the cost of reduced portability. When a new code is written, it is common to find that it does not work as expected or that it crashes prematurely. The code needs to be *debugged.* Debugging involves using *breakpoints* to stop the program flow at a particular location and stepping through the problematic section line-by-line. Optimization is turned off when building the Debug version to maintain consistency between the source code and the binary. This is done by excluding the -O2 flag (or replacing it with -O0). The -g flag is included to write out symbols (variable and function names) into the application file. IDEs typically include some toolbar option to quickly switch between Debug and Release modes. Finally, the C++ codes in this book utilize features added with the 2011 revisions to the language standard called C++11. By now, support for C++11 is ubiquitous, however older compilers require that the code is compiled with the -std=c++11 flag.

The previous example also illustrate how to use the -o flag to specify the output file name. Linux executable do not require extensions, the way .exe is used on Windows. Instead, the file must have the execute permissions set, which is done automatically by the compiler. However, if needed, the flag could also be set using

```
$ chmod u+x cpp_example
```

The code was then launched by prepending the executable file name with ./ characters. By default, the local directory (.) is not part of the executable search path, and the ./cpp_example syntax indicates to launch the file called cpp_example located in the current folder. In the absence of the -o flag, gcc writes the program to a file called a.out (Linux) or a.exe (Windows).

Linux commands can be chained together with &&. The second command starts only if the first one succeeds, which is indicated by its return value. || is used to run another program in case of failure. The prior two lines can be combined into a single one reading

```
$ g++ -O2 cpp_example.cpp -o cpp_example && ./cpp_example
```

Finally, large simulation programs are typically split into multiple files to aid in readability and to reduce compilation time. We discuss the use of build systems in Chapter 8. For now, it is sufficient to know that you can include multiple source files in the call to gcc. These files may require *header files* (discussed shortly) that are located outside the local directory. Additional header search directories are specified with -I as in

```
$ g++ -O2 file1.cpp file2.cpp file3.cpp -I ~/library/inc -o my_app
```

Other commonly encountered flags include -L to specify additional library directories, -l to include precompiled libraries, -D to define macros, -c to only compile the code without linking, **-E** to export the preprocessor output, and **-S** to export the assembly version of the compiled code.

4.1.2   MAIN FUNCTION AND BASIC CODE STRUCTURE

The example code from the prior section implements a function called **main**. While functions are optional in Python or MATLAB, every C/C++ program must define, at the minimum, this single **main** function. This is where the program begins. It does not matter in which file (in multi file programs) **main** appears, but there must be one, and only one, **main** somewhere. The body of the function is placed within a pair of curly braces, **{ }**. Unlike Python, C++ is not white-space conscious and instead braces are used indicate blocks. Semicolons are used to terminate statements. Multiple statements can be placed on the same line. In fact, the entire C++ code could be written as a single line with limited spacing but this would make the code hard to read. Semicolons are not needed after the block-closing right brace, except in the case of structures and classes (more on this later). Flow-control commands, such is loops or conditional statements can execute just a single command, for example

```
if (condition) do_something();
```
or they can execute multiple commands placed inside a brace-delimited block:

```
1    if (condition) {
2        do_A();
3        do_B();
4        ...
5    }
```

The **int** preceding the function name is the *return type*. It specifies that this function returns an integer (a whole number). In the case of **main**, this return value is passed back to the operating system. By convention, the zero value indicates a normal exit. The operating system does not do anything with these values but they are regularly used in *bash scripts* (Bash is the name of a popular Linux console shell). We have in fact already experienced the use of a command return value by utilizing **&&** to chain compilation and program execution. The compiler returns 0 if the compilation is successful, which then leads to the shell launching the second command. The **main** function can also receive optional arguments that capture additional values specified on the command line. The syntax for this alternate version is

```
1    int main(int num_args, char* args[]) {
2        return 0;
3    }
```

The first argument is the number of command line arguments, which includes the executable file name as the first argument. The second argument is an array of C-style strings, defined as character arrays (more on this in the next section), that contain the individual white-space separated arguments. Let's consider this alternate example:

```
1  #include <stdio.h>
2  int main(int num_args, char* args[]) {
3      printf("num_args is %d\n",num_args);
4      for (int i=0; i<num_args; i++)
5          printf("%s\n",args[i]);
6      return 0;
7  }
```

Compiling and running this example (found in `args.cpp`) with some arguments as in

```
$ g++ args.cpp -o args
$ ./args AA 22 CC
```

prints

```
num_args is 4
./args
AA
22
CC
```

Note that all these values are strings. As needed, they can be converted to numeric values using functions such as `atoi`, `atof`, or `stod`.

Here we also demonstrated the use of the `printf` function to produce screen output. This legacy C function uses `%` parameters to format the output string with syntax hopefully already familiar from past work in MATLAB or Python. The signature of this function (its return value and the types of arguments), called the function *prototype*, is found in a file called `stdio.h`. This file is located in some system directory where the compiler was installed. It is part of the C standard library. The `#include` statement loads this file into the program so the compiler can generate the appropriate function call.

4.1.3 VARIABLES

Unlike Python, C++ is a *typed* language. Each variable must be declared before use. The declaration specifies the type of data this variable is to hold. This type cannot subsequently change. We talk more about data types in Section 4.1.7. For now, it is sufficient to know that they fall into three main categories: integers, real values, and logicals. Commonly encountered data types corresponding to these three categories are: `char` and `int` whole number integers, `float` and `double` for single and double precision real numbers, and `bool` true / false logical types. Character strings are stored as arrays of small integers called chars. C++ also gives us the flexibility to define custom container objects built around these basic types. We learn about this topic in Section 4.4. Such a custom data type is used in modern C++ to store character strings using the `string` type.

Variables are declared by prepending the variable name by its type. C and C++ variables and other named objects are case-sensitive: `myVar` is different entity than `MyVar`. Variable names can contain a mix of letters, numbers, and the underscore character, but they cannot start with a number. Multiple variables of the same type can be declared in a single statement by separating the variable names with a comma. In classic C, variable declarations had to be placed at the top of a code block, but this is no longer the case. In fact, C++ standard recommends declaring variables as close to the point of first use as possible. The code below illustrates the declaration and initialization of several variables:

```
1  int main() {
2      int n0 = -123;      // 32-bit integer, historically 16-bit
3      float y0 = 2.1;      // single precision real number
4      double z0 = 3.4;     // double precision real number
5      bool save_data = false; // true / false boolean
6      char sep = ',';      // character value
7      unsigned long int np; // uninitialized unsigned 32-bit integer
8      return 0;            // normal exit
9  }
```

C++ variables are not initialized when declared. They should be assumed to contain "garbage" values that existed in the memory space where the variable happened to be placed. Variables thus need to be initialized prior to use, ideally during declaration. In the above example, the value of **np** is not known, and attempting to use it in an expression leads to an undefined behavior.

We often also need read-only objects for storing values such as physical constants. This is done by including the **const** keyword,

```
const double KB = 1.3806503e-23;          // Boltzmann constant
```

Alternatively, **constexpr** is used to specify expressions which can be evaluated at compilation time. This keyword can lead to additional code optimization since the value may be substituted directly instead of utilizing memory space.

4.1.4 COMMENTS AND PREPROCESSOR DIRECTIVES

The compiler ignores any text starting with **//** until the end of the line. Multiple lines are commented out by enclosing them between **/*** and ***/**. This was in fact the only type of comments originally supported by C, although modern C compilers understand the C++ single line comments as well. One caveat about the multi-line comments is that they cannot be nested. The following leads to syntax error:

```
1  /*
2      /* initialize data */
3      int np = 123;
4  */
```

This issue may be encountered when attempting to comment out a large block of code (such as an entire function) that contain some C-style comments. In such cases, it is customary to utilize *preprocessor directives* to exclude the code, as in

```
1  #if 0
2  ...
3  #endif
```

C or C++ code is actually built using three steps. The code is first run through a *preprocessor* which looks for special directives starting with # signs. We have already encountered one of them, `#include`. That directive is replaced by the contents of the specified file. The resulting output is then run through the *compiler*. The compiler translates each source file individually into an object file. These object files are then *linked* together into the single application. Running the source code through `g++` as we have been doing so far leads to all three steps executed automatically, with the temporary files discarded. We will see later in Chapter 8 how to separate the compilation from linking in order to develop software libraries.

The `#if` directive causes the section of the input file until the following `#elif`, `#else`, or `#endif` to be included only if the specified condition is true. Since zero is always false, by definition, the enclosed section is effectively commented out. This construct can also be used with `#ifdef`, which evaluates to true if the specified macro is defined,

```
1  #ifdef FAST_SOLVER
2  void solver() { /*algorithm 1*/ }
3  #else
4  void solver() { /*algorithm 2*/ }
5  #endif
6  int main() {solve(); return 0;}
```

`#ifndef` tests for undefined expressions. These macros can be defined during program build using the `-D` compiler flag. They can also be specified in project settings of the IDE. Compiling the code using

```
$ g++ -DFAST_SOLVER solver.cpp
```

leads to the program utilizing the, presumably faster, algorithm 1. Otherwise, compiling without this parameter leads to the second algorithm being included. This substitution happens directly on the input *source file*, and the compiler does not even "see" the alternate code. Only one of the solver functions is actually present in the binary output. As such, this approach is useful for excluding code portions that should not be found in the general public release, such as intellectual property owned by a particular customer. Macros can also be defined programmatically using

```
1  #define MY_MACRO
2  #define PI 3.14
3  #define DIST2(x) (x[0]*x[0] + x[1]*x[1])
```

Here we demonstrate three possible uses. The first line simply defines the macro, so that a test within `#ifdef` evaluates to true. The second option leads to any instance of PI in the source file being replaced with the literal string 3.14. Note that a semicolon is not used as otherwise the semicolon would also be substituted. The final option illustrates the possibility to use macros as functions. Here note the presence of the parentheses. They are needed since the expression is substituted directly. Without the parentheses, code such as

```
double z = 2*DIST2(vec);
```

would evaluate to

```
double z = 2*vec[0]*vec[0] + vec[1]*vec[1];
```
which is incorrect.

4.1.5 FUNCTIONS

We already saw how to define the `main` function. Other custom functions are generated with the same syntax. A function definition consists of the return type, the function name, and the list of function arguments specified as a comma-separated list within parentheses. This is followed by the function body. For example:

```
1  double eval(double t, double x) {
2      double z = 2*x;   // some example code
3      return t*t + z;   // return value
4  }
```
We use this function as

```
double val = eval_y(0.2, x);
```
A function can use the return type of `void` to indicate that it does not return a value. C and C++ functions can directly return only a single item.

One peculiarity of C and C++ is that the compiler does not "look ahead". The following code will not compile:

```
1  void run() {
2      double val = eval(0.0, 0.2);
3  }
4  double eval(double t, double x) {
5      /* ... */
6  }
```
The reason is that the `eval` function is introduced only after the attempted use. When the compiler gets to line 2, it does not yet know that `eval` is a function or what arguments it expects. The obvious solution is to move the `eval` definition above `run`. This is not always feasible either due to stylistic choices (organizing the functions in some logical order) or due *circular depen-dence*. This latter issue is encountered if, for this example, `run` needs to call `eval`, but `eval` also needs to call `run`. This difficulty is avoided by splitting the function definition into a separate declaration and implementation. The declaration is provided by the function *prototype*. This is just the first line of the function definition, without the body. Here is an example:

```
1  double eval_data(double x);   // declaration, prototype
2  void run() {
3      double val = eval_data(0.2);
4  }
5  double eval_data(double x) { // implementation
6      /* ... */
7      return some_value;
8  }
```
The prototype is used by the compiler to generate the appropriate function calls using symbolic names. The linker subsequently replaces these links with

the actual instruction jumps to the appropriate location where the function body resides. This is what is meant by "linking". This build approach also dictates that functions that ought to be *inlined* (as suggested by an `inline` keyword) need to have their body placed in the same compilation module where they are being used.

One interesting characteristic of C and C++ is the ability to have multiple functions with the same name. This is known as *function overloading*. These functions need to have different arguments, or at least a sufficiently different return type, so the compiler can distinguish among them.

```
1  double my_fun(double a);
2  double my_fun(int a, double z);
```

Function arguments can also be assigned default values. These arguments need to be placed at the end of the list. For example

```
double solve(double tol, int max_steps = 5000);
```

this function can then be called such as `solve(1e-6)`, in which case the solver runs for up to 5000 iterations (assuming this function implements an iterative solver). Alternatively, we can limit the number of solver steps using `solve(1e-6, 100)`. It is also possible to have functions with a variable argument list. For this, the reader is directed to external references on the language specification, such as Stroustrup 2018.

4.1.6 HEADER FILES

Function prototypes are customarily placed in the already-mentioned header files and are loaded using the `#include` directive. The file name is included between `< >` or `" "`. These those options modify the default search path. The former option is used to load standard and third party library headers from system directories. The second option is used for files that are part of the application being built. In other words, the header files that *you* write are included with

```
#include "Solver.h"
```

Header files are just regular C or C++ style files. As was already mentioned, the `#include` line is replaced by the file's content by the preprocessor. Headers can, and usually do, include other headers. This is needed, to in order to load declaration of custom data types, as discussed later in Section 4.4. We may have a header, called here `Solver.h` that reads

```
1  #include "World.h"
2  void solve(World &world);  // some function that requires World
```

The argument list of the `solve` includes some yet unfamiliar syntax. It is sufficient to realize that this is a prototype of some arbitrary function that requires a definition from `World.h` header.

We can also have another similar file, called let's say `Output.h` that also similarly needs access to `World`. It would read

```
1  #include "World.h"
2  void output(World &world);  // another function requiring World
```

We may then have some other file, let's call it `Main.cpp`, that implements code that requires access to both the solver and output functions. This file would thus use

```
1  #include "Solver.h"
2  #include "Output.h"
```

As both headers themselves include `World.h`, the content of this file is included twice into `Main.cpp`. This could lead to compilation errors due to *duplicate declaration* (i.e. variables or objects declared during the first inclusion are declared again during the second inclusion). This issue is mitigated by wrapping the header content within *header guards*. They utilize preprocessor directives discussed previously in Section 4.1.4 to selectively include the file content only if it has not been included already. They are used as

```
1  #ifndef __SOLVER_H
2  #define __SOLVER_H
3  /* file contents */
4  #endif
```

The first inclusion of the header into the compilation unit leads to the `#ifndef` condition evaluating to true, since the macro `__SOLVER` is (presumably) not yet defined. Subsequent attempts to include this header file result in no code lines actually included, as the `#ifndef` statement now evaluates to false.

4.1.7 DATA TYPES

So far, we have been utilizing data types such as `int` or `double` without providing additional detail about them. Integer and floating point variables come in several variants that alter the range and resolution of values that can be held by the variable. It is important to realize that the computer main random access memory (RAM) does not have different banks for different data types. Instead, RAM is essentially a collection of extremely many 0 or 1 *bits*. These bits are further organized into groups of 8 called *bytes*. Another term encountered in computer storage is *word*. This is a collection of bytes, usually 8 on 64-bit architectures, that the computer uses for memory addresses. The different variable types map to different byte sizes. This size can be sampled using the `sizeof` keyword. The types also control the "meaning" behind the bits. A 32-bit integer and a single precision floating point variable both occupy 4 bytes, but the bit values are interpreted differently.

A *char* stores exactly one byte of information. Each bit represents an increasingly higher power of 2. The actual value of a char is determined by adding the contribution from all bits. This is the basic principle behind *binary mathematics*. A single byte can store any value up to, and including $2^0 + 2^1 + 2^2 + 2^3 + 2^4 + 2^5 + 2^6 + 2^7 = 1 + 2 + 4 + 8 + 16 + 32 + 64 + 128 = 255$. With 8 bits of information, we can store whole numbers from 0 to $2^8 - 1$. An `int` integer used to store 2 bytes (16-bits), but nowadays represents 4-bytes (32-bits). A 32-bit integer thus holds values from 0 to $2^{32} - 1$, or $4,294,967,295$. Historically, we had to use a `long` qualifier to get a 32-bit integer, but these days, there is no difference between an int and a long (or long int). A 64-bit integer is available via long long. A 16-bit integer is obtained using short. The ordering of bytes in memory is known as *endianness*. The 4 bytes making up a 32-bit integer are written out left-to-right in a big endian system, and right-to-left in a little endian system. This is something one needs to pay attention to when saving simulation results using binary representation. A file written on a system of one endianness will require byte-wise rearrangement when loaded on a computer utilizing the other convention.

Negative values are represented by convention by setting the *highest-order bit*. Consider the following set of eight bits:

bit	7	6	5	4	3	2	1	0
value	1	0	1	0	1	1	0	1

Data bits are ordered from right to left, with the right-most value being the least-significant (2^0) bit. This sequence can represent

$$1 \times 2^7 + 0 \times 2^6 + 1 \times 2^5 + 0 \times 2^4 + 1 \times 2^3 + 1 \times 2^2 + 0 \times 2^1 + 1 \times 2^0 = 173 \quad (4.1)$$

But we can also let the 2^7 bit act as the negative sign. Then, this same sequence of binary digits represents

$$(-1) \times \left(2^5 + 2^3 + 2^2 + 2^0\right) = -45 \quad (4.2)$$

A signed 8-bit char can store values $\pm(2^7 - 1)$, or -127 to 127. A 16-bit signed integer has a range of $\pm32,767$. In C++, all variables are by default assumed to be *signed*. The unsigned keyword is used to declare the unsigned variant. This specification changes the way values are written to the screen and can also affects the result of mathematical operations in case of an *overflow*. You will often encounter variables of type size_t. This is just a "nickname" for an unsigned integer. It is often used with functions that require the size of some object, which by definition must be nonnegative.

The bool type holds true or false logicals. These are stored using a single byte. Casting false to an integer returns 0, while a true converts to 1.

4.1.8 STRINGS

Chars are conventionally used to store character strings. Computers have historically relied on a lookup table, called ASCII, to map the numeric value held by the char to the character code. For example, the value of 65 produces the capital letter 'A'. The lower case 'b' is 98, '7' is 55, '+' is 43, space is 32, and '=' is 61. With these codes, the text "A + b = 7" is represented by the following sequence of byte values:

64 32 98 32 61 32 55 0

The final 0 indicates the end of the string. This is at least how strings were store in legacy C codes. A string buffer was allocated such as

char text [10];

This buffer can hold a string containing up to 9 characters along with the single 0-value termination byte. An assignment such as

```
text = "Hello World";   // buffer overrun
```

leads to memory corruption, known as *buffer overrun* since the value being assigned is two characters too long. Nowadays, C++ defines a storage object called **string** to store strings. This object automatically resizes the storage buffer during assignment to avoid these memory errors that tend to plague legacy C applications. Additional character maps, such as Unicode or UTF-8 provide the capacity to capture an expanded character set.

4.1.9 FLOATING POINT MATHEMATICS

Real values are represented using a standard devised by IEEE 754. A 32-bit block of memory is used to capture 3 different pieces of information: a 1 bit negative sign (s), an 8-bit exponent (E), and a 23-bit fraction (or mantissa), M. These three components build the value per

$$\text{value} = (-1)^s \times 2^{(E-127)} \times \left(1 + \sum_{i=1}^{23} M_{23-i} 2^{-i} \right) \tag{4.3}$$

where M_i are the bits of M. C++ offers two main types of floating point data: single precision *float* and double-precision *double*. There is also a *long double* but this one is rarely encountered. Some architectures, such as graphics cards, support half-precision floats. The **double** type follows the same format as described above, except that it uses twice as many bytes: 64 instead of 32. The exponent is now 11 bits long, and the mantissa uses 52 bits. Simulation codes often use double precision in order to avoid numerical round off errors. A single precision floating point number can represent values between approximately 1.175×10^{-38} and 3.403×10^{38} without a loss of precision. This range may be seem sufficient for a typical simulation, but danger lurks beneath!

The smallest value that can be added to a mantissa in a single precision float is 2^{-23} or approximately 1.19×10^{-7}. In practical terms, this quantity, called the *machine epsilon*, is the smallest number that can be added to 1 without being lost,

$$1 + \epsilon > 1 \tag{4.4}$$

This in turn implies that we cannot add two values if they differ by more than seven orders of magnitude. This is not a large difference, especially in numerical simulations. While this limitation could be mitigated by first accumulating the small numbers before adding the resulting sum to the total, such approaches are not always practical. Double precision decreases ϵ to $2^{-52} \approx 2.22 \times 10^{-16}$. The downside is the doubled memory requirement. Some hardware architectures, such as graphics cards, discussed in Chapter 9, were initially designed to work with single or even half precision. Use of double precision leads to a significant decrease in performance. A reduction in performance is also observed on desktop CPUs, if for no reason other than the increased number of bytes that need to be transferred from the memory.

Yet even using double precision, it is important to be aware of *round-off errors*. Computer floating-point arithmetic is not precise. Calculations that are algebraically equivalent may not be so when implemented on a computer. A well-known real world example involves the failure of a Patriot missile to intercept an incoming projectile during the 1991 Gulf War. The interceptor was using a scheme in which the time since launch was computed by a repetitive addition of a small time step - just as we have been doing with the tennis ball example, i.e. something along the lines of

```
1  double time = 0;
2  for (int i=0; i<num_steps; i++)
3      time = time + dt;
```

After a large number of iterations, a small round off error in the addition accumulated to sufficiently alter the total elapsed time. This is because, when implemented on a computer,

$$\sum^{n} \Delta t \neq n\Delta t \tag{4.5}$$

Due to the fast speed of the missile, even a small error in the total elapsed time led to significant difference in the calculated position.

We need to keep in mind the following two rules when working with floating point arithmetic:

1. Use integer addition whenever possible. In the above example, the following holds exactly true

$$\sum^{n} I = nI \tag{4.6}$$

if I is a variable of integer type.

2. Never compare equality of two finite floating point values directly. For example, the following two expression evaluate *mathematically* to the same result:

$$a = -2.3 + 5.3$$
$$b = 0.1 * 30$$

However, there is no guarantee that $a == b$ when this calculation is implemented on a computer. Instead, you should always test the relative difference against some tolerance,

$$\frac{|a - b|}{(|a| + |b|)/2} < \epsilon \tag{4.7}$$

where ϵ is some small number (such as 10^{-7}).

4.1.10 INTEGER DIVISION

There is one more important point to make about data types. C, C++ and other derived languages distinguish between floating point and integer division. The latter happens when both the numerator and the denominator are of integer types. While mathematically, the following three expressions are identical, they produce different results when implemented in C or C++:

```
1   double result;
2   result = 34 / 5;      // 6
3   result = 34.0 / 5;    // 6.8
```

The first division involves both integral nominator and denominator and hence the integer division code is generated. The second contains a real-valued nominator and hence 6.8 is produced.

Important feature of an integer division is that the fractional part is discarded. In the first version, the result is 6, not the nearer 7. The remainder after an integer division can be obtained using the modulus operator %.

```
int rem = 34%5;          // 4 = 34−5*6
```

The modulus operator is often used to implement *strides* in which some operation takes place only every N steps,

```
if (step%N == 0) { /* do something */}
```

Integer division can inadvertently lead to errors. The following code uses arrays and loops, which we have not yet discussed, in an attempt to initialize an n-sized vector to equidistant values in $[-1, 1]$:

```
1   const int n = 10;
2   double x[n];                    // n−sized array of floating point values
3   for (int i=0;i<n;i++)          // loop for i=[0,...,n−1]
4       x[i] = −1.0 + 2*i/(n−1);   // −1 plus value in [0,2]
```

This is however incorrect, as can be proven by outputting the values to the screen:

```
5   for (int i=0;i<n;i++)
6       printf("%.1f ",x[i]);     // one digit after the decimal point
7   printf("\n");                  // new line
```

Running the program, found in **printf.cpp**, leads to

```
-1.0 -1.0 -1.0 -1.0 -1.0 0.0 0.0 0.0 0.0 1.0
```

As you can probably guess, the culprit is the integer division on line 4. To correct it, we need to convert at least one of the four terms to a floating point type. There are many ways, including

```
1   x[i] = −1.0 + 2.0*i/(n−1);        // real value in nominator
2   x[i] = −1.0 + 2*i/(n−1.0);        // real value in denominator
3   x[i] = −1.0 + 2f*i/(n−1);         // f suffix, not recommended
4   x[i] = −1.0 + 2*(double)i/(n−1);  // type cast
```

On the first and second line, we include the .0 fractional part to convert an integer to a double-precision floating point value. The third line accomplishes a similar effect using the f suffix. We need to be careful here, however as this suffix converts the number to the single precision **float**, which may not be desired. Finally, the last line we use (double) to explicitly *cast* the type of i to double precision. Here we use the cast to convert a real value to an integer. The reverse cast, in which we convert a real value incremented by 0.5 to an integer, is used to round to nearest:

```
int rounded = (int)(real_value+0.5);
```

Instead of displaying the values to the screen, we could use fprintf to save data to a file. This function is found in stdio.h just as printf. The example below demonstrates how files are opened, written to, and closed:

```
1  #include <stdio.h>
2  int main() {
3      float x = 1.2;
4      FILE *out = fopen("results.txt","w");  // open for writing
5      fprintf(out,"x = %.2f,  x^2 = %.2f\n",x,x*x);  // write string
6      fclose(out);                            // close the file
7      return 0;
8  }
```

This example uses, text (or ASCII) output format. Files can also be opened for binary output (or input) in which case the individual buffer bytes are written out. Such format can be used to write out restart files, for example. Due to the already discussed endianess issue, nowadays it is customary to convert binary data to a text-like format through *Base64 encoding*.

4.1.11   STREAM INPUT AND OUTPUT

This use of C-style printf and fprintf has gone out of fashion in C++. Modern codes rely on *stream*-based input and output. The benefit of this scheme is that data formatting is automatically based on the variable type (although there are directives to control the width and precision of numeric data, such as setw, setprecision, and scientific). Support for screen output is provided through the <iostream> header. Note the lack of the .h extension. This header defines std::cout and std::cin objects that output to the standard console output (the screen) and read from the standard console input (the keyboard). The following code, found in stream.cpp, illustrates their use:

```
1  #include <iostream>  // support for cout and cin
2  int main() {
3      double val;
4      std::cout<<"Enter a number: ";
5      std::cin>>val;  // read a real value from the keyboard
6      std::cout<<"("<<val<<")^2 = "<<val*val<<std::endl;
7      return 0;
8  }
```

Running the code from the command line, we obtain

```
$ g++ stream.cpp -o stream && ./stream
Enter a number: 2.56
(2.56)^2 = 6.5536
```

The above streams are defined within a *namespace* called std, for "standard library". Namespaces were introduced in C++ to differentiate between functions of identical names belonging to different libraries. Think of them as the last (family) names for the function first names. Since constantly typing std:: can become tiring, we can use

using namespace std;

to make the standard library (STL) objects visible without needing fully qualified names. The **endl** operator inserts a line break and also flushes (writes out) the stream buffer. When writing to a file, we generally prefer to use n to prevent excessive file output as it leads to slower performance. Using n, data is written out only once the buffer fills up or when the file is closed. The **flush** stream member function is used to force file output.

Streams associated with input and output files are opened using data types **ifstream** and **ofstream** found in **fstream** header. C++ file output looks like this:

```
1  #include <fstream>
2  using namespace std;
3  int main() {
4      float x = 1.2;
5      ofstream out("results.txt"); // open file for writing
6      out<<"x = "<<x<<", x^2 = "<<x*x<<"\n"; // write to file
7      return 0;
8  }
```

Note that we no longer need to close the file. It is closed automatically when the **out** variable is destroyed at the conclusion of **main**.

4.1.12 VARIABLE SCOPE

A C or C++ code consists of a series of blocks delimited by { and }. These blocks serve a dual, but related purpose. In the case of loops or conditional statements, they indicate the code to be repeated or executed. They also delineate bodies of functions or custom object definitions. However, they play another important role. Each block defines a new *scope* for declared objects. Imagine that the program starts as a blank piece of paper. This is the *global* scope. Each new block, whether it be associated with a function or a general program statement, is analogous to placing a sticky note on top of the block parent's sticky, or in the case of a top-level function, directly on the original blank canvas. Whenever the compiler encounters a reference to an object, such as a variable, it first checks if it has already been declared in the local scope. If not, it checks the parent scope. This process continues until the global scope. Whenever the block ends, the sticky gets peeled off and tossed into trash, along with any variables that may have been declared within it.

Understanding variable scope is extremely important. It is also a source of confusion to new programmers in C and C++. This confusion is due to the language having no objection to multiple variables sharing the same name, as long as they are declared in different scopes. Take a look at the following example:

```
1  int a = 0;
```

```
2  void F() {
3      int a = 1;
4        {
5          int a = 2;
6        }
7  }
```

Here we have three unique variables, each with their own memory storage location, despite all having the same name a. Let's now review a slightly altered version:

```
1  int a = 0;              // global variable
2
3  void F() {
4      a = 1;              // assign to the global a
5      int a;              // local variable
6      a = 2;              // assign to the local a
7      {
8          int a = 3;      // another local variable
9          cout<<a<<endl;  // prints 3
10      }
11      cout<<a<<endl;     // prints 2
12  }
13
14  void main() {
15      F();
16      cout<<a<<endl;     // prints 1
17  }
```

We begin by declaring a global variable a. The code main function then calls function F. The first line of this function is the assignment a = 1. A variable of this name has not *yet* been declared in the local scope. The nearest higher-up a is the one found in the global scope and hence global::a is set to 1 (note that global::a is not a valid way of referring to this variable in C++, but is used here to distinguish the variables). On line 5, we declare a local variable also called a. The assignment on line 6, despite looking identical to the one line 4, assigns the value to F::a. On the next line, we start a new block. The next declaration and initialization creates and assigns a yet another a, we can call it F::::a. The following cout uses this most local a with value of 3. Subsequently, the code reaches the block end, at which point the variable a created on line 8, F::{}::a is destroyed. The subsequent cout statement uses the variable in its most local scope, which is the a declared on line 5, F::a. Next, the function F() reaches the end of its code block, which causes this a variable to be destroyed. After control returns to main, the cout on line 16 prints 1, as the only a still in existence at that point is the one declared in the global scope on line 1.

Understanding of scope is also important for function calls. Consider the following example:

```
1  void F(int a) {
2      a = 1;
3  }
4
5  void main() {
```

```
6        int  a  =  0;
7        F(a);
8        cout<<a<<endl;
9    }
```

It may seem intuitive that this program prints 1, but this is not the case. Again, despite sharing the same name, the two a variables are unique. The above code is analogous to (note this again is not valid C syntax, so don't try to compile this code):

```
1    void F() {
2        int F::a = main::a;     // copy value of main::a to F::a
3        F::a = 1;
4    }    // F::a is destroyed here
5
6    void main() {
7        int main::a = 0;
8        F();
9        cout<<main::a<<endl;
10   }
```

4.1.13 STATIC VARIABLES

In the prior section we noted that local variables are deleted at the end of a code block, such as a function definition. There is however one exception. Function member variables can be declared with the **static** keyword. Such a variable remains in existence even after the function terminates. Static variables in the essence act as global variables that are accessible only from the declared function. Consider the following code:

```
1    void F() {
2        static int a = 0; // not needed, auto-initialized to zero
3        a++;                    // increment a
4        cout<<a<<endl;
5    }
6
7    void main() {
8        F();
9        F();
10       F();
11   }
```

This code prints

```
1
2
3
```

Such variables were used in legacy C codes when a function needed to "remember" some data, such as a calculation remainder from a prior iteration. Nowadays, the same functionality can be accomplished more elegantly utilizing custom data object member variables, see Section 4.4.

4.1.14   CONDITIONAL STATEMENTS

Conditional statements are introduced using the `if` statement. The test expression can consist of multiple statements grouped with logical *and* and *or* operations. These are written as **&&** and **||**, although the verbose **and** and **or** is also supported. Consider the following example:

```
1    if ((x>=a && x<b) || b>c) {
2      /* do something */
3    }
```

The code executes if either $x \in [a, b)$ or $b > c$. We use **==** to check for equality. Inequality is checked with **!=** and logical expressions can be negated using **!**.

A common mistake involves forgetting the second equal sign in a conditional test, as in:

```
1    int a = 0;
2    if (a = 1) doA();      // incorrectly using = instead of ==
3    else doB();
```

While we would expect the code to call **doB**, it calls **doA**. Instead of comparing the value, the single equal sign assigns 1 to **a**. A logical expression without a conditional expression is interpreted as a boolean check in which any non-zero numerical value evaluates as true. Therefore, the compiler treats the code on line 2 as if consisting of two steps:

```
1    a = 1;
2    if (a!=0) doA();
```

Modern compilers usually produce a warning when encountering this typo.

C also provides *bitwise* AND, OR, and XOR operations denoted using **&**, **|**, and **^**. The resulting value is computed by considering the individual bits. For example, the result of 12 & 9 is 8, since the binary representation of the first operand is **1100**, while the second one reads **1001**. The bitwise AND operation sets the result to 1, if, and only if, both inputs are 1. Thus, the result is **1000**, or 8. Such operations are commonplace in legacy codes which attempt to save memory by utilizing a single integer to store multiple logical states instead of utilizing individual boolean variables. They are then checked using constructs such as `if (flags & 4) { ... }` to perform some action if the 3rd least significant bit (the one corresponding to $2^2 = 4$) is set. Use of the bitwise **&** in conditional statements where **&&** was intended is another source of errors one needs to pay attention to.

Multiple conditional statements can be chained together using `else if` and `else`. Lets say that we want to perform some logic according to a provided operation variable. Valid values are 1, 2, and 3. For any other value, we perform the default action. We implement this logic as:

```
1    if (op==1) { /* do action A */}
2    else if (op==2) { /* do action B */ }
3    else if (op==3) { /* do action C */ }
4    else { /* do default action */ }
```

Comparisons against multiple cases like this are sufficiently common that C++ provides another way of implementing branching using `switch`. This

operand takes as an argument an integer value. The body specifies individual
`case` sections for each possible value. One peculiar feature of this statement is
that code blocks in individual cases do not require curly braces. The code runs
until the end of the entire switch statement, including through the subsequent
cases! To prevent this, we use the `break` keyword to jump out of the `switch`
block. The code in the previous listing can be written as

```
1  switch (op) {
2      case 1: doA(); break;
3      case 2: doB(); break;
4      case 3: doC(); break;
5      default: doDefault();
6  }
```

4.1.15 ENUMERATIONS

One challenge with the prior example is that we need to remember that the
flag value 1 implies algorithm A, 2 is B, and so on. This is where enumerations,
declared using `enum` come in. The following line

```
enum Operation {OP_A, OP_B, OP_C, OTHER};
```
declares a new data type called `Operation` that can have one of the listed
values. These are internally stored as consecutive integers. We assign the value
as

```
Operation op = OP_A;
```
and check against it using

```
1  switch (op) {
2      case OP_A: doA(); break;
3      case OP_B: doB(); break;
4      case OP_C: doC(); break;
5      default: doDefault(); break;
6  }
```
C++ introduced a slight modification to these C-style enumerations. We in-
clude the keyword `class` as in

```
enum class Operation {OP_A, OP_B, OP_C, OTHER};
```
and subsequently use the enum type name when assigning values,

```
Operation op = Operation::OTHER;
```
This syntax helps prevent name clash, if multiple code sections use the same
enumeration keys. It also offers additional functionality such as being able to
define custom values for the keys.

4.1.16 CONDITIONAL OPERATOR

We often need to assign different values to a variable based on a condition.
We may write

```
float x;
if (a == 1) x = 5;
else x = -21;
```

where a == 1 is a placeholder for a generic test expression. C++ allows us to collapse this assignment using a ? : conditional operator:

```
float x = (a == 1) ? 5 : -21;
```

The part before the question mark is the test. If it evaluates to true, then the first value following the question mark is used. Otherwise, the value after the colon is used. We generally avoid the use of this operator in our programs as it tends to lead to a less-readable code. There are however instance when its use is required, such as for conditional setting of reference-type values.

4.1.17 LOOPS

The syntax for the for loop is:

```
for(initializer_list;exit_condition;post-iteration-code)
```

The loop statement consists of three semicolon separated parts. The first part contains code that is executed prior to the loop start. We generally use this section to declare a new local variable to be used as the loop counter. The second section contains some conditional statement. The loop continues to run while this statement is true. The final section is the code to be executed after each iteration. Here we typically increment the counter variable. The statement to execute can be a single instruction or a block of code. Here is a simple example that initializes values of an array to zero:

```
1  double data[size];
2  for (int i=0;i<size;i++) {
3      data[i] = 0;
4  }
```

The i variable exists only within the for loop. The braces are not required if only a single command is being iterated. It should be noted that although the for loop is typically used to repeat some instructions a specified number of times, the syntax is generic-enough to allow termination driven by other conditions. As an example, the following snippet is perfectly acceptable syntax-wise:

```
1  double x = 0.01;
2  for (;x<1.0;) {
3      x *= 1.01;
4  }
```

This code increments a value by 1% on each iteration. It continues for as long as the value is < 1. Note that both the initialization and the post-iteration sections are left blank.

Iterations of this kind are usually written using the while or do-while constructs. The above example can be written as

```
1  double x = 0.01;
2  while (x<1.0) {
3      x *= 1.01;
4  }
```

The do-while is similar except that the condition is checked at the end of the block. In practice this only affects the behavior in cases in which the test expression is false on the first pass. For the following example,

```
1   double x = 1.0;
2   while (x<1.0) {
3       x *= 1.01;
4   }
5   cout<<x<<endl;
6
7   x = 1.0;          // reset
8   do {
9       x *= 1.01;
10  } while (x<1.0);
11  cout<<x<<endl;
```

we obtain 1.0 and 1.01, respectively. On the other hand, identical result is obtain if x is initialized to a number < 1.

A for loop, such as the one used above to initialize the data array, can be rewritten using while as

```
1   // for (int i=0; i<size; i++) { data[i]=0; }
2   int i = 0;
3   while {i<size} {
4       data[i] = 0;
5       i++;
6   }
```

You may have noticed the use of i++ to increment the counter. This expression is a shortcut for i+=1, which itself is a shortcut for i=i+1. There are actually two variants: i++ and ++i. The versions differ in when the value is checked in logical expressions. Consider the following:

```
1   int i=0;
2   while ( (i++)<5 ) cout<<i;
```

The parentheses around i++ are not needed but are added for clarity. This code will output 12345. Now, if the increment is replaced with ++i, the code outputs only 1234. With the second version, the value is incremented prior to being checked for the condition i<5, which makes the loop run for one fewer iteration.

The examples demonstrated so far all utilized loops that execute until the loop-level exit condition is met. It is also possible to terminate a loop from "within" using the break command. For example

```
1   double x = 1.0;
2   while (1) {
3       x *= 1.01;
4       if (x>=1.0) break;   // terminate when x>=1
5   }
```

Here we use while(1) (or alternatively while(true)) to define an *infinite loop*. Now of course, we don't want an actual infinite loop in the code as otherwise the program will never end. We use break coupled with a conditional statement to provide the exit condition.

Another useful keyword is `continue`. It skips the rest of the block and starts the next iteration. Let's say we have an array containing n values and would like to compute the sum of the reciprocals,

$$S = \frac{1}{x_0} + \frac{1}{x_1} + \frac{1}{x_2} + \cdots + \frac{1}{x_{n-1}} \tag{4.8}$$

There is no guarantee that all values are finite. In order to avoid division by zero, we may write

```
1  float x[n];          // some array
2  float rec_sum = 0;   // sum of reciprocals
3  for (int i=0;i<n;i++) {
4      if (x[i]!=0.) {
5          rec_sum += 1/x[i];
6      }
7  }
```

But we can also write

```
1  for (int i=0;i<n;i++) {
2      if (x[i]==0.) continue; // skip this value if zero
3      rec_sum += 1/x[i];
4  }
```

The second approach leads to a cleaner code due to a reduced number of nested blocks.

The C++11 language standard introduced a new type of a *range-based* for loop. It implements the *for-each* loop popularized by languages such as Java or Python. These loops are encountered with standard library storage containers, but you can also use them with your own custom data types as long as you implement few necessary member functions to provide iterator support. For example

```
1  vector<double> data;
2  for (double &d:data)   // range-based loop
3      d = 0;
```

is identical to

```
1  for(int i=0;i<data.size();i++) {
2      data[i] = 0;
3  }
```

Note the use of a reference on line 2 of the range-based loop, &d. Without the reference, the code on line 3 would assign to a temporary loop-local variable d.

4.1.18 ARRAYS AND DYNAMIC MEMORY ALLOCATION

We have already seen that arrays are declared using syntax such as

```
double x[20];   // array of 20 double precision numbers
```
Here x maps to a consecutive memory block large enough to store 20 unique double-precision real values. The individual items are accessed using the array access operator [], such as

```
x[4] = -0.25;
```
As was already mentioned, but bears repeating, C and C++ does *not* contain built-in support for range checking. The following code
```
x[82] = 1.23;
```
will happily compile but will lead to memory corruption as the 1.23 value is stored outside the bounds of the x array. We also need to remember that indexing begins with 0, and hence the valid indexes for this example are 0 through 19.

This is also an example of a *static* declaration. It can only be used when the array size is known at *compile* time. For instance, the following code, which attempts to load the array size from a file, will not compile:

```
1   ifstream in("inputs.txt");   // open file for reading
2   int size;
3   cin>>size;            // read an integer from the file
4   double x[size];       // try to make an array
```
This is because the size of the array depends on whatever value is read in at *run time*. It is not fixed during compilation. However, even if the size were fixed, static declaration is applicable only to relatively small arrays. Program memory is organized into two groups: *stack* and *heap*. The stack is a small buffer that is associated with the program's instruction space. It is used for storing local variables. The actual size is compiler-dependent (and can be changed) but is generally only several megabytes large. With a 1 Mb stack, a double-precision array is limited to only about 130,000 elements. This may seem like a lot, but three-dimensional simulations require much more space.

Much larger arrays can be allocated on the *heap*. This is the main memory, and the size of allocatable arrays is limited by your actual system hardware.[1] This is known as *dynamic memory allocation*. The syntax is

```
1   unsigned int N = 10000;       // or use size_t
2   double *x = new double[N];    // allocate space for N doubles
3   /* use the data here */
4   delete[] x;                    // free memory once no longer needed
```
Note that we use a **new** operator to create the array. This operator asks the operating system to find and reserve a block of contiguous memory sufficiently large to store N double precision values. This block is accessed using its *address*. You can envision the RAM to be a huge one-dimensional byte array, and this address is just the index into this array. The address is stored in a special variable type called a *pointer*, denoted by the asterisk. Here x is a variable of type "pointer to a double-precision data", denoted by **double***. We discuss pointers in more detail in Section 4.3.1. Instead of storing a large array on the stack, we store just the address. The data itself is stored in the main memory. A special **nullptr** value is used to indicate the zero 0x0 (conventionally written in the hexadecimal notation) address, which is used to "invalidate" a pointer.

---

[1] Older 32-bit architectures imposed additional restriction on buffer sizes due to limited addressable space.

Unlike other languages, such as Java, C++ does not implement automatic *garbage collection* for dynamically allocated buffers. It is the responsibility of the programmer to free the memory space once it is no longer needed. This is done on line 4 using the `delete[]` operator. This operator is used with arrays. `delete` without the square brackets is used to free single items, such as instances of custom data objects. A program that repeatedly allocates dynamic memory without properly freeing it will eventually use up all system RAM. This will lead to a system crash, or at least the computer grinding to a halt once the operating system starts using the hard drive as swap space. This is known as a *memory leak*. Some memory leaks are quite difficult to find. In Chapter 8, we introduce a tool called *valgrind* that can be used for this task.

When working with legacy C codes, you will encounter the use of `malloc` and `free`. These two functions serve the same purpose as the `new` and `delete` operators introduced in C++. The argument to `malloc` consists of the number of bytes to allocate. The number of bytes needed for a single variable or a data type is obtained with `sizeof`. The `malloc` function returns the address as a *void pointer*, which is a pointer not associated with any data type. Since C++ is more type sensitive than C, we need to manually convert the pointer to the appropriate type using `(new_type) variable` cast. To illustrate this legacy syntax, the allocation from the listing above would be written as

double *x = (**double***) malloc(N***sizeof**(**double**));   *// free(x);*

Note that here we are using the legacy, C-style cast. C++ provides new type conversion operators given by `static_cast`, `dynamic_cast`, `const_cast`, and `reinterpret_cast`. The first option performs a compile-time type conversion analogous to the legacy C code. Dynamic cast is used as

double *x = **dynamic_cast**<**double***>(malloc(N***sizeof**(**double**)));

It is useful when working with polymorphic objects, discussed later in the chapter, as it allows the program to determine at run time whether some object is of a particular derived type. The operator returns `nullptr` if the conversion fails.

Modern C++ codes tend to avoid the use of *naked pointers* as was the case here, and instead utilize variables of type `vector` to store dynamically allocated arrays. This data type, declared in `<vector>` header is a *templated* (more on templates later) wrapper on top dynamic memory allocation. It is used as

```
1  #include <vector>
2  ...
3  std::vector<double> x(N);   // dynamically allocated N-item double
                                  array
```

Vector provides automatic garbage collection in the sense that the allocated space is automatically freed when the enclosing object goes out of scope. For performance reason, it does *not* perform range checking, and we encounter the same memory corruption issues as with traditional arrays if we are not careful with indexes. Vector, and similar data containers, are introduced in Section 4.4.10.

4.1.19 EXCEPTIONS

There are many instances when a function needs to signal to the caller that an error occurred. Sometimes this signaling can be performed by returning a value outside the expected range. A negative value may indicate an error for a function that nominally returns only zero or positive values. There are instances when this approach does not work. Take for example a storage container that implements some getter for obtaining the reference to one of the items. What value is this getter supposed to return if the input index is out of range? Standard library containers such as **vector** tackle this issue by including a "end" item that is used for this purpose. An alternate approach includes throwing exceptions. Exceptions signal that the algorithm could not continue and essentially gave up. Program control is returned to the calling function, which can *catch* the exception. If the exception is not caught, it propagates up the call stack, and leads to program termination due to an uncaught exception. Exceptions are caught using a *try-block*, as shown below

```
1   #include <iostream>
2   #include <exception>
3   using namespace std;
4
5   void fun(int x) {
6       if (x<0) throw runtime_error("invalid value!");
7   }
8
9   int main() {
10      try {
11          fun(-1);
12      }
13      catch (const std::exception &e) {
14          cerr<<"Error occurred: "<<e.what()<<endl;
15          return -1;
16      }
17
18      return 0;
19  }
```

The block can contain multiple `catch` handlers with the most specific one to the exception type getting executed. If the exception is thrown, we write a message to the standard error stream (which is usually the console but can be redirected) and then terminate the program with a negative return value.

4.2 C++ TENNIS BALL INTEGRATOR

4.2.1 INITIAL VERSION

We will now use the newly acquired understanding of basic C++ to implement a tennis ball integrator analogous to the Python version from Chapter 1. In doing so we will introduce additional concepts, including structures and random number generators. The code from 25 can be converted to legacy C as follows:

```
1   // initial version
2   #include <stdio.h>    // to use printf
3
4   int main() {
5       double x = 0;        // initialize position
6       double y = 2.1;
7       double u = 45;       // initialize velocity
8       double v = 0;
9       double gx = 0;       // initialize acceleration
10      double gy = -9.81;
11      double dt = 0.04;    // assign time step size
12      int ts = 0;          // time step counter
13
14      //show initial values
15      printf("%.2f, %.2f, %.2f, %.2f, %.2f\n",ts*dt,x,y,u,v);
16
17      // rewind by half time step for Leapfrog
18      u -= 0.5*gx*dt;
19      v -= 0.5*gy*dt;
20
21      while (y>0) {        // repeat until ground impact
22          u += gx*dt;      // increment velocity to n+0.5
23          v += gy*dt;
24          x += u*dt;       // increment position
25          y += v*dt;
26          ts++;            // increment time step counter
27          // display current position and velocity
28          printf("%.2f, %.2f, %.2f, %.2f, %.2f\n",ts*dt,x,y,u,v);
29      }
30      return 0;
31  }
```

Here we have included few corrections per the discussion in Sections 1.5.9. Namely, we perform a velocity rewind by half a time step to implement the Leapfrog scheme and utilize integer addition in the computation of elapsed time.

Despite the .cpp extension, this is actual a legacy C code. The code begins by including the stdio.h header needed by the printf function. The main function starts by declaring the required variables and setting their initial values which are output on line 15. On lines 18 and 19, we rewind the velocity by half time step for the Leapfrog scheme discussed in Section 1.5.9. Then on line 21, we jump into the main loop implemented using while. Inside we use the compound += operators to increment velocity and position, and ++ to increment the step counter used to calculate the elapsed time.

Running the program we obtain the following output:

```
$ g++ bouncy1.cpp && ./a.out
0.00, 0.00, 2.10, 45.00, 0.00
0.04, 1.80, 2.09, 45.00, -0.20
...
0.68, 30.60, -0.17, 45.00, -6.47
```

4.2.2 DIFFUSE REFLECTION

Instead of terminating the simulation on ground impact, we can let the tennis ball reflect back as was done in Section 1.5.6. However, we now include a diffuse component. A diffuse reflection is one in which the initial direction is completely forgotten, and the object reflects back such that flux scales with $\cos(\theta)$, where θ is the angle off surface normal. This is how light reflects back from an unpolished surfaces, and also how low-velocity molecules reflect back off surfaces. This is also known as *Lambert's cosine law*. The new direction is sampled per

$$\sin(\theta) = \sqrt{\mathcal{R}_1} \tag{4.9}$$

$$\cos(\theta) = \sqrt{1 - \sin^2(\theta)} \tag{4.10}$$

$$\phi = = 2\pi\mathcal{R}_2 \tag{4.11}$$

$$\hat{e}_n = \cos(\theta) \tag{4.12}$$

$$\hat{e}_{t1} = \sin(\theta)\cos(\phi) \tag{4.13}$$

$$\hat{e}_{t2} = \sin(\theta)\sin(\phi) \tag{4.14}$$

where $\hat{e}_n$ is a unit vector in the surface normal direction (y in our formulation), and $\hat{e}_{t1}$ and $\hat{e}_{t2}$ are the two tangents, here mapping to x and z. The algorithm works by first picking the off-normal angle. The resulting vector is then rotated about the normal vector through ϕ angle sampled from $\phi \in [0, 2\pi)$. Real-world reflections tend to fall somewhere between the specular and diffuse limits.

4.2.3 RANDOM NUMBERS

The two $\mathcal{R}$s in Equations 4.9 and 4.11 are random numbers sampled from uniform distribution in $[0, 1)$. As we have already seen in Chapter 1, numerical simulations utilize so-called random number generators to sample *pseudorandom* numbers. These are consecutive values returned from a long algebraic sequence that to a casual observer appear to be truly random. Early implementations of the C and C++ programming language suffered from the lack of quality random number generators. The generator algorithm was not prescribed by the language standard. Compilers from different vendors used varying algorithms, which led to codes that worked correctly on one platform producing erroneous results when ran on a different system. A common workaround was to include a custom generator by utilizing equations from sources such as Numerical Recipes, Press 2007. This issues was mitigated in the C++11 language update through functions provided by the `<random>` header. The new generators consist of two parts: the random engine, and a distribution function. The engine samples large random integers. The distribution function then remaps these to, let's say, real values uniformly distributed in $[0, 1)$. Mersenne Twister is a popular generator, due to both its speed and a huge period (the size of the sequence before numbers start repeating). These two components are created using

```
1  #include <random>
2  std::mt19937 mt_gen;              //random number generator
3  std::uniform_real_distribution<double> rnd_dist;  //uniform
      distribution
```

A random value is then sampled as shown below

```
double rnd() {return rnd_dist(mt_gen);}
```

This algorithm will return the same sequence every time the code is run due to the generator being initialized with the same default starting seed. We can instead randomize the generator with

```
std::mt19937 mt_gen{random_device()()};   // set starting seed
```

Here **random_device()** instantiates a number generator that produces truly random values by sampling system state parameters, such as the number of running processes or the amount of free memory. Due to the reliance on these queries, this generator is too slow for the general use (typical stochastic simulation may require billions of random values). It is however perfectly usable for seeding another faster generator, such as the Mersenne Twister. The two sets of parentheses are not a typo. The first one provides the (empty) argument list used to construct a temporary object of type **random_device**. The second set then evaluates this object to obtain the starting seed.

4.2.4 UPDATED VERSION

The revised version, found **bouncy2.cpp** implements the above described logic using the code shown below:

```
1  for (int ts=0; ts<num_ts; ts++) {
2      u += gx*dt; // increment velocity to n+0.5
3      v += gy*dt;
4      w += gz*dt;
5
6      x += u*dt;  // increment position
7      y += v*dt;
8      z += w*dt;
9
10     if (y<0) {   // if ground hit
11         // compute velocity magnitude
12         double mag = sqrt(u*u + v*v + w*w);
13         constexpr double a_spec = 0.35; // specularity coefficient
14
15         double mag_new = 0.8*mag; //inelastic reflection
16         cout<<ts<<" "<<mag_new<<endl;
17         if (mag_new<0.05) break;  // terminate once stationary
18
19         // direction for specular reflection
20         double spec_i =  u/mag;
21         double spec_j = -v/mag;   // specular reflection
22         double spec_k =  w/mag;
23
24         // pick new diffuse direction
25         double sin_theta = sqrt(rnd());
26         double cos_theta = sqrt(1-sin_theta*sin_theta);
```

```
27        double phi = 2*PI*rnd();
28
29        double dif_i = sin_theta*cos(phi);   // tangent 1
30        double dif_j = cos_theta;            // normal component
31        double dif_k = sin_theta*sin(phi);   // tangent 2
32
33        // set new velocity
34        u = (a_spec*spec_i + (1-a_spec)*dif_i)*mag_new;
35        v = (a_spec*spec_j + (1-a_spec)*dif_j)*mag_new;
36        w = (a_spec*spec_k + (1-a_spec)*dif_k)*mag_new;
37
38        // move to surface "hack"
39        y = 0;
40    }
41
42   // write out position and velocity
43   out<<ts*dt<<","<<x<<","<<y<<","<<z<<","<<u<<","<<v<<","<<w<<"\n";
44  }
```

The output was modified to write out a `results.csv` text file using C++ `fstream`. Another modification involved converting the code to 3 spatial and 3 velocity components, since the diffuse reflection will no longer be in the plane of the original $x - y$ trajectory. We utilize a α_{spec} specularity coefficient to control the fraction of post-impact direction that retain the specular component. We also utilize a `for` loop to run the code for a prescribed number of iterations, but allow for an early termination once velocity drops below a threshold value. Finally, we should note that the surface impact logic is not completely correct. Instead of blindly setting $y = 0$, the code should compute the actual ground-impact point, and integrate the particle position *and velocity* only up to that location. This is something that is "left as an exercise for the reader".

4.2.5 VISUALIZATION

Unlike Python, C and C++ do not implement any plotting capability. While 2D data can be plotted relatively easily by loading a `.csv` file into a spreadsheet program such as Microsoft Excel, visualizing 3D data is more challenging. One could conceptually write a Python script coupled with a 3D projection Matplotlib plotting mode to generate the plots. However, in our opinion it is easier to utilize a dedicated scientific data visualization tool such as Tecplot, Visit, or Paraview. In our work we mainly utilize the last option. Paraview, available from `paraview.org`, is a graphical front end built on top of the Visualization Toolkit (VTK), see Chapter 8. Figure 4.1 demonstrates the use of Paraview to visualize the trajectory from our example simulation.

If this is your first time using Paraview, you may want to start by navigating to Settings (usually found under the Edit menu bar) and enabling Auto Apply in the General tab. This will avoid needing to press the Apply button everytime you make a change to visualization properties. The settings tab also allows you to change the Color Palette away from the default dark one to one more compatible with saving screenshots for publications. The Colormap

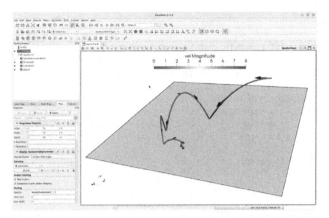

Figure 4.1   Use of Paraview to visualize object trajectory.

Editor, activated from the View menu bar, and other places, allows you change the colormap used for new variables from the default Cool to Warm divergent map to something like the Rainbow Uniform map used in the picture. This colormap attempts to mitigate the artificial contrast issues encountered with the standard rainbow map.

Paraview supports importing data in various file formats, including .csv. Such files open in a Spreadsheet View. You can close this view as we will be using the Render View instead. VTK (and hence Paraview) is built on the concept of a *visualization pipeline* in which the input (the source file in this case) is run through an arbitrary number of filters that manipulate the data. The filter output can act as a source to another filter, or can be rendered to the screen.

The item to be visualized must consist of two parts: geometry and data. The geometry provides the information on the shape of the object being drawn. The data part then provides various scalar or vector properties associated with the nodes and/or cells of the underlying geometry. Despite our .csv file containing columns called x, y, z, Paraview does not "know" that these should be used as point coordinates. Therefore, we need to use a filter such as TableToPointData or TableToStructuredGrid to generate the geometrical description. Filters are added by right clicking on the desired source in the Pipeline Browser and selecting Add Filter. The first of these two filters produces geometry consisting of isolated points. In order to visualize the trajectory, we would like to connect the points with line segments. This is essentially a 1D Cartesian grid with dimensions $N \times 1 \times 1$, where N is the number of samples (table rows). We thus use the latter option. The Whole Extent found in the Properties tab specifies the lower and upper bounds for the i, j, and k indexes. For $N = 1000$, we make sure that the first line reads 0 and 999. Next, we assign the x, y, and z variables to the x, y, z columns.

We can change representation to Wireframe to plot the line. By default, it will be painted using black line but we can instead use of the other variables as the source by selecting it in the Coloring section. We can also increase the Line Width and check the option to Render Lines as Tubes to produce a nicer output.

Depending on the random numbers picked by the simulation, as well as the bounce off coefficients, the resulting simulation may end up with a long "tail" in which the ball is bouncing around in place. We can cut off this section by including the Threshold filter. We want this filter to act on the generated structured grid, and hence we add it by right clicking on the TableToSTructuredGrid filter (filter inputs can also be changed post-addition through this same right-click context menu). We use the Properties section to threshold data for, let's say, $t \in [0, 10]$. Note that when you add the filter, the prior filter output gets hidden. The visibility of individual filters is toggled using the eye icon in the Pipeline Browser.

We would next like to add several arrows to visualize the object's velocity. This can be done using the Glyph filter, which draws small shapes such as arrows or spheres at the nodes of source data. The shapes can be oriented according to a vector data array. At this point, we have three distinct u, v, and w variables, but Paraview does not know that they correspond to the components of a single vector quantity. This is where the Calculator filter comes in. We add it to the output of the Threshold filter, and use the following expression

```
u*iHat + v*jHat + w*kHat
```

The Result Array Name field is set to vel. Under Representation, we also modify the visualization properties to plot the trace (this is essentially the same step as was taken before for the TableToStructuredGrid filter, which is no longer being rendered). We set the Coloring variable to to the Magnitude of the newly computed vel vector. Then we add the Glyph filter, and use vel as both the Orientation Array and the Scale Array. The Glyph Type is set to Arrow and under Masking we select to plot Every Nth Point with stride of 20. Finally, to aid in visualization, a plane with origin $(-1, 0, -4)$ and points $(7, 0, -4)$ and $(-1, 0, 4)$ was added by utilizing the Sources menu. These points correspond to the bottom left, bottom right, and top left corners. The plane is visualized as Surface with Edges.

4.2.6 VTK POLYDATA FORMAT

Use of .csv files for data visualization is not practical and is also of limited use since only Cartesian-like topologies can be converted to renderable shapes. A better approach is to use the native VTK file formats as described in the VTK User's Guide. These files use XML syntax to describe vertex positions, cell connectivity, and to provide data arrays containing values for scalar or vector properties associated with points or cell centers. The two formats of interest are VTKPolyData and VTKImageData. The former encodes surface meshes: geometry composed of polygonal surfaces. This data must be saved in files with `.vtp` extension. The other format, which we use later in this chapter, is used to store data associated with uniform Cartesian grids. These files must utilize the `.vti` extension.

The general syntax of the .vtp file is shown below:

```
1   <?xml version="1.0"?>
2   <VTKFile type="PolyData">
3   <PolyData>
4   <Piece NumberOfPoints="3" NumberOfLines="1" NumberOfPolys="0">
5   <Points>
6   <DataArray Name="pos" NumberOfComponents="3" type="Float64"
        format="ascii">
7   0  2.1  0
8   0.09  2.10196  0
9   0.18  2.1  0
10  </DataArray>
11  </Points>
12  <Lines>
13  <DataArray Name="connectivity" NumberOfComponents="1" type="Int32
        " format="ascii">
14  0  1  2
15  </DataArray>
16  <DataArray Name="offsets" NumberOfComponents="1" type="Int32"
        format="ascii">
17  3
18  </DataArray>
19  </Lines>
20  <PointData>
21  <DataArray Name="vel" NumberOfComponents="3" type="Float64"
        format="ascii">
22  4.5  0.0981  0
23  4.5  −0.0981  0
24  4.5  −0.2943  0
25  </DataArray>
26  <DataArray Name="time" NumberOfComponents="1" type="Float64"
        format="ascii">
27  0  0.02  0.04
28  </DataArray>
29  </PointData>
30  </Piece>
31  </PolyData>
32  </VTKFile>
```

It begins by providing information about the number of different kinds of geometrical entities, such as points, lines, or polygons, that there are in this file. There additional types available, but these are the most common ones encountered in our work. Next the `<points>` section provides the coordinates for all points using $x_0, y_0, z_0, x_1, y_1, z_1, \ldots$ format. The values are provided within a 3 component data array. The `<lines>` section provides the line connectivity (the order of points according to their indexes used for building the line). In this case, the points are just listed in the sequential order. The offsets data array is used indicate the index into the connectivity array where the next line would start, if we had one. The `<PointData>` section lists an arbitrary number of data arrays that provide scalar (single component) or vector (three components) integer or real-valued properties associated with the nodes. With this file we no longer need to use the calculator filter to group the velocity components into a single vector.

4.2.7 STRUCTURES

The challenge associate with outputting the .vtp file is that data is now organized by type, and not by the time step. In other words, we need to have access to all `num_ts` values of position and velocity when writing the output. This necessitates the use of array. We could declare six separate arrays to store the 3 position and 3 velocity components as in

```
1  double *x = new double[num_ts];
2  ...
3  double *w = new double[num_ts];
```

This formulation becomes cumbersome as the number of parameters to track grows, especially as all information about a single item need to be passed to helper functions. For example, using this formulation, the prototype for the `saveVTP` function may read as

void writeVTP (**double** *x, **double** *y, **double** *z, **double** *u, **double** *v, **double** *w, **unsigned int** num_ts, **double** dt);

C and C++ lets us define custom data containers (structures) using the keyword `struct`. We can write

```
1  struct Sample {       // type name
2      double x,y,z;     // position
3      double u,v,w;     // velocity
4  };
```

Here `Sample` is the name of this new container type. Note the semicolon after the declaration. While normally a semicolon is not needed after a closing bracket, it is required after a struct or a class definition. The reason is that a struct can also be defined using a legacy C syntax as

`struct { /* data */ } my_struct;`

in which case the type name comes after the closing bracket.

This new type can then be used to declare variables and arrays using the already familiar syntax:

```
1  Sample smp;                          // single variable of type sample
2  Sample *smp = new Sample[num_ts];    // dynamic array of N samples
```

Member variables are accessed using a dot,

```
1  sample.x = 12.3;    // assign value
2  samples[20].y = 0;  // set y of the 21st element
```

The main loop would then read

```
1  for (int ts=0; ts<num_ts-1; ts++) {
2      smp[ts+1].u = smp[ts].u + gx*dt;      // advance velocity to n+0.5
3      ...
4      smp[ts+1].x = smp[ts].x + smp[ts].u*dt;  // advance position
5      ...
```

The call to the output function is also replaced as

void writeVTP(Sample *smp, **unsigned int** num_ts, **double** dt);

Not only is this more concise, it also leads to a more efficient code due to a reduction in the size of data being passed during the function call. The complete source code for this modified version, including the code for writing out the .vtp file can be found in **bouncy3.cpp**.

Finally, one issue that we need to be aware of when working with structures is that of *circular dependence*. We may define a structure ModA that depends on data of type ModB, but with ModB itself requiring data of type ModA. Due to the top-down compilation process, the following code will lead to errors since ModB is not yet defined at the point of first use:

```
1   struct ModA{
2       ModB *modB;
3   };
4
5   struct ModB{
6       ModA *modA;
7   };
```

We resolve this issue by indicating to the compiler that the name ModB refers to a structure via a *forward declaration*:

```
1   struct ModB;      // forward declaration
2   struct ModA { ... };
3   struct ModB { ... };
```

4.3 POINTERS AND REFERENCES

4.3.1 POINTER ARITHMETIC

We now return to describing additional C++ language concepts. In Section 4.1.18 we saw how to use the **new** operator to dynamically allocate memory. Specifically, we used a construct such as

double *x = **new double**[10];

A static array was previously allocated using

double x[10];

Regardless of the declaration method, individual items in the array were accessed using the [] operator, such as x[0] = 0.0;.

At this point you may be curious about this apparent discrepancy. In the first case, we are declaring a variable called x which is of type *pointer to a double*. Pointers are special variables that hold a memory address - in this case the address returned by the **new** operator. The data at the address held by the pointer is accessed by *dereferencing* the pointer. This is done by prepending the variable name with a * at point of use. Let's consider the following example:

```
1   int A = 0, B = 0;
2   int *p = &A; // p holds address of A
3   *p = 1;      // analogus to A = 1
```

```
4   p = &B;          // p now holds the address of B
5   *p = 2;          // sets B to 2
```

Here we declare two integer variables called A and B and initialize them to zero. We subsequently declare a pointer variable p which is assigned the address (obtained using & operator) of A. We next write 1 to the memory address held by p. Since p is a pointer to a 32-bit integer, the compiler actually generates instructions to set the 4 bytes starting from the given address. Since the address maps to the location of A, the value of this variable changes to 1. Next, we assign the address of B to p. The final line, although being effectively identical to line 3 which at that time updated A, now stores the value of 2 into B, since p now "points" to B (p holds B's address).

One peculiar feature of C and C++ is its support for pointer math. Specifically, pointers can be incremented or decremented by integer values, such as

```
p += 2;          // p is a int* pointer
```

The above expression increments the address stored in p by 2*sizeof(int), or 8 bytes. Now consider a function declaring two integers and a double precision real variable,

```
1   int i = 0, j = 0;   // 2 integers
2   double d = 0;       // double precision values
```

We can declare a pointer p initially pointing to i

```
int *p = &i;     // p points at a
```

We next increment it, which increases the address by 4 bytes (the size of an integer), which most likely corresponds to the address assigned to the variable j (this arrangement is set by the compiler). Therefore, the next expression modifies j.

```
1   p += 1;          // increment address by 4 bytes
2   *p = 7;          // j is modified
```

If we were to increment the pointer again, the address would increase by 4 bytes, which now most likely corresponds to the first half of the 8-byte double d. The next assignment thus corrupts the value stored in that variable,

```
1   p += 1;          // increment p by another 4 bytes
2   *p = 2;          // value in d is corrupted
3   cout<<j<<", "<<d<<endl;  // prints 1, 9.88131e-324
```

Instead of incrementing the pointer step by step, we could also achieve this final outcome using

```
*(p+2) = 2;
```

This may seem like a contrived example to illustrate a peculiarity of the language that you would never encounter in practice. Well, turns out that the array access operator [i] that we are already familiar with is just a shortcut for offsetting and dereferencing a pointer. The following two expressions are identical:

```
1   double *x = new double[10];
2   x[5] = 5;        // access the 6th element
3   *(x+5) = 5;      // another way to access this element
```

As such, it is imperative to pay attention to array indexes as one may easily end up overwriting data outside the allocated space. Depending on the actual memory arrangement, we can end up with corrupted data and normal exit, corrupted data and a "stack smashing" warning, or a premature termination due to a segmentation fault error.

Use of pointers is not limited to the basic data types. Using our `Sample` structure, we can write,

```
1  Sample sample;          // sample object
2  Sample *ptr = &sample;  // pointer to the variable 'sample'
3  (*ptr).u = 2;           // assign value
4  ptr->u = 2;             // another way
```

On line 2, we define a pointer to a structure of type `Sample`. It is initialized with the address of variable `sample`. On line 3, we use the asterisk to dereference the pointer which makes the left side act as a `struct`. The dot is used to access struct members, just as we learned previously. However, pointers to a structure are so common that C and C++ introduced operator `->` to access `struct` (and `class`) members directly from the pointer variable. This is shown on line 4.

4.3.2 REFERENCES

C++ also supports a variable type called *reference*. References are essentially pointers that are automatically de-referenced when used. They are a form of *syntactic sugar*. There is nothing that can be done with references that couldn't be done with pointers; however, the code utilizing references looks cleaner and is less prone to errors. Here is an example:

```
1  int A = 0;       // local variable A
2  int &ref = A;    // ref is a reference to A
3  ref = 1;         // assigns 1 to A
```

A reference variable is defined by prepending the name with an ampersand `&`. Unlike pointers, references cannot be reassigned and thus must be initialized during declaration. The example used previously in which we used the same `p` pointer to modify two separate variables `A` and `B` would not work with a reference. The reference variable `ref` is subsequently used directly without the need to utilize any additional operators. We just need to remember that we are actually modifying the `sample`'s data.

References can also be used with structs (and classes). In that case we just use the regular `.` operator to access member objects,

```
1  Sample &s_ref = sample;  // reference to a struct variable
2  s_ref.u = 2;             // assigns value to a struct component
```

4.3.3 CALL BY REFERENCE

At this point, you may be asking what exactly is the "point" of using pointers or references. After all, we could have easily accomplished the assignment using

```
sample.u = 2;
```

Besides providing access to dynamic memory, pointers and references are primarily used to pass data between different code sections. Consider some function that requires data provided by the `Sample` struct,

```
void doSomething(Sample data) { ... }
```

which would be used as

```
doSomething(sample);      // call the function
```

C and C++ use the *call by value* convention for passing data to functions. As part of the function call, the compiler generates CPU instructions to *copy* the values of the inputs into local variables corresponding to the arguments. We have already introduced this concept during our discussion on variable scope.

We want to reduce the amount of information that needs to be duplicated for performance and memory usage reasons. The `Sample` structure consist of 6 double-precision floats. This adds up to 48 bytes. The size can be checked using `sizeof(Sample)`. Structures will often contain many more member variables.

Given a simulation with millions of objects, similar memory copy introduces a noticeable overhead. However, on a 64-bit architecture, a pointer of any kind is always exactly 8 bytes (64-bits) long. Therefore, instead of passing the large object, we just pass the 8 bytes of memory address. The data is still passed by value, but now only the much smaller address is copied. The function declaration is modified to change the arguments to pointers of type `Sample` as shown below:

```
1  void doSomething(Sample *s) {       // pointer to some sample
2     double r2 = s->x*s->x + s->y*s->y; // use sample's data
3     s->u = 2;                          // assign value to sample
4     ...
5  }
```

Here we included some sample code to illustrate how the data would be used. The function is then called by passing the addresses:

```
doSomething(&sample);      // pass the reference
```

References allow us to accomplish this same thing using a simpler syntax:

```
1  void doSomething(Sample &s) {       // reference to Sample
2     double r2 = s.x*s.x + s.y*s.y;    // use sample's data
3     s.u = 2;                          // assign value to sample
4  }
```

Note that we now utilize the dot . instead of the -> arrow operator since we are working with references. This function is called using

```
doSomething(sample);   // no need to take the address
```

Passing the pointer or a reference to an object is called a *pass by reference*, and is the default in some languages, including Java.

4.3.4 CONSTANT ARGUMENTS

The important caveat is that with a call by reference, the `doSomething` function has access to the variable defined outside its local scope. The function can modify non-local data, as was demonstrated in the prior example. Sometimes this behavior is indeed desired. C and C++ do not allow returning more than one object from a function. Utilizing pointers or references lets us treat some function arguments as additional outputs. This is in fact how functions used to implement distributed processing discussed in Chapter 9 operate. But most of the time, we simply use references and pointers for performance reasons. In that case, in order to ensure that data is not inadvertently modified, we use the **const** parameter to signal to the compiler that the pointer or reference arguments is meant to be read-only. The compiler throws an error if it detects an attempt to write to such a memory address. For example

```
1  void doSomething(const Sample &s) {
2      s.u = 2;  // attempting to assign value to a const reference
3  }
```

leads to

```
error: assignment of read-only reference 's'
```

4.3.5 MULTI-DIMENSIONAL ARRAYS

Unlike many other scientific computing languages, C and C++ do not offer single operation for dynamically allocating multi-dimensional arrays. This does not mean that such arrays cannot be used. Instead, they need to be declared in stages. The same way that

```
double *data;      // pointer to double
```
is a pointer to double precision data,

```
double **data;     // pointer to 'pointer to double'
```
is a pointer to a pointer to double precision data (**double***). Similarly to how we use **new double[n]** to allocate an array of n doubles, we use

```
double **data = new double*[m];
```
to allocate an array of m pointers to doubles. Extending this further to three dimensions

```
double ***data = new double**[l];
```
allocates an array of l 'pointers-to-pointers-to-double' (**double****). The overall strategy for allocating a three-dimensional $n_i \times n_j \times n_k$ array is:

1. Allocate an array of n_i pointers-to-pointers-to-data
2. Assign each of the n_i pointers to point to a new array of n_j pointers-to-data
3. Assign each of the n_j items to point to a new array of n_k data

The implementation looks like this:

```
1  double ***data = new double**[ni];
2  for (int i=0;i<ni;i++) {
3      // each item points to a new nj-sized array of pointers
4      data[i] = new double*[nj];
5      for (int j=0;j<nj;j++) {
6          // each items points to an nk-sized array of doubles
```

```
7        data[i][j] = new double[nk];
8    }
9 }
```

The data is then accessed as follows:

```
data[2][0][7] = 14;
```

The array access operators are processed from left-to-right. The first `data[2]` returns the `double**` pointer that was stored in the `[2]` slot of `data`. The next `[0]` then dereferences this pointer, returning a `double*` pointer. Finally, that pointer is dereferenced, returning the double precision value of interest.

The memory is released by performing deleting all dynamically allocated arrays. We need to begin the deletion with the inner-most items,

```
1 for (int i=0;i<ni;i++) {
2    for (int j=0;j<nj;j++)
3        delete[] data[i][j];  // deallocate double[nk]
4    delete[] data[i];         // deallocate double*[nj]
5 }
6 delete[]   data;             // deallocate double**[ni]
```

What is important to realize here is that the $n_i \times n_j \times n_k$ values stored by this array are *not* located in contiguous memory. Only the final n_k items of each i and j index are stored consecutively. In Chapter 9 we demonstrate that the order with which the 3D array is accessed has a major impact on code performance.

4.3.6   LINKED LISTS

Another use of pointers involves the implementation of *linked lists*. Linked lists are memory data structures consisting of individual "containers" scattered through memory, and chained together with pointers. They can be single-linked, in which case each container stores the address of only the item coming after it. Or they can be double-linked, in which case the pointer to the previous container is also known. Linked lists offer easy removal of single items, but lead to reduced computational performance due to an increase in cache misses. Linked lists also do not provide direct access to an element. Accessing the i-th item requires traversing through $(i-1)$ prior connections. We touch upon these topics in Chapters 5 and 9. Just to illustrate linked lists, consider the following container object:

```
1 struct Cont {
2    int my_data;          // placeholder for container data
3    Cont *next = nullptr;
4 }
```

We use a pointer to store the address of the first item,

```
Cont *containers = nullptr;
```

Adding an item involves allocating a new container and updating the pointers,

```
1 Cont *cont = new Cont();
2 cont->next = containers;  // place in front
3 containers = cont;
```

The list is traversed as

```
1  Cont *cont = containers;   // address of the first item
2  while (cont) {
3     /* do something */
4     cont = cont->next;
5  }
```

The loop terminates once `cont` becomes `nullptr`. A particular item is deleted as

```
1  prev->next = cont->next;
2  delete cont;
```

This requires having a pointer to the previous list item. In the case of a double-linked list, this is `cont->prev` (in which case `cont->next->prev` also needs to be updated). Alternatively, in the iteration loop from the previous listing, we could just maintain this pointer as

```
1  Cont *cont = containers;   // address of the first item
2  Cont *prev = nullptr;
3  while (cont) {
4     /* do something */
5     prev = cont;
6     cont = cont->next;
7  }
```

4.4 OBJECT ORIENTED PROGRAMMING

From linear algebra, we are accustomed to performing math with matrices and vectors. A matrix is essentially a two-dimensional array of coefficients,

$$A_{(r,c)} = A[r][c] \tag{4.15}$$

where r and c are indexes for the row and the column. Utilizing concepts from the prior discussion on multidimensional arrays, we could define a function to allocate a 2D array of matrix coefficients,

```
1  double **alloc2D(int nr, int nc) {
2     double **data = new double[nr];
3     for (int r=0; r<nr; r++)
4        data[r] = new double[nc];
5     return data;
6  }
```

Similarly we can add a function to release the memory once the matrix is no longer needed,

```
1  void free2D(double **data, int nr) {
2     for (int r=0; r<nr; r++)
3        delete[] data[r];
4     delete[] data;
5  }
```

We then use these functions as:

```
1  double **A = alloc2D(nr, nc);
2  A[0][0] = 1.0;   // set coefficients
3  ...
4  free2D(A, nr);
```

But there is a better way. Just as we used a `struct` to declare a `Sample` data type in Section 4.2.7, we can declare a generic `Matrix` storage container. However, instead of using it solely for storing data, we can let it implement built-in functionality by including member functions, called *methods*. Such functions can perform any operation we like, such as data allocation and deletion. They can also provide support for setting data or for performing mathematical operations such as matrix-vector multiplication.

Containers that also perform operations are known as *objects*. Historically, the C language `struct` could only contain data. C++ added a new container of type `class` that supported member functions. This is why C++ was initially called "C with classes". The important characteristic of this approach is realizing that each instance of a class is a separate variable with a different memory location. The member functions operate on the data of the instance from which they are being called. The ability for objects to operate on themselves revolutionized computer programming and led to the birth of *Object Oriented Programming* or OOP. It is not without its cons, as a heavy use of OOP can lead to an overly bloated code. However, with just a basic understanding of OOP, you can create code that is both more readable and also less prone to errors.

Nowadays, a `struct` and a `class` are almost identical. The only difference is in the default access rights. Members of a struct are *public* by default and accessible to the rest of the program. `class` members are *private* and can be accessed only by other class members. In our work, we customarily use the `class` keyword for objects that implement a significant functional logic. The `struct` keyword is reserved for data objects that are primarily data containers with only a limited companion code.

To illustrate a class use, consider this example

```
1   class Matrix {
2   public:           // subsequent members are publicly accessible
3       /* functions */
4       void init(int nr, int nc);
5       void free();
6
7       /* data */
8       double **a = nullptr;  // matrix coefficients
9       int nr = 0;            // matrix dimensions
10      int nc = 0;
11  };
```

On lines 4 and 5 we provide declarations of two functions. The function body can be implemented outside the class declaration by prepending the function name with the class name,

```
1   void Matrix::init(int nr, int nc) {
2       // allocate 2D array
3       a = new double[nr];
4       for (int r=0; r<nr; r++)
5           a[r] = new double[nc];
6
7       // save dimensions
```

```
 8      this->nr = nr; this->nc = nc;
 9  }
10
11  void Matrix::free() {
12      for (int r=0; i<nr; r++)
13          delete[] a[r];
14      delete[] a;
15      a = nullptr;
16  }
```

Function bodies can also be implemented directly in the class declaration. That is in fact the preferred approach for small algorithms that should be inlined. The **this** pointer used in **init** allows us to distinguish the class member variable from the function argument of the same name. This pointer is automatically defined for all classes and contains the memory address of the particular instance. Two objects of the same type, such as

```
1  Matrix A;
2  Matrix B;
```

will contain different address in their **this**, i.e. **&A** and **&B**, respectively. Classes offer an exception to the C++ typical top-down compilation approach in that member functions or variables can be used from code lines above their declaration.

We instantiate and use the matrix object as

```
1  Matrix M;            // new variable of type Matrix
2  M.init(20, 20);      // allocate a 20x20 matrix
3  M.a[i][j] = 2.0;     // set some value
4  M.free();            // release memory
```

4.4.1 CONSTRUCTORS AND DESTRUCTORS

The Matrix class uses the **init** function to allocate memory, and the **free** function to release it once no longer needed. These functions are examples of actions performed during the construction of the variable, and the subsequent destruction once the variable goes out of scope. C++ allows us to define functions that get called automatically during these events. They are called, appropriately, a *constructor* and a *destructor*. They are defined as member functions that do not return a value and have the same name as the class. The name is prepended with a tilde for the destructor. For example, for the **Matrix** object, we have

```
1  class Matrix {
2      public:
3      Matrix();    // constructor
4      ~Matrix();   // destructor
5  };
```

The constructor can take arguments. The destructor cannot have any arguments as there is no way to pass data when the variable is being deleted. With this in mind, we migrate the code previously found in **init** and **free** into the constructor and the destructor. Our class definition now reads:

```
1   class Matrix {
2   public:
3       Matrix(int nr, int nc) {    // constructor
4           // allocate 2D array
5           a = new double[nr];
6           for (int r=0; r<nr; r++)
7               a[r] = new double[nc];
8
9               // save dimensions
10              this->nr = nr; this->nc = nc;
11      }
12
13      ~Matrix() {  // destructor
14          for (int r=0; r<nr; r++) delete[] a[i];
15          delete[] a;
16          a = nullptr;
17      }
18  };
```

The function arguments for the constructor are specified during the variable declaration,

```
1   Matrix M(20, 20);  // new variable of type Matrix
2   M.a[i][j] = 2.0;   // set some value
```

The destructor is called automatically once the containing block exits. The use of a destructor automates dynamic memory deletion, and thus helps to avoid memory leaks. Note that attempting to run

```
Matrix M;
```

which worked perfectly fine in the listing on page 4.4 now leads to a compilation error. The reason is that in the absence of a custom constructor, the compiler automatically adds a *default constructor* that does not take any arguments, i.e.

```
1   class Matrix {
2       Matrix() {}  // default constructor
3   };
```

This constructor is *deleted* when any constructor is declared explicitly. Default constructors can also be deleted using the **delete** keyword as in

```
1   class Matrix {
2       Matrix() = delete;
3   };
```

4.4.2 INITIALIZER LISTS

Besides allocating the memory, our constructor also sets the local **nr** and **nc** variables. This initialization can be accomplished more elegantly with an *initializer list*. It specifies values to use in the construction of class member variables *before* the constructor code runs. The list is specified by placing a : after the constructor arguments, and consists of comma separated pairs of **member_var{value}** pairs,

```
1   class Matrix {
2       Matrix (int nr, int nc) : nr{nr}, nc{nc} {
```

```
3        /* rest of the code here */
4
5    };
6    ...
7    const int nr;
8    const int nc;
9 };
```

The compiler automatically distinguishes the member **nr** from the **nr** value passed as the function argument. In this example we also added the **const** keyword to make the variables read only. The constructor is the only function allowed to assign values to constant members. Once they are set, they cannot be modified. Constant members are particularly useful for storing dimensions of dynamically allocated arrays.

The initializer list is also the only way to set reference member variables. Let's say we have another class called **Solver**, which needs to keep track of the matrix to be used in the algorithm. We would define:

```
1 class Solver {
2    Solver (Matrix &mat) : A{mat} {}    // store the reference
3    Matrix &A;    // reference to a matrix object
4 };
```

which would then be used as in

```
1 Matrix M;
2 Solver solver(M);    // pass a reference to M
```

Initializer lists are also used for *constructor chaining*. We can define multiple constructors with different argument types. For instance, let's say that we would also like to set all matrix coefficients to some value on initialization. We can add another constructor,

```
1 Matrix (int nr, int nc, double val) {
2    /* allocation code from before */
3
4    for (int r=0; r<nr; r++)   // set value
5        for (int c=0; c<nc; c++) data[r][c] = val;
6 }
```

The first part of this function is identical to the already defined constructor. Instead of duplicating the code, we just add a call to it using the initializer list:

```
1 Matrix (int nr, int nc, double val): Matrix(nr,nc) {
2    for (int r=0; r<nr; r++)   // set value
3        for (int c=0; c<nc; c++) data[r][c] = val;
4 }
```

We can then instantiate a matrix variable using either **Matrix A(nr,nc);** or **MatrixA(nr,nc,val);**.

4.4.3 ACCESS CONTROL

At this point, there is nothing preventing the programmer from assigning data to matrix coordinates outside the allocated range, i.e.

```
1  Matrix M(7,5);          // a 7x5 matrix
2  M.a[9][2] = 0;          // memory corruption
```

As was already alluded to, structs and classes can control which data can be accessed publicly. This is done using keywords `public:`, `protected:`, and `private:`. The specified access type remains active until another access control section is encountered, i.e.

```
1  class Matrix {
2  public:
3      /* public members */
4  private:
5      /* private members */
6  public:
7      /* more public members */
8  };
```

The first option specifies members that can be accessed from the rest of the code. Data can be made read-only by adding the `const` or `constexpr` (for values known at compile time) keywords. Variables and functions declared within a `protected` or a `private` block are *not* accessible from outside the class. As discussed later in 4.4.8, classes can inherit from their parents. The difference between these two keywords only arises in the derived objects. Protected data can be accessed by the child class, while private cannot. In other words, access to a private data is limited to the particular class in which it was defined.

With this in mind, we can move the coefficient array to a protected block. This prevents direct access to the array from the rest of the code. We then define a *setter* and a *getter* function for setting and retrieving data. Routing data access through these functions makes it possible to add range checking,

```
1  class Matrix {
2  public:
3      ...
4      void set(int r, int c, double v) {  // setter
5          if (isValid(r, c)) a[r][c]=v;
6      }
7      double get(int r, int c) {          // getter
8          if (isValid(r, c)) return a[r][c];
9          return 0;    // default value
10     }
11     const int nr, nc;   // dimensions
12
13 protected:
14     bool isValid(int r, int c) {
15         return (r>=0 && r<nr && c>=0 && c<nc);
16     }
17     double **a;    // coefficient array
18 };
```

The `isValid` functions checks if $r \in [0, n_r)$ and $c \in [0, n_c)$. We also demonstrate the use of a boolean expression directly in the `return` statement. This is identical to

```
1  if (r>=0 && r<nr && c>=0 && c<nc) return true; else return false;
```

Since performing these validations leads to additional CPU cycles, we could use preprocessor directives to include them only in the debug version:

```
1  double get(int r, int c) {
2  #ifdef NDEBUG
3      if isValid(r,c)
4  #endif
5      return a[r][c];
6  }
```

The preprocessor includes the block only if the NDEBUG macro is defined. This macro (along with _DEBUG) is set automatically by many IDEs for the Debug build. When built under Release mode, the compiler (which runs after the preprocessor), would only see

```
1  double get(int r, int c) {
2      return a[r][c];
3  }
```

Similar debug-only checks can be accomplished using the **assert** expression,

```
1  #include <assert.h>
2  double get(int r, int c) {
3      assert(r>=0 && r<nr && c>=0 && c<nc);
4      return a[r][c];
5  }
```

This test is included only when NDEBUG is defined. If the assertion expression evaluates to false, the program terminates with an *assertion error*.

4.4.4 FRIENDS

Instances may arise in which we need to grant access to class' private or protected data to a non-related class ObjA. Such a class can be denoted a "friend" by including

friend class ObjA;

inside the class granting accesses. Similar syntax is also available for non-member functions.

4.4.5 OPERATOR OVERLOADING

Given the migration of the coefficient array a to the protected section, each of the following two lines leads a compilation error:

```
1  A.a[2][3] = 55; v // set value
2  cout<<A.a[0][1];  // get value
```

We need to route the data access through the setter and the getter:

```
1  A.set(2,3,55);
2  cout<<A.get(0,1);
```

This may get a bit cumbersome. One strength of C++ (at least in our opinion) is its support for operator overloading. Custom operator algorithms are implemented using functions with names **operatorX** where X is the operator being overloaded. For example, a new definition for addition would be specified using `operator+`. While we can define custom code, we cannot change the number of expected arguments. Addition is a binary operation, and thus requires two arguments. Similarly, the unary array access operator `[]` needs to take just a single argument, and as such, it is not possible to write code that would let one use `[i,j]` for data access. Such a syntax is however possible with the function call operator `()`, which can take an arbitrary number of arguments. Instead of writing

```
1  class Matrix {
2      ...
3      double get (int r, int c) {return a[r][c];}
4  };
```

we can include

```
1  class Matrix {
2      ...
3      double operator() (int r, int c) const {return a[r][c];}
4  };
```

With this definition, we *get* the value of coefficient $a_{5,3}$ with

```
cout <<A(5,3);
```

The `const` keyword after the function definition signals to the compiler that the function does not modify any class data and can thus be used with constant references and temporary objects. This operator returns a *copy* of the coefficient value, and thus is not compatible with the use on the left side of an expression. The following will not work given the definition so far

```
A(5,3) = 55.0;
```

In order to set a value, we need a *reference* to the coefficient storage location in the `Matrix::a` array. We thus add another `()` operator as:

```
1  class Matrix {
2      ...
3      double operator() (int r, int c) const {return a[r][c];}  //get
4      double& operator() (int r, int c) {return a[r][c];}        //set
5  };
```

Please note that the second operator returns a reference, allowing it to be used in expressions that need to modify (set) the coefficient. With these two definitions, we can now write code such as

```
1  A(5,3) = 55.0; // set value
2  cout <<A(0,1); // get value
```

The first operator version is now no longer strictly necessary since the reference can be used on both sides of an expression. However, the second, non constant version cannot be used to *read* values neither from temporary objects nor from const references. As such, it is recommended to implement both types. While the function body looks identical, the produced code is not the same. In the first version, we return a *copy* of the value held in `a[r][c]`. In the second version, we return the actual pointer (via a reference) to this object, `&a[r][c]`.

Operators can also be defined outside a class definition. Such an approach is used when defining operators not associated with any particular instance. For example, we can implement matrix-vector multiplication using

```
1  vector<double> operator*(const Matrix &A,
2                           const vector<double> &vec) {
3    ...
4  }
```

where **vector<double>** is a 1D array of double precision values. We cover this standard library storage container in Section 4.4.10.

Another commonly encountered case involves the overload of boolean operators ! and **bool** (the latter is defined using **operator bool()**). These operators can be used to check the "validity" of the object using syntax such as **if (A)** .... File input and output stream handlers, **ifstream** and **ofstream** use this method to indicate whether the file was open successfully. The stream objects also use the « bit shift operator for data output. This operator nominally shifts integer values by the specified number of bits to the left, which is analogous to multiplication by 2^n, where n is operand. The original usage is generally found only in high performance codes utilizing integer mathematics. We can overload this operator to add output support for our custom objects using

```
1  std::ostream& operator<<(std::ostream &out, const vec3& v) {
2    out<<v[0]<<", "<<v[1]<<", "<<v[2];
3    return out;
4  }
```

Here we are assuming that **vec3** is a custom data type for three-dimensional vectors. With this definition, one may write out (let's say) velocity components with

```
1  vec3 velocity;
2  cout<<"Velocity is "<<velocity<<endl;
```

This will generate output such as

```
Velocity is 11.11, 22.22, 33.33
```

4.4.6 COPY AND MOVE

Another operator of interest is the assignment = operator. It let's us write code such as

```
1  Matrix A, B;
2  A = B;
```

This is called the *copy* operation. The compiler automatically implements this operator if not explicitly specified. However, the implicit definition performs a *shallow copy* in which pointers are copied without duplicating the actual memory block they refer to. Hence, after a shallow copy,

```
1  B(1,2) = 3;
2  cout<<A(1,2);  // prints 3
```

Both matrices share the coefficient array and as such are essentially the same matrix due to `A::a` and `B::a` storing the same memory addresses. In order to properly duplicate the matrix, we need to implement a *deep copy* as follows:

```
1   class Matrix {
2      Matrix& operator =(const Matrix &other) {
3         assert(nr==other.nr && nc==other.nc);  // dims must match
4         for (int r=0;r<nr;r++)
5            for (int c=0;c<nc;c++) a[r][c] = other.a[r][c]; // or
                   other(r,c);
6      }
7      return (*this);  // return reference to us to allow chaining
8   };
```

This operator takes in a single argument, which is a constant reference to the "other" matrix. The `const` keyword is again used to indicate to the compiler that we are performing a read-only access, and can thus be called with temporary objects or with constant references.

Whenever we define an assignment operator, we should also define a custom *copy constructor*. This is a special constructor that creates a new object that is a copy of another one. A copy constructor is defined as follows:

```
1   class Matrix {
2      Matrix (const ref &o)  :  Matrix(o.nr,o.nc) {  // copy
             constructor
3         for (int r=0;r<nr;r++)
4            for (int c=0;c<nc;c++) a[r][c] = o(r,c)
5      }
6   }
```

We first use constructor chaining to allocate space for a $n_r \times n_c$ matrix. Next we copy the data from the "other" matrix o to us, just as was done for the assignment operator.

Let's say that you want to write a function that generates and initializes a matrix to be used for a simulation code. You could write the following:

```
1   Matrix buildMatrix(int nr, int nc) {
2      Matrix M(nr,nc);  // make a new matrix
3      for (int r=0;r<nr;r++)
4         for (int c=0;c<nc;c++) { /* set coefficients */ }
5
6      return M;
7   }
```

which would be called as

```
Matrix A = buildMatrix(nr,nc);
```

This code starts off by allocating a new matrix inside the `buildMatrix` function. The `return` statement at the end returns a *copy* of this matrix. Therefore, space for the secondary matrix A needs to be allocated first, and then data from M is copied to A. After this, the M matrix is destroyed.

This copy is completely unnecessary. Instead of copying all data from M to A, and subsequently deleting M, we can simply *move* the data over. A move let's us "steal" data from a temporary object. It is up to the compiler to decide whether a move or a copy is used, but we can hint that we want to move the data by enclosing the object in `std::move` as in

```
return std::move(A);
```

The move operation is used with references to *r-values*. In typical assignment, A = M, the expression on the left is called the *l-value*. We can only assign to objects with a non-ephemeral storage. For instance, in the expression

```
Matrix A = Matrix(nr,nc);
```

we create a temporary **Matrix** object on the right side. This object is not associated with any variable name, and exists only for the brief duration of this assignment operation. This is an *r-value*. Such an object can appear on the right side of an expression. However, they are not allowed on the left side. The following expression would not make sense:

```
1  Matrix(nr,nc) = A;
```

since there is no actual named object to which the data assignment should be made. Up to this point, we have been using references to l-value objects. Turns out that C++ also supports so-called *r-value references* denoted by **&&** that can bind to temporary objects. Move operations are defined for references of this type, since the purpose of a move is to work with temporary objects that will be deleted anyway. The definition is as follows:

```
1   class Matrix {
2      Matrix (Matrix &&other): Matrix{other.nr, other.nc) { // move
          constructor
3         a = other.a;     // copy over the pointer
4         other.a = nullptr;  // invalidate other's pointer
5      }
6
7      void operator= (Matrix &&other) { // move assignment
8         assert (nr==o.nr && nc==o.nc);
9         a = other.a;
10        other.a = nullptr;   // invalidate other's pointer
11     }
12  }
```

The reader is suggested to review Stroustrup 2018 for more information on these concepts.

4.4.7 STATIC MEMBER FUNCTIONS AND NAMESPACES

Class member functions can be declared *static* to associate them with the class type, but not a particular instance. Such functions are used to implement generic functionality that is related to the topic of the class but does not utilize any member variables of a particular instance. For example, we may want the Matrix class to implement a function that instantiates and initializes a new identity matrix. We may write the code such as

```
1   class Matrix {
2      public:
3      static Matrix makeIdentity(size_t nr) { ... }
4   }
```

This function would subsequently be used as

```
Matrix I = Matrix::makeIdentity(40);
```
Such "utility" functions can alternatively be combined into namespaces,

```
1  namespace MatrixUtils {
2    Matrix makeIdentity(size_t nr);
3    ...
4  };
```

with implementation and use through `MatrixUtils::makeIdentity`. Namespaces are also useful for organizing physical constants.

4.4.8 INHERITANCE AND POLYMORPHISM

Let's consider an algorithm for performing matrix-vector multiplication $\vec{b} = A\vec{x}$. Each term of $\vec{b}$ is given by

$$b_r = \sum_k^{n_c} a_{r,k} x_k \tag{4.16}$$

Given our matrix class definition, we may be inclined to implement an operator for this operation that reads

```
1  vector<double> operator*(const Matrix &A, const vector<double> &x
   ) {
2    vector<double> b(A.nr);  // vector to store results
3    for (int r=0;r<A.nr;r++) {
4      double sum = 0;
5      for (int c=0;c<A.nc;c++) sum += A(r,c)*x[c]
6      b[r] = sum;
7    }
8    return b;
9  }
```

It is important to realize that the loop on line 5 is computing a dot product between the matrix row r and the vector $\vec{x}$. Thus, the algorithm can be generalized as

```
1  vector<double>operator*(const Matrix &A, const vector<double>&x){
2    vector<double> b(A.nr);
3    for (int r=0;r<A.nr;r++) {
4      b[r] = A.dotRow(r,x);
5    }
6    return b;
7  }
```

with the dot product implemented as a member of function of Matrix

```
1  double Matrix::dotRow(int r, const vector<double>&x) const {
2    double sum = 0;
3    for (int c=0; c<nc; c++) sum += a[r][c]*x[c];
4    return sum;
5  }
```

This implementation assumes that **A** is a dense matrix. However, as we have seen in Chapter 2, discretization of engineering governing equations tends to lead to sparse matrices with only several non-zero terms per row. Sparse matrices can be further divided into subcategories such as *banded*, in which the non-zero coefficients are limited to several diagonal bands, *Toeplitz*, which besides being banded, contain identical values on each row, and *identity*, which contain only ones on the diagonal. Of course, we can also have a generic sparse matrix.

The matrix-vector multiplication algorithm does not care about the type of matrix that is being used. It simply needs the value of the dot product. This dot product will be computed differently for each matrix type. For example, for an identity matrix, it is equal to b_r. A sparse matrix needs to multiply just the columns for which the coefficients are known to be non-zero. This behavior can be accomplished by utilizing *inheritance*. C++ classes can derive from a parent class, in which case they inherit the functions and variables declared in the parent. This is written as

```
1   class A { double x; };
2   class B : public { double y;};
```

An instance of B class type now contains two member variables: x and y. The `public` keyword indicates that A members should retain their original access rights. Otherwise, public members become protected.

Similarly, member functions from the base class A are also inherited. This is where things get really interesting. As we know by now, C++ allows us to overload functions. As such, we can let the base class act as an "interface" that provides the signature of some generic functionality that we want implemented, such as the row-vector dot product. We then declare multiple derived data objects with each implementing this functionality in a way that is the most appropriate to that particular object. Other code that needs this functionality can be written to utilize objects of the *base* type without caring to know just what type of a derived type it is dealing with. This is known as *polymorphism*.

Polymorphism is an extremely powerful concept. It helps you write code that is more modular, easier to read, while also simpler to maintain. The basic premise is that given

```
1   class SparseMatrix : Matrix { ... };
2   SparseMatrix sparse_mat;   // instance of the derived class
```

we can write

```
Matrix &mat = sparse_mat;   // base-type reference to derived type
```

If this is not immediately obvious, we are assigning a reference to a derived object to a reference variable of the base type. The rest of the code can then use the `mat` reference without needing to know just what type of matrix it is. It just knows that the object referenced by `mat` implements all functionality specified in the base class declaration. This same outcome is also accomplished using pointers, i.e.

```
1   class IdentityMatrix : Matrix { ... }; // another derived type
2   void doSomething(Matrix *mat);   // function using a Mat pointer
3   IdentityMatrix identity_mat;   // instance of identity matrix
4
```

```
5    Matrix *mat = &sparse_mat;
6    doSomething(mat);       // perform some action with sparse matrix
7    doSomething(&identy_matrix); // now use an identity matrix
```

In this example, we introduced a generic function that needs a `Matrix` pointer. We subsequently call it with instances of two different derived types.

Returning to our matrix-vector multiplication example, polymorphism implies that we can define a custom `dotRow` function for each matrix type. We can then write

```
1    vector<double> res1 = sparse_mat*b;
2    vector<double> res2 = identity_mat*b;
```

Here the same `operator*` function gets used as long as the left-hand operand is of type that derives from the base `Matrix` type. There is however one wrinkle. Function calls are "fixed" at compile time by the linker adding a jump instruction to a particular function body. However, in this case, the multiplication operator needs to call a different `dotRow` depending on the actual type of matrix reference. The linker cannot determine this at build time. Therefore, we need to signal to the linker that the actual function call lookup needs to be delegated to run time. This is done by flagging the function in the base class (through which we access the derived functionality) as *virtual*. Consider the following:

```
1    class Base{
2        void sayHello() {cout<<"I am Base"<<endl;}
3    };
4
5    class Derived : public Base {
6        void sayHello() {cout<<"I am Derived"<<endl;}
7    };
8
9    Derived derived;
10   Base &ref = derived;
11   ref.name();  // prints "I am Base"
```

Calling `ref.sayHello` prints I am Base, despite `ref` reference being initialized with an object of the Derived type. This is because the linker created a call to `Base::sayHello` during the code build. But now, if we mark the function as virtual:

```
1    class Base{
2        virtual void sayHello() { ... };
3    };
```

we obtain the desired I am derived.

In this particular example, the base class provided a body for this function. Quite often we want the base class to simply act as an interface that only describes the required functionality, but does not actually implement it. This is done by making the function *pure virtual*. It is denoted by adding = 0 after the declaration. Our base Matrix type can now read

```
1    class Matrix {
2        public:
3        Matrix (int nr): nr{nr} {}
```

```
4    virtual double dotRow(int r, const std::vector<double>&x)
         const = 0;
5    virtual double& operator()(int i,int j) =0;
6    virtual double operator()(int i, int j) const =0;
7
8    virtual ~Matrix() {}    // virtual destructor
9
10   const int nr;    // number of rows
11 };
```

A class containing a pure virtual function cannot be instantiated. Such a class is called an *abstract class*. It can only be used a reference type through which concrete implementations get accessed. The following will not compile

```
Matrix M(nr);  // error, Matrix is an abstract class
```
However, using the type as a reference to a concrete implementation is acceptable, as in

```
1  DenseMatrix dense_mat(nr,nc);
2  Matrix &A = dense_mat;
```

We also added support for overloaded operators, as the implementation of these is also matrix-type specific. Also notice that the destructor is marked as virtual in the base class. A virtual destructor ensures that the correct derived destructor is called. Otherwise, the destructor from the base class would be executed. This would lead to an incomplete data cleanup, since any dynamically allocated resources from the derived class would not be released. We then define a derived class, such as one implementing a five banded sparse matrix as

```
1  class FiveBandMatrix : public Matrix {
2      public:
3      FiveBandMatrix(int nr, int di) : Matrix(nr), di{di} { ... }
4      ~FiveBandMatrix() { ... }
5
6      double& operator()(int r, int c)  {
7          if (c-r==-di) return a[r];
8          else if (c-r==-1) return b[r];
9          else if (c-r==0) return this->c[r];
10         else if (c-r==1) return d[r];
11         else if (c-r==di) return e[r];
12         else throw std::runtime_error("Unsupported operation: " +
                 std::to_string(r) + " " + std::to_string(c));
13     }
14
15     double operator()(int r, int c) const { /* same as above */}
16
17     double dotRow(int r, const std::vector<double>&x) const {
18         double sum = 0;
19         if (a[r] && r-di>=0) sum += a[r]*x[r-di];
20         if (b[r] && r-1>=0) sum += b[r]*x[r-1];
21         sum += c[r]*x[r];
22         if (d[r] && r+1<nr) sum += d[r]*x[r+1];
23         if (e[r] && r+di<nr) sum += e[r]*x[r+di];
24         return sum;
25     }
```

```
26
27   protected :
28       double *a = nullptr ;    // i−di
29       double *b = nullptr ;    // i−1
30       double *c = nullptr ;    // main diagonal
31       double *d = nullptr ;    // i+1
32       double *e = nullptr ;    // i+di
33       const int di ;           // offset from a to b
34   };
```

This matrix is used for the heat equation solver in Section 4.5.

4.4.9 TEMPLATES

The matrix we have defined so far stores double precision data. Let's say we also need a version that stores single precision floats and another one for integers. We could implement unique versions such as

```
1    class FMatrix {
2    public :
3        FMatrix () {
4            a = new float *[ ni ];
5            for (int j=0;j<nj;j++) a[j] = new float[nj];
6        }
7        float operator(int i , int j) {return a[i][j];}
8    protected :
9        float **a;
10   };
11
12   class IMatrix {
13   public :
14       IMatrix () {
15           a = new int *[ ni ];
16           for (int j=0;j<nj;j++) a[j] = new int[nj];
17       }
18       int operator(int i , int j) {return a[i][j];}
19   protected :
20       int **a;
21   };
```

These classes would be identical, except for the difference in their data types. We can avoid such code duplication using *templates*. A template defines a keyword that is substituted at *compile time*. As such, templates have no impact on code performance. They do however require that the implementation is available to the compiler, which in practical terms necessitates placing all templated code in the header files. A template version could be defined as follows:

```
1    template<typename K>
2    class Matrix {
3    public :
4        Matrix () {
5            a = new K*[ ni ];
6            for (int j=0;j<nj;j++) a[j] = new K[nj];
7        }
8
```

```
9      K operator(int i, int j) {return a[i][j];}
10   protected:
11     K **a;
12   }
```

We tell the compiler that K is a stand-in for some *type name*, such as a basic data type like an integer or a custom class. We instantiate the object by specifying the type in the declaration:

```
1   Matrix<float> fmatrix;  // matrix with floats
2   Matrix<int> imatrix;    // matrix with ints
3   Matrix<vec3> vec3_matrix; // matrix with a custom type vec3
```

The template argument is part of the type definition and hence is also used when specifying function arguments:

void doOp(Matrix<**double**> &A) {...}

Since continuously writing out these definitions can become tiresome, we can define "nicknames" for custom types. This is done with the **using** keyword,

```
1   using DMatrix = Matrix<double>;
2   using FMatrix = Matrix<float>;
```

This allows us to write

DMatrix A(ni, nj);

and

void doOp(DMatrix &A) {...}

It is also possible to use templates to substitute values. For example, an object that stores a prescribed number of items of arbitrary type can be declared as

```
1   template<typename K, int S>
2   struct MyCont {
3    K data[S];   // statically allocated array of S items of type K
4   }
```

Then, writing MyCont<string,10> would give us an array of 10 strings, while MyCont<float,5> provides an array of 5 floats.

4.4.10 STORAGE CONTAINERS

The concepts discussed so far are utilized in storage containers that are part of the standard library since the C++98 revision. These include **vector**, linked **list**, **map**, and **set**. Additional containers such as **unordered_map** were introduced in C++11. A **vector** is a wrapper on top of a standard array that allows the array to grow dynamically. The array can be expanded from empty using **push_back** or **emplace_back**. The first version requires an actual object of the desired type to be specified, while the second one constructs the item in place, possibly leading to a faster performance. The vector contains internal *capacity* which is the size of the allocated array. Once the capacity is exceeded, the push operation results in the allocation of a new array that is larger than the original according to an implementation-specific growth policy. The old data is then copied into the new buffer and the old array is deleted. If we know ahead of time how big the array we will need to be, we can avoid these unnecessary memory copies using **reserve**. Consider the following code:

```
1  #include<vector>
2  vector<double> vals;         // empty array
3  for (int i=0;i<20;i++) {
4      vals.emplace_back(i);    // add a new entry
5      cout<<vals.size<<", "<<vals.capacity<<endl;
6  }
```

Adding

```
vals.reserve(20);
```

above the **for** loop eliminates the memory copies. We can also initialize the vector to start off with a certain number of default items. For example

```
1  vector<double> vals(20, -1);  // 20 doubles initialized to -1
2  for (int i=0;i<20;i++) vals[i] = 0;   // modify values
```

The **vector** class overloads the array access [] operator and is thus accessed using familiar notation.

These storage containers are compatible with the new *range-based* for loops. The above assignment could be rewritten as

```
for (double &d: vals) d = 0;
```

Note the use of a reference in the for loop to allow modifying the data.

These range-based loops take advantage of *iterators*. An iterator is a class that points to a particular item and implements a custom increment operator to advance to the next element. It also overloads the * dereferencing operator to access the data at the iterator position. A for-each loop is identical to the following code

```
1  vector<double>::iterator it = vals.begin();  // first element
2  while (it!=vals.end()) {   // until we reach the end
3      *it = 42;       // assign some value to the current element
4      it++;           // advance the iterator to the next element
5  }
```

The **list** defines a *linked list*, which we have already seen in Section 4.3.6. We use a list as follows:

```
1  #include<list>
2  list<double> vals;
3  for (int i=0;i<5;i++) vals.push_back(i);
4  for (double val:vals) cout<<val<<" ";   //0 1 2 3 4
5
6  // erase the second element
7  list<double>::iterator it = vals.begin();
8  advance(it,2);
9  vals.erase(it);
10 for (double val:vals) cout<<val<<" ";   //0 1 3 4
11 }
```

A **map** is another powerful container. It is used to create a "dictionary" mapping data of one type to another. Here is an example of mapping strings to integers:

```
1  #include <map>
2  map<string, int> my_map;
3  my_map["abc"] = 42;
4  cout<<"The code for 'abc' is "<<my_map["abc"];
```

Finally, a **set** defines a container for unique values. Consider ·

```
1  #include <set>
2  set<int> data;
3  data.insert(5);
4  data.insert(2);
5  data.insert(5);
6  data.insert(7);
7  for (int d:data) cout<<d<<" ";
```

will print **5 2 7** despite 5 being inserted twice. The map and set containers utilize < operator to compare the keys. They can be used with arbitrary custom types, as long as the type implements a custom comparison operator. Alternatively, a comparator object can be provided during instantiation.

4.4.11 SMART POINTERS

Polymorphism is particularly useful for generating code that is ambiguous to the actual implemented type. For example, we can store an array of matrixes using

```
vector<Matrix *> mats;
```

and subsequently populate it as

```
1  mats.push_back(new DenseMatrix(...));
2  mats.push_back(new IdentityMatrix(...));
```

and so on. Pointers are needed since derived objects can be accessed through the base class only using pointers or references; however, the STL containers do not support the storage of references. With this approach we need to remember to delete the allocated objects once no longer needed. This can be avoided by utilizing **smart pointers**. These are small classes that wrap a pointer and also implement *reference counting*. Once all objects utilizing the pointer go out of scope, the class destructor frees the pointer. There are several kinds of smart pointers, with **unique_ptr** being the simplest. The stored pointer can be used by only a single object without being able to be shared. We use it as follows:

```
1  #include <memory>
2  vector<unique_ptr<Matrix>> mats;
3  mats.push_back(new DenseMatrix(...));
4  mats.push_back(new IdentityMatrix(...));
5  for (auto mat:mats) mat->solve();
```

There is no longer the need to free the data. Here we also demonstrate the use of **auto** variables. This keyword lets the compiler choose the appropriate type, which in this case is **unique_ptr<Material>**. Here we also demonstrate the **auto** variable type which let's the compiler to automatically pick the correct type. This type is commonly utilized with templated standard library types that would otherwise require a lengthy descriptor.

4.4.12 LAMBDA FUNCTIONS AND FUNCTORS

Let's say that your code needs to compute a generic math operation on a vector of data. Using enums we could write

```
1  enum class OpType {SQUARE, SQRT, ABS};
2  void performOp (double *x, int ni, OpType op_type) {
3      if (op_type==OpType::SQUARE) {
4          for (int i=0;i<ni;i++) x[i] = x[i]*x[i];
5      } else if (op_type==OpType::SQRT) {
6          for (int i=0;i<ni;i++) x[i] = sqrt(x[i]);
7      } else if (op_type==OpType::ABS) {
8          for (int i=0;i<ni;i++) x[i] = abs(x[i]);
9      }
10 }
```

which you would then call as in performOp(x,ni,OpType::SQRT). But there is an alternate way. Functions can receive other functions as arguments. We can define

```
1  void performOp (double *x, int ni, double op(double)) {
2      for (int i=0;i<ni;i++) x[i] = op(x[i]);
3  }
```

We then define the operator function as

```
double square(double x) {return x*x;}
```

which would be used in the calls as

```
performOp(x,ni,square);
```

The sole purpose of the **square** function is to implement the operation to be used by **performOp**. This function will never be called from anywhere else in the program. It may thus be preferred to pass the algorithm logic in directly utilizing an anonymous lambda function. Lambda functions are defined using the following syntax:

```
[capture_list] (arguments) -> return_type { body }
```

The capture list, arguments, and the return type are all optional. The capture list lists variables existing in the higher scope that should be accessible to the lambda function. The return type is automatically deduced by the compiler and needs to be specified only if it would be ambiguous. With a lambda function, our code now reads

```
performOp(x,ni,[](double x) {return x*x;});
```

This example works well if the function does not require any external data. But let's say that instead of performing a simple math operation, the operation needs to return a value from a look-up table, which takes some time to build. We don't want having to have to rebuild the LUT on each call. While we could utilize a global variable to hold the table, this is not recommended. The better option is to encapsulate the data with the processing function. We can define a custom class that overloads the () function call operator (note, we have done this previously with the Matrix class, but there we used it to access data). Consider

```
1   class LookUp {
2      LookUp() { buildLUT();} // constructor
3      double operator() (double x) {return interpolate(x);}
4
5   protected:
6      vector<double> values; // some data
7      double interpolate(x); // interpolates from the LUT
8   };
```

We can now write the following:

```
1   LookUp lookup; // this builds the LUT
2   lookup(x);
```

The overloaded function call operator let's this class behave as a function which makes it compatible with the argument list of **performOp**. Class objects that can be used as functions are called *functors*. The C++ standard library implements many common algorithms (such as sorting) in the <algorithm> header that depend on specification of custom comparison operators via one of these methods. We use this concept in our codes to wrap the random number generator and distribution into a class that is subsequently used just as **rnd()** function found in Section 4.2.3,

```
1   class Rnd {
2   public:
3      //constructor: set initial random seed and distribution limits
4      Rnd(): mt_gen{std::random_device()()}, rnd_dist{0,1.0} {}
5      double operator() () {return rnd_dist(mt_gen);} // in [0,1)
6   protected:
7      std::mt19937 mt_gen;                              // generator
8      std::uniform_real_distribution<double> rnd_dist; // distrib.
9   };
10  Rnd rnd; // instantiate
```

This approach allows us to easily create an array of random number generators to be used by parallel algorithms utilizing multithreading, see Chapter 9.

4.5 HEAT EQUATION SOLVER IN C++

We demonstrate these advanced concepts through a C++ implementation of the steady-state 2D heat equation solver. The full source code is found in the **ch4/heat2** subdirectory. The code is divided into several source and header files, as is customary in larger simulation codes. It can be built using

```
$ g++ -O2 *.cpp -o heat2d
```

We also included a **run.sh** *bash script file* that performs this compilation and, assuming successful outcome, executes the binary. It reads as

```
#!/bin/bash
g++ -O2 *.cpp -o heat2d
if [ $? -eq 0 ]; then
   ./heat2d
fi
```

The first line is called a *shebang* and indicates the interpreter to use. The $? found on line 3 captures the return value of the previously executed program. If it is zero, implying no error returned by gcc, then the freshly compiled program is executed. This same logic could also be implemented more concisely by chaining the commands together using &.

4.5.1 STORAGE OBJECTS

We now briefly describe the code. Due to space constraints, only the relevant portions are included in print. The data.h header defines several storage containers. All relevant logic is implemented directly in the class definitions and as such there is no companion .cpp file. The first storage object is a structured called World. It simply collects information about the computational domain such as the number of nodes and cell spacing. These are set by the constructor,

```
1   struct World {
2      World(int ni, int nj, double dx, double dy):
3         ni{ni},nj{nj},nu{ni*nj},dx{dx},dy{dy} {}
4
5      int U(int i, int j) {return j*ni+i;}
6
7      const int ni;
8      const double dx;
9      ...
10  };
```

The U member function converts a 2D (i, j) index into a consecutive "unknown" index per

$$u = j \cdot n_i + i \qquad (4.17)$$

We also define a FiveBandMatrix for storing a sparse five-band matrix, utilizing concepts described previously in Section 4.4.8. It derives from base abstract class Matrix and provides implementations for the dotRow function, and the const and non-const variants of the () operator. We then define a Field object to store a generic two-dimensional array. Here we similarly implement a custom data access operator. This object is defined using a template argument,

```
1   // generic 2D data container
2   template<typename T>
3   class _Field {
4   public:
5      _Field(World &world, T def=0) : world{world} { ... }
6      ~_Field() { ... }
7
8      T& operator()(int i,int j) {return data[i][j];}
9      T operator()(int i,int j) const {return data[i][j];}
10     World &world;        // world reference
11
12  protected:
13     T **data;            // 2D array
14  };
```

Note that no bounds checking is performed. We also define names for a double and a bool variant, as well as a double vector

```
1  using Field = _Field<double>;
2  using FieldB = _Field<bool>;
3  using dvector = std::vector<double>;
```

4.5.2 MATRIX SOLVER

A Gauss-Seidel solver is found in `solver.cpp` with a function prototype in `solver.h`. The relevant part is shown below:

```
1   #include "solver.h"
2
3   void gsSolve(Matrix &A, dvector &b, Field &x2d) {
4       size_t nu = b.size();
5       World &world = x2d.world;
6
7       if (nu !=A.nr) throw runtime_error("Mismatched dimensions");
8
9       // flatten the x vector
10      dvector x(nu);
11      for (int j=0;j<world.nj;j++)
12          for (int i=0;i<world.ni;i++) x[world.U(i,j)] = x2d(i,j);
13
14      // solver loop
15      for (int it=0; it <10000; it++) {
16          for (int r=0; r<A.nr; r++) {
17              double dot_nomd = A.dotRow(r,x) - A(r,r)*x[r];
18              double g = (b[r] - dot_nomd)/A(r,r);
19              x[r] = x[r] + 1.4*(g-x[r]);            // SOR
20          }
21
22          // residue check every 25 iterations
23          if (it%25==0) {
24              double sum=0;
25              for (int r=0; r<A.nr; r++) {
26                  double R = b[r] - A.dotRow(r,x);
27                  sum += R*R;
28              }
29              ...
30          }
31      } // it
32
33      // unpack 1D solution
34      for (int j=0; j<world.nj; j++)
35          for (int i=0; i<world.ni; i++) x2d(i,j) = x[world.U(i,j)];
36  }
```

The main feature to notice is that this function operates with the base `Matrix` type. As such, the solver is agnostic of the matrix type, and could very well be used with a dense or a general sparse matrix instead. It uses the `dotRow` function, implementation of which is specific to each matrix type, to evaluate the new estimate for b_j

$$x_r = \left[b_r - \left(\sum_{k}^{n_c} a_{r,k} x_k - a_{r,r} x_r \right) \right] / a_{r,r} \qquad (4.18)$$

The term in the parentheses is a dot product of r-th matrix row with the solution vector $\vec{x}$ but with the contribution from the main diagonal excluded. The residue check is executed every 25 iterations. It uses this same dot product function to evaluate $R_r = b_r - \sum_{k}^{n_c} a_{r,k} x_k$, which is accumulated for the L2 norm calculation.

The location to hold the solution is passed into the solver as a reference to a two-dimensional field. Since the solver itself operates on a 1D vector, this 2D data is first "flattened". We copy the data instead of just starting with an empty array so that previously computed results can be used for the initial guess in simulations consisting of multiple time steps. At the conclusion of solver iteration, the 1D vector is "inflated" (or unpacked or reshaped) back into the 2D variant. Here we utilize the $u(i, j)$ function to perform this mapping.

4.5.3 VTK IMAGEDATA OUTPUT

Next, output.cpp provides the file output support. The function prototype is declared in the output.h header file within a namespace,

```
1  #ifndef __OUTPUT_H
2  #define __OUTPUT_H
3
4  #include <map>
5  #include <string>
6  #include "data.h"
7
8  namespace Output {
9      void saveVTI(std::map<std::string, Field*> fields, FieldB &
           fixed);
10 }
11 #endif
```

This construct allows us to organize several related functions into a common group. The function receives a *map* of string - double precision 2D field pairs that captures the data fields to be exported. The string is used to provide the name of this property. The "fixed" field flags Dirichlet nodes. It represents another type of data that may need to be exported, to support, for instance, thresholding of the displayed data.

The data is stored using VTK's ImageData file format. These files, which must utilize the .vti extension, capture 3D Cartesian grids. Since the geometry of a Cartesian grid is fully prescribed by the grid's origin, cell spacing, and the number of grid nodes in each direction, there is no need to output node vertices or connectivity info.

```
1  <VTKFile type="ImageData">
```

```
2   <ImageData WholeExtent="0 200 0 200 0 0" Origin="0 0 0" Spacing="
        0.1 0.1 0">
3   <Piece Extent="0 200 0 200 0 0">
4   <PointData>
5   <DataArray Name="T (K)" NumberOfComponents="1" format="ascii"
        type="Float64">
6   ...
7   </DataArray>
8   <DataArray ... > ... </DataArray>
9   </PointData>
10  </Piece>
11  </ImageData>
12  </VTKFile>
```

Here we have a $201 \times 201 \times 1$ two-dimensional grid. The spacing in z is not relevant due to the lack of cells in that direction. The `<Piece>` section specifies which part of the whole domain this file provides. While here we provide data for the entire extent - and hence the `Piece` section could in fact be omitted - Paraview allows the grid to be split among multiple files. This is useful when visualizing results from large parallel simulations in which each processor outputs data just for the domain it worked on. In fact, Paraview even supports a truly parallel visualization in which large input files remain stored on separate computers. The rest of the file contains data arrays for node-centered `<PointData>` and/or cell-centered `<CellData>`. For cell data, it is imperative to remember that there is one fewer cell in each dimension than we have nodes. The data is output by iterating over the provided map,

```
1   for (std::pair<string, Field*> pair: fields) {
2       Field &F = *pair.second;
3       out<<"<DataArray Name=\""<<pair.first<<"\" NumberOfComponents
            =\"1\" format=\"ascii\" type=\"Float64\">\n";
4       for (int j=0;j<world.nj;j++)
5           for (int i=0;i<world.ni;i++) out<<F(i,j)<<" ";
6       out<<"\n</DataArray>\n";
7   }
```

This map-based approach allows this function to output arbitrarily many fields. They just need to be populated into the map by the calling function. The data is written out following the same ordering used by $u(i,j)$ function, i.e. all data for a single i index are written out before moving on to the next j index. Here we also demonstrate iteration over a map. Each entry maps to a `std::pair` object. This object contains two member variables called **first** and **second** which correspond to the key and the value, respectively. The results are stored in a file called `heat2d.vti`. Had we implemented an unsteady heat equation solver, we could use stream formatters such as `setw` and `setfill` to write results at distinct time steps to files with names such as `heat2d-00100.vti`, `heat2d-00200.vti`, and so on. Paraview automatically groups files with a common prefix into an animation group.

Figure 4.2 shows one possible visualization of the produced results. Here we used the WarpByScalar filter to turn the 2D data into a 3D representation. The temperature data along the warped surface is visualized using the Blue to Orange divergent heat map with color discretization turned off to obtain

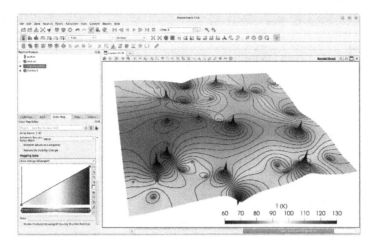

Figure 4.2 Heat equation solution visualized in Paraview.

smoothly varying colors. Image quality is improved by enabling ray tracing. We also use the Contour filter to display contour lines. When used on a 3D source data, this filter generates isosurfaces. Other filters of interest for volumetric data include Slice and Clip. the first option generates a planar (by default) cut when applied to 3D data, and a line when applied to 2D data. Values along the line can be visualized in the Line Chart view mode. The Clip filter is used to hide part of the data so that we can look "inside" an object. All filters offer several visual representations under the filter's properties tab. Surface shows the external surface of the rendered object. This is in fact what is used in this Figure. Surface with Edges adds the cell borders, which is useful when wanting to illustrate the mesh resolution. Wireframe shows just the cell edges. Slice can be used with 2D data. It defaults to plotting values using flat shading, with the same color used for the entire cell (this is in fact how cell-centered data are visualized). It also turns off lightning effects. Finally, the Volume option uses volume rendering to plot three-dimensional data as a colored "cloud". This option is particularly useful when visualizing gas plumes.

4.5.4 MAIN DRIVER

Finally, all these functions are called from `heat2d.cpp`,

```
1  // random number generator
2  class Rnd { ... };
3  Rnd rnd;    // instantiate
4
5  int main() {
6      World world(201,201,0.1,0.1);
7      Field T(world);
8      FieldB fixed(world);
9
```

```
10      // pick random heat sources
11      for (int s=0;s<40;s++) {
12          int i = 1+rnd()*(world.ni−2);
13          int j = 1+rnd()*(world.nj−2);
14          fixed(i,j) = true;
15      }
16
17      // coefficient matrix and forcing vector
18      FiveBandMatrix A(world.nu, world.ni);
19      dvector b(world.nu);
20
21      // Finite Difference of Laplace equation
22      for (int j=0; j<world.nj; j++)
23          for (int i=0; i<world.ni; i++) {
24              size_t u = world.U(i,j);
25              if (fixed(i,j)) {A(u,u)=1; b[u] = 50+100*rnd();
26                              continue;}
27
28              if (i==0) {A(u,u)=1; A(u,u+1)=−1; b[u]=0;}
29              else if (i==world.ni−1) {A(u,u)=1; A(u,u−1)=−1; b[u]=0;}
30              else { /*standard stencil*/
31                  ...
32              }
33          }
34
35      gsSolve(A, b, T);            // call the solver
36
37      // build output map and write file
38      map<string,Field*> fields;
39      fields["T (K)"] = &T;
40      Output::saveVTI(fields, fixed);
41
42      cout<<"Done!"<<endl;
43      return 0;
44  }
```

We begin by instantiating the world object for a 201×201 node grid. Note that this container does not actually allocate any data. It only stores these domain dimension. We then instantiate a double-precision data field to store node-centered temperature values, and another boolean field to store the location of the internal Dirichlet (fixed) nodes. We then use an instance of the Rnd class to sample 40 random nodes that are flagged as fixed. Note that we use this class as a function, as described in Section 4.4.12. We next instantiate a variable to hold the coefficient matrix, and a 1D vector to store the forcing vector $\vec{b}$. We then loop over all mesh nodes and set the appropriate **A** matrix coefficients according to the Finite Difference discretization. Fixed nodes have just one set on the main diagonal and b set to a random value. We then call the gsSolve function. The address of the T field is added to a map of output fields. Finally, the data is stored by calling the saveVTI function with the fully qualified name including the namespace. And that's it. You are now experts in the use of C++ for numerical analysis!

5 Kinetic Methods

This and the following chapter introduce methods relevant to simulations of gas dynamics. Specifically, in this chapter, we cover particle-based, Lagrangian approaches. Then in the following chapter, we introduce methods applicable to mesh-based, Eulerian integration schemes. Although we focus on gas modeling, the methods introduced are relevant to other disciplines. We develop codes simulating free molecular, collisional, and low-density plasma flow past an infinitely long cylinder. The topics covered here are applicable to fracture mechanics, planetary orbits or solar system formations, traffic flows, or even the spread of contaminants or aerosols.

5.1 INTRODUCTION

On the microscopic level, gases are simply atoms, molecules, and in the case of plasmas, ions and electrons colliding with each other while being affected by external forces such as gravity or electromagnetics. Conceptually we could simulate gas flows by integrating equations of motion for all particles using schemes demonstrated in Chapter 1. This is known as a Lagrangian (or particle-based) approach and is the algorithm used in Molecular Dynamics and N-Body simulation codes. A simple calculation quickly reveals that simulating every single real atom is not feasible for macroscopic engineering problems. For an ideal gas, pressure p is related to the number density n and temperature T via the ideal gas law, $p = nk_bT$, where $k_b \approx 1.3806 \times 10^{-23}$ J/K is the Boltzmann constant. Using sea-level atmospheric conditions of $p = 1$ atm (101,325 Pa) and $T = 20°$C (293.15 K), we can see that there are about 2.5×10^{25} molecules per cubic meter. Using just the minimum six single-precision floating point numbers to capture each particle's position and velocity, we would need over 5×10^{14} Tb of memory per cubic meter just to store the particle data. Further, assuming that it takes a computer 10 ns to integrate the position and velocity of a single particle, advancing the system through a single integration step would take over 7 billion years. Even at the $p = 10^{-6}$ Torr (1 Torr is 133.3 Pa) pressures found in vacuum chambers, the memory and run-time requirements are staggering at over 700,000 Tb of RAM and 10.2 years.

5.1.1 THERMALIZATION

Luckily, it turns out that modeling every single particle is not necessary. Imagine that you have a perfectly elastic box into which you somehow loaded two discrete molecular populations: population A with all molecules moving with speed v_A, and population B with all molecules moving with speed v_B. We can let $f(\vec{x}, \vec{v}, t)$ be a function that describes the probability of finding a molecule

DOI: 10.1201/9781003132233-5

with velocity $\vec{v}$ at time t at position $\vec{x}$. This function is known as the *velocity distribution function*, or VDF for short. If we consider the entire box as a single control volume (so the position is irrelevant), and if we only care about the speed $v = |\vec{v}|$, then the initial speed distribution can be described using

$$f(v, t = 0) = \begin{cases} f_A & ; v = v_A \\ f_B & ; v = v_B \\ 0 & ; \text{otherwise} \end{cases} \tag{5.1}$$

After a sufficiently long time, you return to the box and characterize the speeds of all molecules. To your surprise, you find that instead of recovering the two original peaks, the two populations have merged into a single continuous distribution given by

$$f(v)_M = 4\pi \left(\frac{m}{2\pi kT} \right)^{3/2} (v - v_d)^2 \exp \left(-\frac{(v - v_d)^2}{v_{th}} \right) \tag{5.2}$$

Here v_d is a parameter called drift velocity and $v_{th} = \sqrt{2kT/m}$ is the thermal velocity. For this particular example, $v_d = 0$, but it is included here for generality. By plotting the relationship for different values of v_{th}, we can observe that thermal velocity, and hence T, affects the width of the distribution. In the limit of $T \to 0$, all velocities collapse to v_d.

Equation 5.2 is known as the Maxwell-Boltzmann velocity distribution function. Statistical mechanics tells us that this is the highest entropy state that a molecular system will naturally evolve to, as noted in Vincenti and Kruger 2002 among others. A gas population with velocity described by this distribution is said to be thermalized. Local collisions still occur and lead to the transfer of momentum between individual molecular pairs. However, the global distribution of velocities no longer changes.

5.1.2 FLOW PROPERTIES

At this point, you may be wondering why this is relevant. When it comes to simulating gas flows, we are generally interested in resolving the temporal and spatial variation in macroscopic properties such as gas density, mean velocity, and temperature. It turns out that these properties are related to so-called *moments* of the velocity distribution function, i.e., $n \sim \int_v f(v) dv$, $v \sim \int_v f(v) v dv$, $E \sim \int_v f(v) v^2 dv$ where the integration is over the velocity space. The evolution of the VDF is governed by the Boltzmann equation, see Section 6.5. For the case of $f = f_M$, these moments reduce to the familiar Navier-Stokes equations that form the foundation of Computational Fluid Dynamics (CFD). N-S equations are partial-differential equations (PDEs) that are solved using Eulerian, mesh-based approaches introduced in Chapter 6. These equations can alternatively be derived by considering mass, momentum, and energy conservation in a small volume due to the flux of these properties across the volume bounding surface. The requirement for thermalization may not be explicitly stated in the derivation but is implied. Our definition of temperature inherently assumes the Maxwellian VDF (i.e. temperature is a parameter of Equation 5.2). Similarly, the value of mean velocity is related to the shape of the distribution. In the case of the initial dual peak state in Equation 5.1, there is a clear distinction between the mean ($v_A < v_{mean} < v_B$) and the most probably speed $v_{mp} = v_A$ if $f_A > f_B$ else v_B. While a difference between the mean and the most probable speed also exists for f_M, the discrepancy there is not as pronounced.

5.1.3 KNUDSEN NUMBER

The take away message from the prior paragraph is that for a thermalized population, it is possible to compute the macroscopic properties of interest by integrating governing partial differential equations on a computational grid. This allows us to forget about the microscopic nature of gas flow, and ignore the massive computational requirements that would be inherent in attempting to model all individual particles. However, this simplification requires that the population is indeed thermalized. In the above formulation, we stated that we had to wait "sufficiently long" before returning to the box. In this context, a sufficiently long time implies that enough time has passed for all molecules to undergo multiple momentum exchange collisions. The number of collisions per second is given by collision frequency ν

$$\nu = n\overline{\sigma v} \tag{5.3}$$

where n is the gas number density and σ is a term called collision cross-section. Imagine that we have a molecular beam of constant flux Γ_0 passing through a slab filled with atoms. The concentration of atoms in the slab is given by n. If σ is the cross-sectional area in which the incident molecule collides with the target (i.e. $\pi(r_1 + r_2)^2$ for a hard sphere model), then the probability of that molecule undergoing a collision in a distance dx is $P = n\sigma dx$. The incident flux reduces according to

$$\frac{d\Gamma}{dx} = -\Gamma n\sigma dx \tag{5.4}$$

which can be integrated to yield

$$\Gamma = \Gamma_0 \exp(-n\sigma x) \tag{5.5}$$

or

$$\Gamma = \Gamma_0 \exp\left(\frac{-x}{\lambda_m}\right) \tag{5.6}$$

where $\lambda_m = 1/(n\sigma)$ is called the mean free path. It corresponds to the distance over which the incidence flux drops by a factor of e. In practice, λ_m is assumed to indicate the average distance traveled by molecules between collisions.

The mean free path can be compared to some characteristic length. This ratio, given by

$$K_n = \lambda_m / L \tag{5.7}$$

is known as the Knudsen number. Let's consider gas confined within an impermeable container, such as the previously introduced box. We let L correspond to the container size. If the distance between collisions is much smaller than the size of the box, then $l_m \ll L$ and $K_n \ll 1$. This flow is collision dominated. Wall interactions play a negligible role since the velocity with which the molecule bounces off after a wall collision is quickly altered by interactions with other molecules. The velocity distribution function can be expected to rapidly thermalize. This flow is said to be in *continuum*. Gases in a continuum are referred to as fluids, since due to frequent collisions, they act as a single entity. On the other hand, if $K_n \gg 1$, the likelihood for two molecules colliding as they cross the enclosure is negligible. In this case, the dynamics is dominated by wall impacts and intermolecular collisions can be ignored. Each atom acts as an individual entity that is not affected by other particles. Instead, it is the interaction with walls that alters the velocity distribution since real-world wall collisions are neither specular nor elastic. This state is known as the *free molecular flow*. This is the state that is encountered in much of the space environment or during component testing in vacuum chambers, assuming sufficient pumping capacity. Finally, we can have a state for which $K_n \approx 1$. In this case, a molecule undergoes only a few collisions between wall impacts. Molecular collisions are not frequent enough to ensure continuum, but are also not so rare that they can be disregarded. This state is known as *rarefied flow* and is encountered in plumes of spacecraft plasma thrusters or during the initial entry into planetary atmospheres.

5.1.4 KINETIC METHODS

Unlike for continuum flows with $K_n \ll 1$, rarefied gases and gases in the free molecular flow state are not collisional enough to assume thermalization. Methods based on the solution of the Navier-Stokes equations (i.e. CFD) are not applicable. Instead, we need to self-consistently resolve the evolution of the VDF. The VDF is subsequently used to extract macroscopic flow properties of interest (perhaps with an assumption about the meaning of temperature). As we have already seen, it is not possible to directly model every single atom in a real-world system. Turns out that is not necessary. Equations of motions tell us that the displacement of a particle is due to its velocity, and that the velocity, in the absence of collisions, is changed only due to the action of forces, i.e.

$$d\vec{x} = \vec{v}dt \tag{5.8}$$

$$d\vec{v} = \left(\vec{F}/m \right) dt \tag{5.9}$$

where $\vec{F} = \vec{F}(\vec{x}, \vec{x}, t)$. All particles sharing the same velocity phase-space coordinate $(\vec{x}, \vec{v})$ encounter the same velocity change $d\vec{v}$, and the same spatial displacement $d\vec{x}$. The only exception are collisions which randomly scatter velocities of a subset of this population. We can thus let our code simulate some selected VDF "parcels" (referred to as simulation particles). As particles move about the domain, they advect their contribution to the velocity distribution function to new locations. The resulting local VDF can then be utilized to compute the local macroscopic flow properties. Methods that resolve the VDF self-consistently are referred to as *kinetic*.

5.2 EULERIAN AND LAGRANGIAN FORMULATION

In general, fluid dynamics are solved using the *Eulerian* approach, in which governing equations that control the evolution of properties of interest are solved on a stationary grid. Gas flows that require the kinetic treatment (i.e. rarefied and free molecular flows) are modeled using the *Lagrangian* approach, in which some parcels of information (such as particles) migrate through the domain. We can visualize the difference between Eulerian and Lagrangian methods with an example. Let's say that you wanted to model gas expansion into an initially empty chamber. The chamber is connected to a large pressurized tank but is also continuously pumped with two pumps. With the Eulerian approach, we utilize one or more governing equations that control the evolution of properties of interest. Here for simplicity, let's assume that molecular motion is controlled by diffusion alone, so that the unsteady diffusion equation (discussed in more detail in the following chapter),

$$\frac{dn}{dt} = D\nabla^2 n \tag{5.10}$$

provides a valid description for the evolution of number density. The parameter D is the diffusion coefficient.

To actually solve the system, we first discretize the computational volume into a mesh as shown. We utilize methods introduced in Chapter 2 and discussed further in Chapter 6 to numerically integrate the governing equation. As an example, we may implement the Forward Time Centered Space (FTCS) method to rewrite the governing equation as

$$n_{i,j}^{k+1} = n_{i,j}^k + \Delta t D \left(\frac{n_{i-1,j}^k - 2n_{i,j}^k + n_{i+1,j}^k}{\Delta^2 x} + \frac{n_{i1,j-1}^k - 2n_{i,j}^k + n_{i,j+1}^k}{\Delta^2 y} \right) \tag{5.11}$$

The subscript k corresponds to the time step. We begin the simulation with n initialized to some starting value, such as $n = 0$ (complete vacuum) everywhere. Along the boundaries, we specify other relationships to capture the physical behavior at that location. We can let $n = n_{in}$ be fixed along the nodes corresponding to the tank inlet. This boundary assumes that the tank is sufficiently large so that gas concentration variation is insignificant during the time considered by the simulation. Similarly, we can assume perfect pumps such that $n = 0$ along the pump interface. These are examples of *Dirichlet* boundary conditions. We also need to make the walls impermeable. This condition implies that there should be no flow across the wall, and hence gas

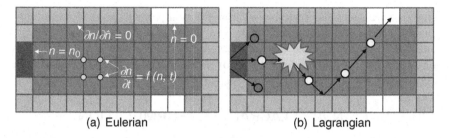

(a) Eulerian   (b) Lagrangian

Figure 5.1   Comparison of gas modeling approaches.

concentration should not be varying in the direction normal to the wall. In other words $\partial n/\partial \hat{n} = 0$. Here $\hat{n}$ is the normal vector. This is an example of a *Neumann* boundary condition. Equation 5.11 is then repeatedly evaluated subject to the specified boundary conditions to march the solution forward in time. At each time step k, we have a value for gas number density at each grid node, $n_{i,j}^k$.

With the particle approach, we similarly start with an empty discretized domain. However, instead of specifying boundary conditions for density, we begin by injecting particles into the domain from the region corresponding to the tank inlet as shown in Figure 5.1(b). The number of particles to inject at each time step is derived from the desired mass flow rate. The initial velocity of each particle is sampled (usually using a stochastic, random-number based approach) from a prescribed velocity distribution function. The velocity and position of each particle are then advanced by integrating the equations of motion through a time step Δt. After each move, we check the particle for collisions with solid objects. This specifically involves testing if the line segment $\vec{x}^k \rightarrow \vec{x}^{k+1}$ had crossed an object interface. This may involve performing line-triangle intersection checks against a surface mesh. If a surface impact is detected, we subsequently apply some wall model algorithms. Possible surface interaction algorithms include reflecting the molecule back to the domain, having the molecule "stick" to the surface (and thus be removed from the simulation), or in the case of energetic impacts, emitting new species from the surface due to sputtering or secondary electron emissions. In this example, molecules impacting the wall are reflected, however those impacting the pump are removed from the simulation. Molecules can also collide with each other. It is in fact these collisions that lead to the isotropization of pressure that enables the flow to diffuse into low-pressure regions. At each time step, the simulation consists of a large number of particles with unique positions and velocities. These particle properties can be used to calculate mesh-averaged gas densities, mean velocities, and temperatures.

Assuming correspondence between the collisional algorithm and the diffusion coefficient D as well as other boundary conditions, we can expect an agreement in the mesh-averaged densities computed from these two approaches. The primary difference is that, unlike the fluid method, the kinetic treatment does not make any assumptions about the velocity distribution within the gas region, and allows it to be computed self- consistently. It could conceptually be applied across the entire spectrum of Knudsen numbers. The continuum fluid approach is applicable only to $K_n \ll 1$. However, this is not quite true in practice. Later in this chapter we will discuss methods for simulating collisions. As the Knudsen number decreases, the number of collisions increases rapidly and the kinetic treatment becomes impractical. Kinetic methods are thus realistically suitable only to free molecular flows and rarefied gases. For continuum flows, fluid approaches produce the solution in a fraction of the time of a kinetic code. We now introduce computational approaches for performing particle-based simulations. We start by developing code modeling collisionless free-molecular flow of neutral molecules around a solid object. Here we learn how to compute the mesh-averaged properties from particle data. We then add collisions using the Direct Simulation Monte Carlo (DSMC) approach. We close the chapter by introducing the electrostatic particle in cell (ES-PIC) method for simulating plasma flows. Here we also introduce the Monte Carlo Collisions (MCC) method for modeling collision between populations at different concentrations.

5.3   REDUCED DIMENSIONALITY

Many engineering problems exhibit some spatial symmetry or periodicity. These features allow us to develop codes that operate on a smaller domain, or that consider only one or two spatial dimensions. This subsequently leads to a faster simulation run time, or alternatively, the ability to utilize a finer computational mesh using available resources.

As an example, let's consider the case of uniform flow past a sphere simulated on a Cartesian domain as shown in Figure 5.2(a), with the flow injected in the normal direction from the face indicated by the gridlines. Cutting the domain along the vertical plane leads to two halves that are a mirror image of each other. The same is true for the horizontal plane. We can thus simulate just one-quarter of the domain and subsequently rebuild the entire volume by employing reflection. This symmetry can be introduced by incorporating a zero Neumann boundary condition $\partial()/\partial \hat{n} = 0$ along the plane of symmetry for any scalar property. For vector quantities, we require zero component in the direction normal to the boundary. In the case of velocity, this implies zero net flow as there are as many fluid parcels moving "to the left" as there are "to the right" and hence no net flow across the reflection plane. Particles crossing the symmetry boundary are specularly reflected back to the computational domain to simulate identical particles entering from the virtual mirrored space.

Next, let's say that instead of a sphere we have a long cylinder with the axis in the horizontal direction. This scenario is depicted in Figure 5.2(b). We can slice the domain along an arbitrary vertical plane and obtain the identical

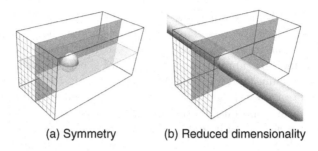

(a) Symmetry   (b) Reduced dimensionality

Figure 5.2 Examples of symmetric and two dimensional simulation domains.

geometrical representation: a rectangular domain enclosing a circular disc. The uniformity of the ambient flow implies that these cutting planes share the same boundary condition. As long as we stay away from the cylinder ends, we can expect the solution to remain invariant with the plane position along the cylinder axis. Assuming the long axis is in the z direction, we have

$$\frac{\partial()}{\partial z} = 0 \tag{5.12}$$

where $()$ is any property of interest. This is an example of a two-dimensional XY planar problem. These codes allow us to describe the computational domain using just a two-dimensional grid. The discretized equations that need to be solved are simplified since the derivative terms in the deleted direction disappear. Let's assume that the 3D domain uses $N_3 = n_i \times n_j \times n_k$ nodes (matrix rows). By eliminating the k component, the matrix reduces to $N_2 = n_i \times n_j \times 1$ rows. Except for perfectly local equations, solver convergence tends to scale as best with $N \log(N)$ (sub-linear scaling). We thus reduce simulation time by approximately $n_k \log(n_k)$ times. For a $41 \times 41 \times 41$ 3D domain, this results in a $61\times$ speed up. It is important to take advantage of reduced dimensionality whenever possible. 2D approximations are often utilized to obtain ballpark estimates of 3D problems before dedicating computational resources to the "for the record" 3D simulation.

Reduced dimensionality is also encountered in cylindrical configurations. Let's consider steady laminar flow[1] through a nozzle. Here we can imagine that any plane aligned with the nozzle axis will contain the same flow field regardless of its azimuthal orientation. In this case

$$\frac{\partial()}{\partial \theta} = 0 \tag{5.13}$$

This is an example of an axisymmetric (RZ) flow. Flow in cylindrical geometries however sometimes develop vortices or flow features in the azimuthal

---

[1]Note that unsteady flows such as simple vortex shedding rapidly break geometric symmetries. Special care must be taken to ensure reduced dimensional solutions do not artificially inhibit natural symmetry breaking in such cases. The behavior of 3D turbulent flows are also fundamentally distinct from 2D turbulence.

direction. Such features cannot be captured by the RZ code. In that case, we may need to consider azimuthal $R\theta$ domain. It may even be possible to string together several $R\theta$ planes connected via a simplified one-dimensional (1D) model for the z direction. One-dimensional codes are also commonly encountered when modeling an environment bounded by two parallel plates. Finally, other configurations exhibit spatial periodicity with the domain wrapping around on itself. An example of a periodic system is the tokamak, which is a toroidal vacuum chamber that uses magnetic fields to confine high density plasma for fusion research. Periodic boundaries are modeled by treating the two opposite faces as the same surface.

5.4   FREE MOLECULAR FLOW

Let's now return to the example of uniform flow past an infinitely long cylinder from Figure 5.2(b). Here we assume that the gas is composed of atomic oxygen of sufficiently low density such that collisions can be ignored ($K_n \gg 1$). The overall modeling approach is relatively simple. We begin with an empty domain. We subsequently iterate through many time steps. At each step, we first introduce new simulation particles from the $x = 0$ edge. Particle velocities and positions are integrated through the Δt time step. Particles impacting the cylinder are diffusely reflected back. Periodically we also compute mesh-averaged macroscopic gas properties and save them for visualization. Once the simulation reaches a steady state, here implying that particles have had a chance to fully traverse the domain such that $\dot{m}_{in} = \dot{m}_{out}$, we begin averaging the results to reduce statistical noise arising from the use of random numbers in the particle injection algorithm. Finally, at the conclusion of the main loop, we perform close-out activities, such as saving the final set of results. This algorithm is summarized below:

```
initialize
for ts in num_ts do
    sample particles
    integrate velocities
    integrate positions
    check for boundaries
    periodically save results
    if steady state then
        sample data for averaging
    end if
end for
save final results
```

The following sections discuss these steps in detail. Due to space constraints, it is not feasible to include the entire source code here. Instead, only the relevant sections are included in print. It is suggested that the reader follows along using the full source code available in the chapter's FMF code subdirectory. We begin by describing the storage containers. We then discuss the algorithm for moving particles. The section is closed with a review of averaging strategies for computing macroscopic gas properties.

5.4.1 PARTICLE STORAGE

Particle codes require two types of storage containers: one for storing the particles, and one for storing the volumetric mesh data. The volume mesh is not strictly needed for integrating particles in free molecular flows; however, it is needed for computing macroscopic flow fields for visualization.

Each simulation particle needs to capture discrete position and velocity components, (x, y, z) and (u, v, w). While it is possible to store the values using arrays of real values, our recommendation is to group information about a single particle into a custom data structure such as

```
1  struct Particle {
2     double3 pos;
3     double3 vel;
4  };
```

where double3 is an object storing a double[3] array while also implementing math operations via C++ overloaded operators. This example shows just the minimum required data. Often we will need additional fields such as a tracking id, or rotational and vibrational energies.

Particles of a single molecular type are stored within a *Species* class. This object encapsulates all common properties such as molecular mass, charge, and macroparticle weight, i.e.

```
1  class Species {
2  public:
3     vector<Particle> particles;   // particle array
4     double mpw;          // macroparticle weight
5     double mass;
6     size_t np;           // number of valid particles
7  };
```

As we have already seen, it is not computationally feasible to model all real particles. Instead, we use some relatively small number of simulation particles to statistically represent a system containing many more physical atoms or molecules. The number of real particles represented by each simulation particle is known as *macroparticle weight*,

$$w_{mp} = \frac{N_{real}}{N_{sim}} \tag{5.14}$$

This parameter can be uniform for all particles, or at least for all particles of each individual material species as is the case here. Alternatively, unique value can be defined for each particle, in which case the mpw variable would be found in the Particle data structure. There are benefits and drawbacks to both approaches. The uniform weight reduces memory requirements. It is also required by some collisional schemes, such as the standard Direct Simulation Monte Carlo algorithm, discussed next. The uniform weight approach however makes it difficult to resolve the high velocity "tail" of the distribution function as well as flows with a large dynamic range of densities. This tail contains very few particles compared to the bulk population, however they are of critical importance to processes such as ionization or surface erosion. Using constant weight, it is unlikely that many, if any, particles get sampled here. The use of uniform weighing then leads to an artificial quenching of these processes. The sampling can be improved by utilizing variable weight. Particles are sampled such that a prescribed number is loaded in each velocity bin. The weight is used to correlate the sampled number to the actual count of real particles with those velocities. The challenge with the variable weight approach is that these inelastic processes lead to energy loss. The initially energetic low weight particles become low-energy, low-weight particles that intermingle with the rest of the bulk population represented by particles with a much higher w_{mp}. These low weight particles have a negligible impact on the bulk population, but still require CPU processing and memory storage. These low weight particles thus need to be periodically merged into fewer, higher weight particles. Care needs to be taken in the implementation of the merge scheme to conserve properties of the velocity distribution function, see Gonoskov 2022 for more information.

Before continuing, it is important to make sure that you understand that simulation particles are not physically "larger" or "heavier" than real particles. Instead, we should think of them as a stack of w_{mp} particles sharing the same velocity $\vec{v}$ located at the same physical location $\vec{x}$. Since all these particles are acted on by the same force, they all receive the same acceleration and hence the same velocity change. The particles end up with identical new velocity and hence move in unison. As such there is no need to simulate every single particle. We simulate just a single representative from each "stack". acceleration is calculated using the physical properties (mass, radius, charge, and so on) of a single *physical* particles. However, when computing macroscopic properties, such as gas number density or surface flux, we need to remember that we actually have a stack of particles, and thus multiply the contribution from a single particle by w_{mp} in order ot conserve system mass.

5.4.2 COMPUTATIONAL DOMAIN

The computational domain is usually described by a volume mesh. Both structured and unstructured meshes are commonplace. The benefit of a structured, Cartesian mesh is the ease in converting physical $\vec{x}$ position to the mesh "logical coordinate" $\vec{l}$ encoding the containing cell and the relative position within the cell. For a 3D Cartesian mesh, this is simply the real-valued (i, j, k) index. It is computed per

$$\vec{l} = \frac{\vec{x} - \vec{x}_0}{\Delta h} \tag{5.15}$$

The integer part of each $\vec{l}$ component corresponds to the cell index in the applicable direction. The fractional part indicates the relative placement in the cell. For example, $\vec{l} = (12.3, 1.8, 4.6)$ indicates a point located in the cell with origin at node index $(12, 1, 4)$, and located $(30\%, 80\%, 60\%)$ of the way toward the diagonally opposite node $(13, 2, 5)$. Such a direct mapping does not exist for unstructured meshes. Instead, a computationally intensive search algorithm needs to be utilized to locate the cell containing a particular $\vec{x}$. On the other hand, unstructured meshes simplify the resolution of embedded boundaries. Such boundaries can be resolved on a Cartesian mesh only by flagging nodes (or cells) internal to the geometry. This results in smooth objects degenerating into a sugarcubed (or staircased) representation. Possible work around include the use of adaptive mesh refinement or the use of cut cells. This limitations is applicable only to mesh-based data, such as the plasma potential introduced in Section 5.6. Particle-surface impacts are resolved independently using a surface mesh or an analytical description of geometries.

The examples in this chapter utilize the Cartesian approach for simplicity. We use a `World` container to store information about the mesh. An incomplete version of this data structure is listed below. The World class stores the mesh origin $\vec{x}_0$, the cell spacing $\Delta \vec{h}$, as well as the number of nodes for each dimension. World also stores objects of type `Field`. This is another custom container wrapping a 2D array, and also implementing helper functions for setting values or for performing arithmetic operations with the help of overloaded operators. For now, we use just a single field to store node volumes. In the latter sections, the storage is expanded to include quantities such as charge density and plasma potential. The `ni` and `nj` properties are exposed in the public section of the class as various parts of the code need ready access to mesh dimensions. The fields are marked constant (and thus set in the constructor) since mesh dimensions are tied to the allocated field memory buffers and thus need to remain invariant.

```
1   class World {
2   public:
3       World(int ni, int nj);   // constructor
4       const int ni, nj;        //number of nodes
5       const int2 nn;           //another way to access node counts
6       Field node_vol;          //node volumes
7
8   protected:
9       double2 x0;              //mesh origin
10      double2 dh;              //cell spacing
11  };
```

5.4.3 INITIALIZATION

The particle and mesh objects are set up during code initialization. This code is shown below.

```
1   int main(int argc, char *args[]) {
2       // initialize domain
```

```
3     World  world(161,121);
4     world.setExtents({0,-0.15},{0.4,0.15});
5     world.setTime(5e-8,4000);
6
7     // set  objects
8     double phi_circle = -10;   //set  default
9     world.addCircle({0.15,0},0.03,phi_circle);
10
11    // set  up  particle  species
12    vector<Species> species;
13    species.push_back(Species("O", 16*AMU, 0*QE, 1e15, world));
```

The **setExtents** function is used to store positions of the two diagonal corners of the computational mesh, and to also save the length of each edge. This function also calls a function responsible for computing node volumes,

```
1     void  World::computeNodeVolumes()  {
2       for  (int  i=0;i<ni;i++)
3         for  (int  j=0;j<nj;j++)  {
4           double V = dh[0]*dh[1];        //default  volume
5           if  (i==0 || i==ni-1) V*=0.5;  //half per  boundary
6           if  (j==0 || j==nj-1) V*=0.5;
7           node_vol[i][j] = V;
8         }
9     }
```

Node volumes are utilized in the computation of gas number densities. A "node volume" in this context refers to the region from which particle data is being accumulated for this subsequent computation. In the case of non-boundary nodes, the node volume is identical to the cell volume. However, along domain boundaries, only the fraction of the cell located inside the domain can contribute particles. This effectively means that the volume is halved for each boundary Cartesian dimension. Nodes along a boundary face, but away from the edges and corners, have their volume halved. Nodes along edges (and thus two boundary faces) have the volume reduced by a factor of four. Finally, corner nodes have the volume reduced eight-fold.

The **addCircle** function merely records the provided centroid and radius. These are then used by an **inCircle** function to classify a location as internal or external to the geometry,

```
1     bool  World::inCircle(const  double3  &x)  {
2       double2  x2 {x[0],x[1]};
3       double2  r = x2-circle_x0;   //ray  to  x
4       double r_mag2 = (r[0]*r[0] + r[1]*r[1]);
5       if  (r_mag2<=circle_rad2) return  true;
6       return  false;
7     }
```

In a later version of the code, this function additionally flags mesh nodes internal to the geometry. We also record information about the desired time step size Δt and the desired number of time steps. Finally, we add our material species. We are simulating oxygen atoms with a mass of 16 atomic mass units, zero charge, and $w_{mp} = 10^{15}$. The species are stored within a vector. The rest of the code is written to operate on a species vector. As such, the code can be oblivious to the actual number of material types. Expanding the simulation to include additional materials then simply involves adding another entry into the vector during initialization. A similar approach is taken for particle sources, discussed next.

5.4.4 SOURCES

Particles are injected into the simulation domain by sampling *sources*. The actual meaning of a "source" depends on the physical problem being simulated. Sources can be volumetric, in which case they load particles in a prescribed volume. They can also be surface-based, as is the case here. We are interested in simulating a flow of uniform density molecules entering the computational domain through the x^- face. In other words, we need a source that injects particles along the $x = 0$ boundary such that the local density matches a prescribed value. Similarly, we require the source the produce some prescribed velocity distribution function. In this case, we desire to resolve a drifting Maxwelian VDF. In the limit of $T \to 0$, such a model becomes a *cold beam source* with all particles injected with the same velocity. For a flow normal to the x^- face, we then have $u = u_\perp$ with $v = 0$ and $w = 0$. Here $u_\perp$ is the drift velocity. Mass conservation tells us that

$$\dot{N} = n u_\perp A \tag{5.16}$$

Here $\dot{N}$ is the particle injection rate and A is the source area. For two-dimensional problem we assume unit depth. $\dot{N}$ is related to mass flow rate through $\dot{m} = m\dot{N}$. The number of real particles entering the domain in a time interval Δt is $N = \dot{N}\Delta t$. The number of simulation particles to inject can then be calculated using

$$N_{sim,f} = N/w_{mp} \tag{5.17}$$

This relationship yields a fractional value. The loading algorithm works with whole particles and hence $N_{sim,f}$ needs to be converted to an integer. Rounding may seem like a viable option, but it leads to an under- or over-prediction of the desired flow rate since the value is adjusted in the same direction at each time step. Instead, we can utilize a random number to floor some value while ceiling others,

$$N_{sim} = \lfloor N_{sim,f} + \mathcal{R} \rfloor \tag{5.18}$$

where $\lfloor \rfloor$ is the truncation operation obtained by performing an integer cast in C++. $\mathcal{R}$ is a random number sampled from the uniform distribution. Consider a case with $N/w_{wp} = 12.8$. We can expect that, on average, out of 10 random numbers, 8 will be larger than 0.2. We have 8 instance where $12 + \mathcal{R} \geq 13$, and 2 instances for which $12 + \mathcal{R} < 13$. The first 8 values get truncated to 13, while the other two become 12. The average is the desired 12.8. Alternatively, we can keep track of the fractional reminder using a static (or a class member) variable f_{rem}, which is initialized to 0. We calculate the integer number of particles from

$$N_{sim} = \lfloor N_{sim,f} + f_{rem} \rfloor \qquad (5.19)$$

The f_{rem} variable is then updated per

$$f_{rem} = f_{rem} + N_{sim,f} - N_{sim} \qquad (5.20)$$

Note that while these methods approximate correct average fluxes, they do have potential to introduce artificial correlations in time and artificially low variance for low flux surfaces. A better option to mitigate these numerical artifacts is injection of the number of particles specified by the inverse Poisson cumulative distribution function (CDF) at the average $\lambda = N/w_{mp}$ rate, $N_{sim,f} = \texttt{PoisINV}(u, \lambda)$, where u is a uniform random sample from the CDF using an efficient implementation such as Giles 2016.

Our example code defines sources as objects derived from a base `Source` class. This class specifies the interface that the derived classes need to implement. Specifically, the derived classes are expected to provide a `sample()` function that performs the appropriate source injection:

```
1  class Source {
2  public:
3      virtual void sample() {};
4      virtual ~Source() {};   // for destruction through base class
5  };
```

An example of a derived class is `ColdBeamSource`, given below

```
1  class ColdBeamSource: public Source {
2  public:
3      ColdBeamSource(Species &species, World &world, double v_drift,
4                     double den) : sp{species}, world{world},
5                     v_drift{v_drift}, den{den} {}
6
7      void sample();   // generate particles
8
9  protected:
10     Species &sp;      //reference to the injected species
11     World &world;     //reference to world
12     double v_drift;   //mean drift velocity
13     double den;       //injection density
14 };
```

This interface-based approach makes it possible to use a vector to store all sources, as shown below:

```
1  const double nda = 1e22;                    //mean atom density
2  vector<unique_ptr<Source>> sources;
3  sources.emplace_back(new ColdBeamSource(atoms, world, 5000, nda));
```

The rest of the code can then remain agnostic of the injection sources and sample particles by iterating over this vector as can be seen in the listing for the code main loop in the following section.

The particle loading algorithm is listed below. This algorithm uses the random-number based approach from Equation 5.18.

```
1   void ColdBeamSource :: sample () {
2       double3 dh = world . getDh ();
3       double3 x0 = world . getX0 ();
4
5       //area of the XY plane , A=Lx*Ly
6       double Ly = dh [1] * ( world . nj −1);
7       double A = Ly * 1;
8
9       //compute number of real particles to generate
10      double num_real = den * v_drift * A * world . getDt ();
11
12      //number of simulation particles
13      int num_sim = ( int ) ( num_real / sp . mpw0+rnd ());
14
15      //inject particles
16      for ( int i =0; i <num_sim ; i ++) {
17          double3 pos {x0 [0] , x0 [1]+rnd () * Ly , 0};
18          double3 vel {v_drift ,0 ,0};
19          sp . addParticle ( pos , vel );
20      }
21  }
```

As can be seen on line 18, all particles are loaded with the same velocity. This cold-beam approach is used here to illustrate the impact of collisions and electrostatic attraction. Simulating a real gas flow with a finite temperature requires implementing a "warm beam" source in which the velocity is sampled, for example, as

```
1   double3 vel ;
2   do {
3       vel = sp . sampleIsotropicVel (T);
4       vel += vel0 ;               // add drift component
5   } while ( dot ( vel , vel0 ) <0);  // pick again if pointing backwards
```

Here we use a helper function to sample thermal velocity component. This function first selects a random point on a unit sphere to obtain the direction,

$$a \equiv \cos(\phi) = \mathcal{R}_1 \tag{5.21}$$

$$\theta = 2\pi \mathcal{R}_2 \tag{5.22}$$

$$\hat{n}_i = \sqrt{1 - a^2} \cos(\theta) \tag{5.23}$$

$$\hat{n}_j = \sqrt{1 - a^2} \sin(\theta) \tag{5.24}$$

$$\hat{n}_k = a \tag{5.25}$$

This direction is next scaled by random speed sampled from the Maxwellian distribution function. While several approaches exist, one popular method is based on first sampling one-dimensional velocity using an algorithm of Birdsall and Langdon 1991 as

$$v_M = v_{th} \left(\sum_{i=1}^{M} R_i - \frac{M}{2} \right) \left(\frac{M}{12} \right)^{-1/2} \tag{5.26}$$

where v_{th} is the thermal velocity. For $M = 3$, the above equation reduces to

$$v_M = v_{th} \left(R_1 + R_2 + R_3 - 1.5\right) \tag{5.27}$$

While equation 5.27 can adequately resolve the bulk population, the reliance on only three random numbers leads to an artificial cut off of the high-velocity tail, since $v_M \in \pm 1.5 v_{th}$.

5.4.5 MAIN LOOP

The simulation main loop is shown below:

```
1   while(world.advanceTime()) {
2       //inject particles
3       for (auto &source:sources)
4           source->sample();
5
6       //move particles
7       for (Species &sp:species) {
8           sp.advance();
9           sp.computeNumberDensity();
10          sp.sampleMoments();
11      }
12
13      // check for steady state
14      world.checkSteadyState(species);
15
16      //screen and file output
17      Output::screenOutput(world,species);
18      Output::diagOutput(world,species);
19
20      //periodically write out results
21      if (world.getTs()%100==0 || world.isLastTimeStep()) {
22          Output::fields(world, species);
23          Output::particles(world, species,10000);
24      }
25  }
```

Each time step begins with new macroparticles injected into the computational domain by sampling "sources". This step is implemented by using the C++ range-style for loop to iterate over all items stored in the **sources** vector and calling the object's **sample** function. Since this function is marked as **virtual** in the base class, the appropriate code is executed at run time depending on the type of the derived object. We next loop over all particles species. A similar polymorphic approach could be taken here if we were interested in simulating material species modeled via different numerical approach. In that case, the **advance** function may lead to particles being moved for kinetic species, and fluid equation integrated for continuum materials. We don't do that here, and all materials are handled via the particle approach. After the particle positions are advanced, we call another function to compute mesh-averaged number density. We also sample velocity moments which will be used to generate macroscopic flow results.

Next, a call is made to a function that checks whether the simulation has reached the steady state. Steady state is used to mark the start of data averaging in order to reduce numerical noise. We then include several calls to functions responsible for generating output. There are different types of output that particle codes generally produce. First, there is the output to the console (or some diagnostics window in the case of a GUI application) that provides the immediate visual feedback to the end user. In our example, this output includes the current time step and the number of particles of each material species. We can also include additional data, relevant to codes discussed in the following sections, such as the number of particle collisions or the range of plasma potential. However, since the amount of screen real-estate is limited, we typically won't be able to include all important diagnostics without generating messy output spanning multiple lines. Therefore, we also output to a file. This data may be formatted as a comma-separated table, with a new row added for each new time step. Alternatively, it can be in the form of a textual log. The first variant is generally more useful as it allows us to graph in real-time, while the simulation is running, the variation in quantities such as particle count or the total kinetic energy to gauge how close the simulation is to reaching the steady state. We also periodically save simulation results. These results come in several types. First, we have data tied to the volume mesh, such as the spatially varying macroscopic flow density, mean velocity, or temperature. The actual format of the output file will depend on the utilized visualization tool. Our example stores this information as a Paraview-compatible VTK image data (.vti) file as defined in the VTK User's Guide, Kitware Inc. 2021.

We may also have results that map to the surface geometry. In the case of a tessellated model, this output will involve exporting a surface mesh along with any associated cell or node-centered quantities using a polygonal data format, such as VTK's .vtp. This is not done here as the code uses an analytical representation of the geometry. Finally, we may also want to export data associated with the simulation particles themselves. This generally involves exporting particle positions and velocities, along with other relevant data such as mass, charge, radius, or a tracking id. Due to the usually large number of particles, only a small subset is exported. Exporting the same particles each time (by the code keeping track of ids to export and adding new entries as tracked particles get removed from the simulation) allows us to create particle traces illustrating the paths particles take as they traverses the computational domain. Such traces are particularly informative when analyzing mass transfer involving gray-body reflections. Particles are also saved in the .vtp format. These output files are typically quite large, and we output them only at some prescribed time step interval. Saving the results in files with a common prefix, such as `fields_0000.vti`, `fields_0100.vti`, `fields_02000.vti` instructs Paraview to group the files into an animation sequence. The danger with this approach is that if the prescribed time step "skip" is not large enough, you may find – and speaking form experience here – that a long simulation failed to produce the needed results due to the hard drive running out of space. It may be thus be preferred to keep overwriting the same output file with the latest data.

5.4.6 PARTICLE PUSH

The integration of particle velocities and positions (commonly referred to as a particle push) is handled by the **advance** function. Fundamentally, this function implements the following algorithm

```
1  for part in particles:
2      part.vel += (F/m)*dt
3      part.pos += part.vel*dt
```

where F/m is the acceleration acting on the particle. In general, this algorithm needs to be supplemented with a boundary check in which we test the particle for surface impacts and for leaving the computational domain. Particles bouncing off a surface, or being reflected by a symmetric domain boundary need to be advanced through the remainder of the time step. This remaining push may involve additional surface impacts. These multiple bounces can be handled by wrapping the position advance in a **while** loop that continues until the particle position is advanced through the entire time step. The code is given below.

```
1  void Species::advance() {
2      // loop over all particles
3      for (Particle &part: particles) {
4          // increment particle's dt by world dt
5          part.dt += world.getDt();
6
7          double3 F{0,0,0}; // no force
8
9          // update velocity from dv/dt=F/m
10         part.vel += (F/mass)*part.dt;
11
12         // keep iterating while time remains and particle is alive
13         while (part.dt>0 && part.mpw>0) {
14             double3 pos_old = part.pos;
15             part.pos += part.vel*part.dt;
16
17             // did this particle leave the domain?
18             if (!world.inBounds(part.pos)) {
19                 part.mpw = 0; //kill the particle
20             }
21             //check for particle hitting the object
22             else if (world.inCircle(part.pos) && !world.inCircle(
                   pos_old)) {
23                 double tp = world.lineCircleIntersect(pos_old, part.
                       pos);
24                 double dt_rem = (1-tp)*part.dt;
25                 part.dt -= dt_rem;
26
27                 //move particle *almost* to the surface
28                 part.pos = pos_old + 0.999*tp*(part.pos-pos_old);
29                 double v_mag1 = mag(part.vel); //pre-impact speed
```

```
30              if (charge==0)  // neutrals
31              part.vel = sampleReflectedVelocity(part.pos,v_mag1);
32              else { /*ions*/
33                  part.mpw = 0;  // kill the particles
34              }
35              continue;
36          }
37          part.dt = 0;  // this particle completed the whole step
38      } // time step loop
39  } // loop over particles
```

We begin by incrementing the time step through which the particle is to be integrated by the system Δt. In this example, each particle stores its own Δt field. In memory-sensitive codes, it may be preferred to just utilize a local variable here. Line 7 includes a placeholder code for setting the force acting on the particle. As there is no force in this initial simulation, this vector variable is set to zero. Particle velocity is then updated on line 10. We then enter the `while` loop on line 13. Line 15 involves advancing the particle position through the entire remaining Δt but not before the old position is saved. We next need to perform the boundary check. Since this code includes only open boundaries, any particle that is no longer in the computational domain is removed from the simulation. This is accomplished by marking the particle for removal by setting its weight to zero on line 19.

The conditional statement on line 22 checks for particles impacting the solid surface. In general, this check should happen before the test for volume bounds to catch particles impacting surfaces near domain boundaries. But since the sphere is located far from boundaries in this simulation, it is more efficient to first check for particles leaving the domain before jumping into the more computationally expensive calculation of geometry impact testing. In a general code operated on a tessellated surface geometry, this test would involve line-triangle intersection checks. Here we use the `inCircle` function that analytically determines if a point is inside the object. We next compute the intersection point by calling `lineCircleIntersect` function. This function determines the parametric position t_p at which

$$| [\vec{x}_1 + t_p(\vec{x}_2 - \vec{x}_1)] - \vec{x}_c| = r \tag{5.28}$$

where $\vec{x}_1$, $\vec{x}_2$, and $\vec{x}_c$ are the old and the new position of the particle, and the circle centroid, respectively. This value gives us the fraction of the time step the particle used up to reach the surface. The time step remaining though which the particle still needs to move through post impact is given by

$$\Delta t_{rem} = (1 - t_p)\Delta t \tag{5.29}$$

We use the t_p parameter to update the particle position. This is done on line 28. However, instead of placing the particle at the surface, we place it just short of the impact location. This is done for numerical reasons. Computer floating point math may lead to difficulties in classifying particle position as internal or external to the surface, especially when dealing with tessellated geometries.

The utilized approach forces the particle to remain in the gas region, thus helping to reduce non-physical "leaks". Here we also need to implement the surface interaction physics. Two processes are illustrated here. Neutrals (atoms or molecules) bounce off the surface by having their velocity set to a new velocity with direction following the cosine law of the surface normal. This logic is implemented by the `sampleReflectedVelocity` function. Ions, on the other hand (although this simulation does not actually include any ions), are "killed off". This behavior approximates surface neutralization in simulations in which the concentration of the resulting neutral species is not of importance.

5.4.7 PARTICLE REMOVAL

The particle push loop leaves some particles marked for removal. The actual removal algorithm depends on the storage scheme used for the particle list. As can be seen from the `Species` class, particles are stored as `vector`, which is a wrapper on a contiguous array. Arrays are preferred over linked-lists due to their faster memory throughput due to increased data locality. On the other hand, a naive implementation of particle removal would call for copying down all particles past the one being removed to fill the "hole",

```
1    for ( size_t i=i_remove; i<N-1; i++)
2        particles [i] = particles [i+1];
```

The overhead of such a copy quickly eliminates any benefit offered by the contiguous data storage even if `memcpy` were to be used instead of the `for` loop. Since the actual ordering of particles in the array is not important, we eliminate dead particles with a backfill. The removed particle is replaced by the last item in the array. Consider the case illustrated in Figure 5.3. Here a particle at index [1] is replaced by the particle previously at index [6]. The particle array size is now decremented from 7 to 6. Next, let's say that the particle at index [3] also needs to be removed. This particle is replaced with the content of the item at the end of the list given by index [5]. After this copy, the particle array size is decremented again such that the size of the particle array is 5. It needs to be noted that we don't typically resize the allocated memory space, since the available free space at the end of the buffer will likely be taken up by new particles injected at the next time step. Instead, we just keep track of the active number of particles along with the buffer capacity. The `vector::resize()` function performs this desired array size reduction without deallocating the memory space. This algorithm is implemented as shown below:

```
1    // perform a particle removal step
2    size_t np = particles.size();
3    for ( size_t p=0;p<np;p++) {
4        if (particles [p].mpw>0) continue;   // ignore live particles
5        particles [p] = particles [np-1];    // fill the hole
6        np--;    // reduce count of valid elements
7        p--;     // decrement p so this position gets checked again
8    }
```

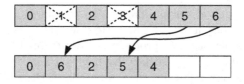

Figure 5.3   Removed particles are replaced with particles from the end of the array.

```
9   //now  delete  particles[np:end]
10  particles.erase(particles.begin()+np,particles.end());
```

5.4.8 STEADY STATE

While this is not always the case, many gas kinetic studies start with a transient phase during which the computational domain is filled with particles and the flow field is established. This transient phase is then followed by the steady state, during which the flow field remains time invariant. The stochastic nature of particle simulations however implies that there will be some small variation in flow properties between time steps due to the numerical noise. This noise can be removed by performing data averaging once the steady state is achieved. The actual process for computing the flow data is discussed in the following section. Before we start the data averaging, we need to know whether the simulation has reached the steady state. This can be done by implementing a check that compares a time step to time step variation in global properties such as the total mass, momentum, and energy to a threshold value. This check is performed by a **steadyState** function found in **World** that is tied to the **advance** function. The function iterates over the particle species and uses species member functions to accumulate total particle mass, momentum, and energy,

$$m = \sum_s \sum_p m_{s,p} \tag{5.30}$$

$$p = \sum_s \sum_p m_{s,p} v_{s,p} \tag{5.31}$$

$$e = \sum_s \sum_p m_{s,p} v_{s,p}^2 \tag{5.32}$$

where $m_{s,p}$ is the mass of the p-th particle of species s. Steady state is set once

$$\left| \frac{\psi^k - \psi^{k-1}}{\psi^k} \right| \leq \epsilon_{tol} \tag{5.33}$$

is true for $\psi \in [m, p, e]$.

5.4.9 MESH-AVERAGED PROPERTIES

Up to this point, we have discussed algorithms for working directly with the particles. We are generally interested in the macroscopic flow properties such as density, stream velocity, or temperature. These properties are computed by averaging particle positions and velocities in a cell. Number density is simply the number of molecules per unit volume. We could begin by overlying a mesh over the computational domain and in each cell, and computing density from

$$n = \frac{N w_{mp}}{\Delta V} \tag{5.34}$$

The formulation in Equation 5.34 is an example of *0th order weighting*. However, this approach leads to a sharp change in properties as particles move from cell to cell. The 1st order (linear weighing) scheme offers a good compromise between speed and the ability to capture smooth variation in data. When coupled with field solvers, additional care must be taken to ensure consistency between how cell properties are computed and applied to particle forcing as described in Birdsall and Langdon 1991. With this 1st order approach, we distribute particle properties to the nodes of the surrounding cell according to the position in the cell. The closer a particle is located to a node, the greater fraction of that particle's mass, momentum, or energy is deposited to that node. The interpolation weights are set from the partial volumes (or areas in the case of a 2D code) of the subcells formed by utilizing the particle's position as one of the nodes. Consider the image in Figure 5.4 that illustrates the area-based weighing for a two-dimensional Cartesian grid. The cell is subdivided into four quadrants according to the particle position. The relative area of a quadrant, A_k/A_{tot} provides the interpolation weight used to deposit data onto the diagonally opposite node. Imagine grabbing the particle and moving it toward node $(i+1,j)$. As you do this, the area shaded in white grows larger and larger, until finally encompassing the entire cell once the particle position coincides with the node. The entirety of particle data will then be deposited onto this single node. The interpolation weights are given by

$$w_{i,j} = (1 - di)(1 - dj) \tag{5.35}$$
$$w_{i+1,j} = (di)(1 - dj) \tag{5.36}$$
$$w_{i+1,j+1} = (di)(dj) \tag{5.37}$$
$$w_{i,j+1} = (1 - di)(dj) \tag{5.38}$$

This interpolation algorithm is implemented in `Field::scatter` function,

```
1   void scatter(const double2 &lc, T value) {
2       int i = (int)lc[0];
3       double di = lc[0] - i;
4       int j = (int)lc[1];
5       double dj = lc[1] - j;
6
7       data[i][j] += (T)value*(1-di)*(1-dj);
```

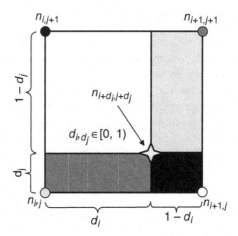

Figure 5.4 Visualization of mesh to particle interpolation.

```
8      data[i+1][j]   += (T)value*(di)*(1-dj);
9      data[i+1][j+1] += (T)value*(di)*(dj);
10     data[i][j+1]   += (T)value*(1-di)*(dj);
11 }
```

This function uses the provided logical coordinate (fractional cell index) to first determine the cell containing the particle by utilizing the whole part of each coordinate. This value is set by casting the real value to an integer, which discards the fractional part. The fractional part is set by subtracting the starting real value from the just computed whole part. The fractional parts for each dimension are used to compute the weights per the equations above.

Stream (or drift) velocity is computed similarly to number density by scattering particle velocities to the grid. We then have

$$\bar{u} = \frac{\sum_p w_p \vec{u}_p}{\sum_p w_p} \tag{5.39}$$

where the denominator is the sum of particle weights on each node. This is the same quantity used in the computation of number density.

We are usually also interested in the gas temperature. For a Maxwellian fluid, $KE = (1/2)kT$ for each degree of freedom. Hence

$$\frac{1}{2}m(u^2 + v^2 + w^2) = \frac{3}{2}kT \tag{5.40}$$

This equation is valid only for gas at rest (no drift) since monoergetic gas with $u = v = 0, w = w_0$ has zero temperature, regardless of the magnitude of w_0. The mean bulk velocity thus first needs to be subtracted

$$\frac{1}{2}m(\Delta^2 u + \Delta^2 v + \Delta^2 w) = \frac{3}{2}kT \qquad ; \Delta u = u - u_{mean} \tag{5.41}$$

Temperature could then be computed per

$$T \sim \frac{\sum_p w_p (\Delta v)_p^2}{\sum_p w_p} \tag{5.42}$$

To avoid having to sweep through all the particles a second time after computing the mean velocity, the linearity of the moment integrals enables calculation of T from first and second moments accumulated simultaneously as described next. Considering the three spatial directions separately, we want to evaluate

$$\overline{vv}_i \equiv \frac{\sum_p w_p \left(v_i^p - \bar{v}_i\right)^2}{\sum_p w_p} \tag{5.43}$$

Expanding the squared term

$$\overline{vv}_i \equiv \frac{\sum_p w_p \left[(v_i^p)^2 - 2\bar{v}_i v_i^p + (\bar{v}_i)^2\right]}{\sum_p w_p} \tag{5.44}$$

Separate into three terms

$$\overline{vv}_i \equiv \frac{\sum_p w_p \left(v_i^p\right)^2}{\sum_p w_p} - 2\bar{v}_i \frac{\sum_p w_p v_i^p}{\sum_p w_p} + (\bar{v}_i)^2 \frac{\sum_p w_p}{\sum_p w_p} \tag{5.45}$$

The middle fraction is simply $\bar{v}_i$ hence

$$\overline{vv}_i \equiv \frac{\sum_p w_p \left(v_i^p\right)^2}{\sum_p w_p} - 2(\bar{v}_i)^2 + (\bar{v}_i)^2 \tag{5.46}$$

or

$$\overline{vv}_i \equiv \frac{\sum_p w_p \left(v_i^p\right)^2}{\sum w_p} - (\bar{v}_i)^2 \tag{5.47}$$

This is *average of squared velocity* minus *average velocity, squared*. Temperature is then given by

$$T = \frac{m}{3k} \left(\overline{uu} + \overline{vv} + \overline{ww}\right) \qquad ; \overline{uu} = \overline{u^2} - (\bar{u})^2 \tag{5.48}$$

The above relationships suggest that macroscopic flow properties can be obtained through a sampling of *velocity moments*, v^0, v, and v^2. This sampling is performed by the `Species::sampleMoments` function,

```
1  void Species::sampleMoments() {
2      for (Particle &part:particles) {
3          double2 lc = world.XtoL(part.pos);
4          n_sum.scatter(lc, part.mpw);
5          nv_sum.scatter(lc, part.mpw*part.vel);
6          nuu_sum.scatter(lc, part.mpw*part.vel[0]*part.vel[0]);
7          nvv_sum.scatter(lc, part.mpw*part.vel[1]*part.vel[1]);
8          nww_sum.scatter(lc, part.mpw*part.vel[2]*part.vel[2]);
9      }
10     num_samples++;
11 }
```

This function loops through all particles of the species, and for each, uses the field scatter function to accumulate velocity moments scaled by the macroparticle weight. We also keep track of the number of samples, as this value is needed to compute the density. We generally sample for many time steps in order to reduce noise. Sampling can be optionally reset after each file output if some transient behavior is expected, or can continue indefinitely from the time when steady state is reached. The reset is accomplished with

```
1  void Species :: clearSamples () {
2      n_sum = 0; nv_sum = 0; nuu_sum = 0;
3      nvv_sum = 0; nww_sum = 0; num_samples = 0;
4  }
```

This function utilizes the overloaded assignment operator to assign an entire field to zero. The accumulated samples are subsequently used by Species::computeGasProperties to set the flow fields,

```
1  void Species :: computeGasProperties () {
2      den_ave = n_sum/(num_samples*world.node_vol);
3      vel = nv_sum/n_sum;   //stream velocity
4
5      for (int i=0;i<world.ni;i++)
6          for (int j=0;j<world.nj;j++) {
7          double count = n_sum(i,j);
8          if (count<=0) {T[i][j] = 0; continue;}
9
10         double u_ave = vel(i,j)[0];
11         double v_ave = vel(i,j)[1];
12         double w_ave = vel(i,j)[2];
13         double u2_ave = nuu_sum(i,j)/count;
14         double v2_ave = nvv_sum(i,j)/count;
15         double w2_ave = nww_sum(i,j)/count;
16
17         double uu = u2_ave − u_ave*u_ave;
18         double vv = v2_ave − v_ave*v_ave;
19         double ww = w2_ave − w_ave*w_ave;
20         T[i][j] = mass/(2*Const::K)*(uu+vv+ww);
21         }
22  }
```

Note that we start with the computation of number density on line 2. The average number of real particles, per node, given by the total count divided by the number of samples, is scaled by the node volume to obtain the density. The node volumes are set during the World initialization. We then compute the average low velocity by dividing the accumulated $\sum_p w_p \vec{v}_p$ sum by the accumulated macroparticle weight count $\sum_p w_p$. Both of these operations utilize the overloaded division operator to perform this operation on the entire field. The calculation of temperature is implemented directly by looping over all data to avoid the need to create additional fields holding the squared quantities. We loop over all nodes, and on each, set the temperature per Equation 5.48.

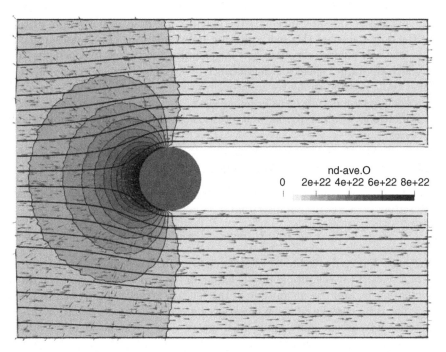

Figure 5.5   Free molecular flow simulation of a 2D uniform flow past a solid cylinder.

5.4.10   RESULTS

Figure 5.5 plots the flow field obtained after 6000 time steps with slightly over 4000 steps used for steady-state sampling. The contour visualizes number density while the darker streamlines indicate flow velocity directions. The reflection of neutrals off the sphere is readily apparent by the increased number density along the upstream facing half. We can also observe that this reflection is limited to the $\pm 90°$ off flow direction, as expected. The number density behind the sphere is seen to be reduced to zero. This reduction arises from the shadowing of the incident beam by the sphere.

5.5   COLLISIONS

The plot in Figure 5.5 contains some unusual features. First, the zero-density wake behind the sphere appears contrary to nature's propensity to relax systems toward smooth equilibrium states. Secondly, while we see an increase in density in front of the cylinder, the reflected molecules do not interact with the incoming flow. Both of these observations arise from the lack of collisions. Two popular algorithms for modeling collisions include Monte Carlo Collisions (MCC), see Birdsall 1991, and Direct Simulation Monte Carlo (DSMC), see Bird 1994. The first algorithm simulates collisions between a source particle and a target gas cloud. Since only the macroscopic flow properties are utilized for the target species, this method is not nominally momentum-conserving. It is on the other hand very fast and is generally acceptable for configurations in which the target species is significantly denser than the source population. The DSMC method simulates collisions between particle pairs. It is applicable to modeling momentum transfer in like gases. Its downside lies in the increased computational complexity, as well as the lack of support for variable-weight particles.

5.5.1 MONTE CARLO COLLISIONS

The MCC algorithm involves looping through the particles, and for each, computing collision probability from

$$P = 1 - \exp(-n_a \sigma |v_r| \Delta t) \tag{5.49}$$

The value P is compared to a random number, with the collision taking place if $P > \mathcal{R}$. Note that the probability is based on the number density of the target population n_a and the magnitude of the relative velocity during the collision, $|v_r|$. The term σ is the collision cross-section, with analytical forms for various material interaction pairs found in literature. As such, MCC approximates collisions of a source particle with a target cloud. There is no specific target particle to participate in the collision and receive the momentum transferred from the source. MCC is applicable to interaction between a trace and dense population where the lack of momentum conservation has a negligible impact on the target. In the plasma propulsion community, MCC is commonly used for charge exchange (CEX) collisions arising between fast plume ions and non-ionized neutral propellant that slowly diffuses out of the thruster. This algorithm can be written as follows:

```
1   for (Particle &part: ions.particles) {
2       // evaluate neutral density at particle position
3       double3 lc = world.XtoL(part.pos);
4       double na_p = neutrals.nd.gather(lc);
5
6       // also gather target drift velocity and temperature
7       double va_p = neutrals.vel.gather(lc);
8       double Ta_p = neutrals.T.gather(lc);
9
10      // generate a virtual target particle
11      double3 vel_a = sampleMaxwellianVelocity(va_p,Ta_p);
12
13      // compute relative velocity
14      double3 v_rel = part.vel - vel_a;
15      double v_rel_mag = mag(v_rel);
16
17      // compute cross section
18      double sigma = sigmaCEX(v_rel_mag);
19
20      // compute collision probability
21      double P = 1 - exp(-na_p*sigma*v_rel_mag*dt);
```

```
22      if (P>rnd()) {
23          // replace ion velocity with the virtual neutral velocity
24          part.vel = vel_a;
25      }
26  }
```

We begin by looping over all particles making up the source material. For each, we evaluate the target density. Here we are assuming that the density of the target neutrals is stored in a `Field` object that allows the value to be interpolated to an arbitrary logical coordinate `lc`. This interpolation is performed using the `gather` function, which collects data from the surrounding grid nodes using the same weighing scheme as `scatter`. Next, on lines 7 and 8 we similarly sample the target gas drift velocity and temperature. These values help us to construct a random virtual target particle. We next compute the magnitude of the relative velocity and use it to evaluate the cross section σ utilizing some, often experimentally derived, fit. Next, collision probability is computed per Equation 5.49. For the charge exchange collision, the colliding particle has its velocity set to that of the virtual particle to simulate the atom converting to the ion. In this classical implementation, we simply discard the fast atom. We assume that the that the target density is sufficiently high, and the collision rate infrequent, that this omission leads to a negligible impact on momentum conservation.

5.5.2 DIRECT SIMULATION MONTE CARLO

The MCC algorithm cannot be used to simulate the momentum exchange (billiard-ball like) collisions between molecules of the same material. But it is precisely this interaction that is needed to address the zero-density wake behind the cylinder. This is where the Direct Simulation Monte Carlo (DSMC) method comes in.

DSMC actually collides two simulation particles and as such, momentum can be conserved. The MCC approach is oblivious to the implementation details of the target population. The neutrals could be simulated as particles or fluid. This is no longer the case with DSMC. As it collides two individual particles, this method is applicable only to simulations in which both the source and the target species are represented using particles. DSMC is also significantly more computationally expensive than MCC. Therefore, it is common to encounter a mix of MCC and DSMC algorithms implemented in the same numerical simulation code. MCC is used for interactions between rarefied and dense species, while DSMC is used for the remaining momentum exchange collisions.

For two particles to collide, they need to be located in close vicinity to each other. To avoid a N^2 search, we start by sorting particles into cells. The cells should ideally be sized according to the collision mean free path so that all particles within a single cell can be treated as possible collision partners. Even utilizing only particles from the same cell, simulating collisions between all possible pairs would be computationally expensive for a large number of particles. Bird's *No Time Counter* (NTC) scheme provides an efficient alternative in which we first calculate the number of collision test groups in each cell,

$$N_g = \frac{1}{2} N_1 N_2 w_{mp} (\sigma v_r)_{max} \frac{\Delta t}{\Delta V} \tag{5.50}$$

Here N_i is the number of particles of species i located in the cell, w_{mp} is the macroparticle weight, $(\sigma v_r)_{max}$ is a parameter, and ΔV is the cell volume. The real-valued N_g is converted to an integer by n = (int)(N_g+rnd()) or by keeping track of the remainder in each cell,

```
1  int n = (int)(N_g+rem[c]);
2  rem[c] = N_g-n;
```

We subsequently select two random particles from the cell particle list for every test group. The relative velocity, $\vec{v}_r = \vec{v}_1 - \vec{v}_2$ is used to evaluate the cross-section, $\sigma = \sigma(v_r)$. Collision probability is given by

$$P = \frac{\sigma v_r}{(\sigma v_r)_{max}} \tag{5.51}$$

The $(\sigma v_r)_{max}$ parameter is updated after each time step according to the values encountered during processing. This parameter scales the number of collision to drive the ratio between the number of test groups and actual collisions towards unity. The initial value should be set by the user such that $N_g \geq 1$ in the cells where collisions are expected to take place. The probability of collision between two particles depends on their relative velocity. This behavior is frequently simulated using a *variable hard sphere* (VHS) or *variable soft sphere* (VSS) models that are tuned to capture effective viscosity and diffusive dependencies on temperature. Both assume that the molecules interact as spheres of radius changing according to the relative velocity. In other words, the collision cross-section is not constant. However, in all the hard sphere models, we assume that the scattering direction is isotropic in the collision plane. The VHS cross-section is given by Bird as

$$d = d_{ref} \sqrt{\frac{[2kT_{ref}/(m_r v_r^2)]^{\omega - 0.5}}{\Gamma(2.5 - \omega)}} \tag{5.52}$$

$$\sigma = \pi d^2 \tag{5.53}$$

where

$$m_r = \frac{m_1 m_2}{m_1 + m_2} \tag{5.54}$$

is the reduced mass. Parameters d_{ref} and ω for many common gases are given in Appendix A of Bird 1994.

Just as with MCC, a collision is assumed to take place if $P > \mathcal{R}$, where $\mathcal{R} \in [0,1)$ is a random number. Consider the collision between two billiard ball-like molecules. From momentum conservation

$$m_1 \vec{v}_1 + m_2 \vec{v}_2 = m_1 \vec{v}_1^* + m_2 \vec{v}_2^* = (m_1 + m_2) \vec{v}_m \tag{5.55}$$

where $\vec{v}_m$ is the *center of mass* velocity and v^* is the post-collision velocity. In order to conserve energy, the relative velocity magnitude needs to stay constant during the collision, $|\vec{v}_r^*| = |\vec{v}_r|$. Then

$$\vec{v}_1^* = \vec{v}_m + \frac{m_2}{m_1 + m_2}\vec{v}_r^* \tag{5.56}$$

$$\vec{v}_2^* = \vec{v}_m - \frac{m_1}{m_1 + m_2}\vec{v}_r^* \tag{5.57}$$

All that remains is picking the new direction for $\vec{c}_r$. In the hard sphere model, we begin by sampling cosine of a random elevation angle,

$$\cos\chi = 2\mathcal{R}_1 - 1 \tag{5.58}$$

When then select random azimuth angle per

$$\epsilon = 2\pi\mathcal{R}_2 \tag{5.59}$$

Velocity components are given by

$$\vec{v}_r^* = <v_r\cos\chi, v_r\sin\chi\cos\epsilon, v_r\sin\chi\sin\epsilon> \tag{5.60}$$

The DSMC algorithm is implemented as a class derived from a base **Interaction** object that enforces the existence of an **apply** function,

```
1  class Interaction {
2  public:
3     virtual void apply(double dt) = 0;
4     virtual ~Interaction() {}
5  };
```

This generalization makes it possible, just as was the case with particle sources, to include collisions, and other material interactions in a way that makes the simulation main loop agnostic of the actual model. We add the following initialization prior to the start of simulation main loop:

```
1  //setup material interactions
2  vector<unique_ptr<Interaction>> interactions;
3  interactions.emplace_back(new DSMC_MEX(atoms, world));
```

We then modify the main loop to contain:

```
1  //perform material interactions
2  for (auto &interaction:interactions)
3     interaction->apply(world.getDt());
```

The DSMC class is defined as follows. The collision cross-section parameters are hard-coded for Nitrogen.

```
1  class DSMC_MEX: public Interaction {
2  public:
3     DSMC_MEX(Species &species, World &world) : species{species},
          world{world} {
4        mr = species.mass*species.mass/(species.mass+species.mass);
5        c[0] = 2*152e-12;      //diameter
6        c[1] = 0.77;           // omega
```

```
7          c[2] = 2*Const::K*273.15/mr;   //parameter at 273.15 K
8          c[3] = std::tgamma(2.5-c[1]);  //Gamma(5/2-omega)
9      }
10     void apply(double dt);
11 protected:
12     void collide(double3 &vel1, double3 &vel2, double mass1,
                double mass2);
13     double evalSigma(double g_rel) {
14         return Const::PI*c[0]*c[0]*pow(c[2]/(g_rel*g_rel),c[1]-0.5)
                /c[3];
15     }
16
17     double sigma_cr_max = 1e-14;  //some initial value
18     double mr;
19     double c[4];
20     Species &species;
21     World &world;
22 };
```

The collision algorithm is implemented in the **apply** function. It begins
by sorting particles into cells. This sort utilizes a dynamically allocated array
of **vector<Particle*>** items. The array size matches the number of cells.
The vector then stores pointers to the particles located within that cell. This
allocation and sort is performed with the following code:

```
1  void DSMC_MEX::apply(double dt) {
2      // first we need to sort particles to cells
3      vector<Particle*> *parts_in_cell;
4      int n_cells = (world.ni-1)*(world.nj-1);
5      parts_in_cell = new vector<Particle*> [n_cells];
6
7      // sort particles to cells
8      for (Particle &part:species.particles) {
9          int c = world.XtoC(part.pos);
10         parts_in_cell[c].push_back(&part);
11     }
```

Next, after setting some common parameters such as the cell volume and
particle weight, the code jumps into the loop over cells. In each cell, the
number of collision test groups is first computed per Equation 5.50. This is
shown below:

```
1      double sigma_cr_max_temp = 0;  // reset for max computation
2      double dV = world.getCellVolume();  // internal cell volume
3      double Fn = species.mpw0;       // mpw, Bird's notation
4      int num_cols=0;  // reset collision counter
5
6      // now perform collisions
7      for (int c=0;c<n_cells;c++) {
8          vector<Particle*> &parts = parts_in_cell[c];
9          int np = parts.size();
10         if (np<2) continue;
11
12         // compute number of groups according to NTC
13         double ng_f = 0.5*np*np*Fn*sigma_cr_max*dt/dV;
14         int ng = (int)(ng_f+0.5);   // number of groups, round
```

The code next loops over the test groups. For each, two unique particles are selected from the list of cell particles. Their relative velocity is used to compute the cross section σ, and to compute the collision probability per Equation 5.51. The probability is compared to a random number. If the collision is found to happen, the code calls the collision handler in `collide` to perform the actual interaction. Velocities of the two colliding particles are passed to this function by reference, which allows the `collide` function to directly modify them. Finally, at the conclusion of loop over the cells, we clean up the dynamically allocated particle sort array, and also update the $(\sigma v_r)_{max}$ value.

```cpp
// assumes at least two particles per cell
for (int g=0;g<ng;g++) {
    int p1, p2;
    p1 = (int)(rnd()*np);   // index of first particle

    do {
        p2 = (int)(rnd()*np);
    } while (p2==p1);

    // compute relative velocity
    double3 cr_vec = parts[p1]->vel - parts[p2]->vel;
    double cr = mag(cr_vec);

    // evaluate cross section
    double sigma = evalSigma(cr);

    // eval sigma_cr
    double sigma_cr=sigma*cr;

    // update sigma_cr_max
    if (sigma_cr>sigma_cr_max_temp)
        sigma_cr_max_temp=sigma_cr;

    // collision probability
    double P=sigma_cr/sigma_cr_max;

    // did the collision occur?
    if (P>rnd()) {
        num_cols++;
        collide(parts[p1]->vel, parts[p2]->vel,
                species.mass, species.mass);
    }
}

delete[] parts_in_cell;

if (num_cols) {
    sigma_cr_max = sigma_cr_max_temp;
}
}
```

The collision logic is implemented as shown below:

```cpp
void DSMC_MEX::collide(double3 &vel1, double3 &vel2,
                       double mass1, double mass2) {
    double3 cm = (mass1*vel1 + mass2*vel2)/(mass1+mass2);
```

```
4
5      // relative velocity, magnitude remains constant through the
          collision
6      double3 cr = vel1 - vel2;
7      double cr_mag = mag(cr);
8
9      // pick two random angles, per Bird's VHS method
10     double cos_chi = 2*rnd()-1;
11     double sin_chi = sqrt(1-cos_chi*cos_chi);
12     double eps = 2*Const::PI*rnd();
13
14     // perform rotation
15     cr[0] = cr_mag*cos_chi;
16     cr[1] = cr_mag*sin_chi*cos(eps);
17     cr[2] = cr_mag*sin_chi*sin(eps);
18
19     // post collision velocities
20     vel1 = cm + mass2/(mass1+mass2)*cr;
21     vel2 = cm - mass1/(mass1+mass2)*cr;
22  }
```

Addition of this relatively simple algorithm modifies the results from Figure 5.5 to the ones seen in Figure 5.6. The molecules impacting and reflecting from the sphere become, due to collisions, entrained in the incoming free stream. This leads to the formation of the characteristic bow shock. Furthermore, downstream of the sphere, collisions introduce diffusion that fills the void in the wake.

5.6 PARTICLE IN CELL

We have now seen how to incorporate collisions in particle methods. Kinetic methods are also attractive for simulations that involve body forces. Let's consider plasmas, which are ionized gases consisting of neutral atoms, ions, and enough electrons to balance the space charge. The dynamics of the charged particles is affected by electric and magnetic fields. These fields can be applied externally or can arise within the plasma due to charge separation and current flows. Dense plasmas are simulated using the magnetohydrodynamics (MHD) method, which is the CFD analogue for non-neutral flows. As we know by now, fluid methods inherently assume that the gas species are thermalized. While it may be acceptable to assume thermalization for electrons even at low number densities, this assumption does not hold for ions. Many plasma processes of interest are strong functions of the velocity distribution function since reactions such as sputtering, ionization, or chemical reactions tend to require some minimum threshold energy to be viable. Even if we start with a thermalized population, these reactions deplete the energetic tail of the distribution function. This depletion is not easily captured using fluid methods; however, it can be resolved self-consistently using particle approaches. The particle-based approach for modeling plasmas is known as Particle in Cell (PIC), see Birdsall and Langdon 1991. The section below introduces the basics of this method. Additional implementation details are available in our

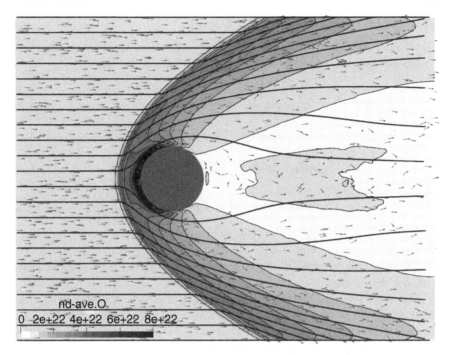

Figure 5.6 DSMC simulation of a 2D uniform flow past a solid cylinder.

recent book, Brieda 2019.

5.6.1 LORENTZ FORCE

The primary difference between PIC and the neutral gas algorithms developed
so far arises from the need to incorporate electromagnetic effects in equations
of motion. The acceleration experienced by a charged particle is given by the
Lorentz force,

$$\frac{d\vec{v}}{dt} = \frac{q}{m}\left(\vec{E} + \vec{v} \times \vec{B}\right) \tag{5.61}$$

The $\vec{E}$ and $\vec{B}$ terms are the electric and the magnetic fields. The presence
of the $\vec{v}$ term on the right-hand side requires careful use of a corrective algorithm (such as the Boris method, see the previously mentioned references
for implementation details) to avoid non-physical energy gain. However, if the
magnetic field is zero, as we assume in our example, acceleration reduces to
$\vec{F} = q\vec{E}$. Particle velocities and positions can then be integrated by utilizing
the Leapfrog method as,

$$\vec{v}^{k+0.5} = \vec{v}^{k-0.5} + q\vec{E}^k \Delta t \tag{5.62}$$

$$\vec{x}^{k+1} = \vec{x}^k + \vec{v}^{k+0.5} \Delta t \tag{5.63}$$

5.6.2 ELECTROSTATICS

The electric field arises from the four governing equations of electromagnetics known as Maxwell's equations. They consist of the following:

$$\text{Gauss' law:} \quad \nabla \cdot \vec{E} = \frac{\rho}{\epsilon_0} \tag{5.64}$$

$$\text{Gauss' law for magnetism:} \quad \nabla \cdot \vec{B} = 0 \tag{5.65}$$

$$\text{Faraday's law:} \quad \nabla \times \vec{E} = -\frac{\partial \vec{B}}{\partial t} \tag{5.66}$$

$$\text{Ampere's law:} \quad \nabla \times \vec{B} = \mu_0 \left(\vec{j} + \epsilon_0 \frac{\partial \vec{E}}{\partial t} \right) \tag{5.67}$$

Here $\rho \equiv \sum_s q_s n_s$ is the *charge density*, and $\vec{j} \equiv \sum_s q_s n_s \vec{v}_s$ is the *current density*. For two species ion-electron plasma containing only singly-charged ions, $\rho = e(n_i - n_e)$.

Faraday's law tells us that if $\partial \vec{B}/\partial t = 0$, $\nabla \times \vec{E} = 0$. A vector field with a zero curl is said to be irrotational. Such a vector field can be defined in terms of a scalar potential ϕ as

$$\vec{E} = -\nabla \phi \tag{5.68}$$

This $\partial \vec{B}/\partial t = 0$ condition is known as the *electrostatic assumption*. Since Ampere's law indicates that magnetic field arises from currents, this condition also implies that the current density is low enough for the self-induced magnetic field to be negligible.

The above expression for $\vec{E}$ can next be substituted into Gauss' Law:

$$\nabla \cdot (-\nabla \phi) = \frac{\rho}{\epsilon_0} \tag{5.69}$$

or

$$\nabla^2 \phi = -\frac{\rho}{\epsilon_0} \tag{5.70}$$

This is the Poisson's equation. It has the same form as the steady-state heat equation covered in Section 2.6.2, and can thus be solved using the solver already developed. An alternate approach, based on time integration of Faraday's and Ampere's equations is needed for electromagnetic problems for which $\partial B/\partial t \neq 0$. This approach is introduced in Section 6.6.4.

5.6.3 INTEGRATION ALGORITHM

The resulting Electrostatic Particle in Cell (ES-PIC) method is given by the following algorithm:

1. Divide computational domain into volume cells.
2. Inject simulation particles as needed.

3. Interpolate particle positions to the volume mesh to compute charge density $\rho = \sum_s q_s n_s$.
4. Use ρ to compute ϕ by solving the Poisson's equation $\nabla^2 \phi = -\rho/\epsilon_0$.
5. Use ϕ to compute $\vec{E} = -\nabla\phi$.
6. Accelerate particles according to $d\vec{v}/dt = q(\vec{E} + \vec{v} \times \vec{B})$, with $\vec{E}$ and $\vec{B}$ interpolated to each particle.
7. Optionally include collisions.
8. Push particles to new positions per $d\vec{x} = \vec{v}\Delta t$.
9. Output results, as needed.
10. Return to Step 2 until exit criterion met.

The name of the method arises from the coupling between particle and volume data. Particles are used to obtain mesh-averaged charge density in step 3. The computation then proceeds on the mesh, resulting in the calculation of electric field components on grid nodes (or cell-centers, according to implementation). Particle velocities are then integrated using field values interpolated from field values in the surrounding cell. Particle positions are advanced in the already familiar way.

5.6.4 HYBRID MODELING

There is one wrinkle. An electron is $\approx 1823\times$ lighter than a proton and about $243,000\times$ lighter than Xenon used in plasma propulsion. Kinetic and potential energy of a charged particle is related through

$$\frac{1}{2}mv^2 = |q\Delta V| \tag{5.71}$$

where ΔV is a potential difference. Particle originally at rest will accelerate to

$$v = \sqrt{\frac{2|q\Delta V|}{m}} \tag{5.72}$$

after falling down a ΔV potential "hill". The final velocity of two particles is related by

$$\frac{v_2}{v_1} = \sqrt{\left|\frac{q_1 m_2}{q_2 m_1}\right|} \tag{5.73}$$

Using this relationship, we see that for the same applied potential voltage, electrons reach velocities nearly $500\times$ faster than Xenon ions.

Integration of Equation 5.61 requires that particles do not move more than one cell at a time, as otherwise the intermediate values of the electric field get skipped over. The integration time step Δt is controlled by the fastest moving material species. Directly simulating electron dynamics requires using time steps several hundreds or thousands of times smaller than what is needed for ions. In engineering applications, it is the dynamics of the heavy particles that is of interest. Running the code at electron time scales leads to excessively long run times. A 300× increase in the number of time steps means that a code that would complete in a day will require nearly a year to do so. Such long run times are incompatible with production timelines.

Considering the frame of reference of ions, electrons can be assumed to respond instantaneously to any disturbance. Therefore, we can assume that electrons act as a background neutralizing fluid that achieves quasi-steady state between each ion integration step. This approximation makes it possible to reduce the electron momentum equation

$$m_e n_e \left[\frac{\partial \vec{v}_e}{\partial t} + (\vec{v}_e \cdot \nabla) \vec{v}_e \right] = -e n_e \vec{E} - \nabla p_e - m_e n_e (\vec{v}_i - \vec{v}_e) \nu_{ei} \qquad (5.74)$$

to a form that balances contribution from the hydrodynamic pressure and finally yielding (the derivation details are not important here)

$$n_e = n_0 \exp \left[\frac{e(\phi - \phi_0)}{k_B T_e} \right] \qquad (5.75)$$

This expression, known as *Boltzmann relationship*, can then be substituted into the Poisson's equation,

$$\nabla^2 \phi = -\frac{e}{\epsilon_0} \left[Z_i n_i - n_0 \exp \left(\frac{e(\phi - \phi_0)}{k T_{e,0}} \right) \right] \qquad (5.76)$$

which has the form

$$\nabla^2 \phi = b(\phi) \qquad (5.77)$$

This non-linear equation needs to be solved using approaches such as the "poor man's Gauss-Seidel" or Newton-Raphson wrapper on top of a linear PCG (or other) solver as was discussed in Section 3.8.2. The use of Equation 5.76 leads to a *hybrid PIC*, in which some species (namely the ions and neutrals) are modeled as particles, while other species (such as electrons) are modeled as fluid. This approach is commonly found in engineering plasma simulations.

5.6.5 SUGARCUBING

An example PIC code can be found in the `PIC` subdirectory. During initialization, we call the already seen `addCircle` and a new `addInlet` function,

```
1  double phi_circle = -10;
2  world.addCircle({0.15,0},0.03,phi_circle);
3  world.addInlet();
```

The specified ϕ_{circle} value is used to fix potential in the region corresponding to the circle. This is done by flagging the nodes internal to the object as Dirichlet. Doing so leads to a degenerate, sugarcubed representation of the circle as visualized in Figure 5.7. This inability to represent smooth conics is a disadvantage of using Cartesian grids for PIC simulation. On the other

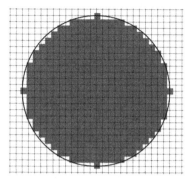

Figure 5.7 Sugarcubed representation of a conical object.

hand, the classification of particles to cells can be performed more rapidly than on unstructured grids. Such grids also require the use of the Finite Element or Finite Volume Methods for the field solver. The fixed, Dirichlet nodes are flagged using the `object_id` field. Instead of just storing the circle centroid and radius, the `addCircle` function now loops through all nodes, and uses the `inCircle` function to classify the node as internal or external. Internal nodes get their `object_id` set to a non-zero value. These nodes are subsequently skipped over by the potential solver.

5.6.6 INITIALIZATION

This example code simulates the neutral species already seen in the previous two sections, as well as a new ion species, added via

```
1  species.push_back(Species("O+", 16*AMU, 1*QE, 1e2, world));
```
Ions are injected into the simulation through a new cold beam source,
```
1  const double ndi = 1e10;    //mean ion density
2  sources.emplace_back(new ColdBeamSource(ions,world,5000,ndi));
```
Additionally, we instantiate a Poisson solver object and use it to solve the initial plasma potential. This solver utilizes the Gauss-Seidel method to solve Equation 5.76. This resulting field is used by particle sources to rewind velocities for the Leapfrog method.

5.6.7 IMPLEMENTATION

The newly added sections of the main loop are listed below:

```
1  while(world.advanceTime()) {
2      /* push particles, compute number density, etc. */
3
4      // compute charge density
5      world.computeChargeDensity(species);
6
7      //update potential
8      solver.solve();
```

```
9
10      //obtain electric field
11      solver.computeEF();
12
13      /* screen and file output */
14    }
```

The `computeChargeDensity` function uses the already computed species-specific number densities to set ρ,

```
1  void World::computeChargeDensity(vector<Species> &species) {
2      rho = 0;
3      for (Species &sp:species) {
4          if (sp.charge==0) continue;    //don't bother with neutrals
5          rho += sp.charge*sp.den;
6      }
7  }
```

The ϕ computed by calling `solver.solve()` is then numerically differenced to set $\vec{E} = -\nabla\phi$. Subset of this code is shown below:

```
1  void PotentialSolver::computeEF() {
2      /* ... */
3      for (int i=0;i<world.ni;i++)
4          for (int j=0;j<world.nj;j++) {
5              /*x component*/
6              if (i==0)
7                  ef[i][j][0] = -(-3*phi[i][j]+4*phi[i+1][j]-phi[i+2][
                       j])/(2*dx); //forward
8              else if (i==world.ni-1)
9                  ef[i][j][0] = -(phi[i-2][j]-4*phi[i-1][j]+3*phi[i][j
                       ])/(2*dx); //backward
10             else
11                 ef[i][j][0] = -(phi[i+1][j] - phi[i-1][j])/(2*dx);
                       //central
```

Another change that is required involves incorporating the Lorentz force in the particle pusher. This requires only a minimal change to replace the placeholder $\vec{F} = 0$ vector with a value computed by interpolating the mesh-based electric field onto the particle position. Here we use the gather function,

```
1  // electric field at particle position
2  double3 ef_part = world.ef.gather(lc);
3
4  // update velocity from F=qE
5  part.vel += ef_part*(part.dt*charge/mass);
```

The surface impact handler is also modified to model surface neutralization. It is accomplished by removing the source particle of the ion species and injecting new particles into the neutral species. The number of particles to inject is based on the ratio of macroparticle weights,

$$N_{inject} = \left\lfloor \frac{w_{mp,ion}}{w_{mp,atom}} + \mathcal{R} \right\rfloor \tag{5.78}$$

The implementation is shown below:

```
1  if (charge==0) // neutrals
```

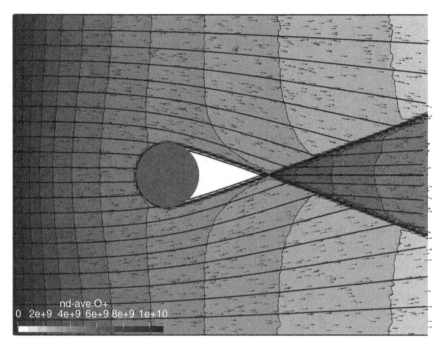

Figure 5.8 ES-PIC simulation of ion flow past a charged cylinder.

```
2       part.vel = sampleReflectedVelocity(part.pos,v_mag1);
3   else { // ions
4       double mpw_ratio = this->mpw0/atoms.mpw0;
5       part.mpw = 0; // kill source particle
6       //inject neutrals
7       int mp_create = (int)(mpw_ratio+rnd());
8       for (int i=0;i<mp_create;i++) {
9           double3 vel = sampleReflectedVelocity(part.pos,v_mag1);
10          atoms.addParticle(part.pos,vel);
11      }
12  }
```

Due to the massive difference in macroparticle weights ($w_{mp,ion} = 10^2$, $w_{mp,atom} = 10^{15}$) in this example simulation, it is unlikely that this surface neutralization algorithm will ever create any new particles. This large difference was required in order to simulate sufficiently high neutral density to demonstrate the formation of a bow shock.

The resulting ion density profile as well as ion velocity streamlines can be seen in Figure 5.8. The fixed negative potential on the sphere leads to a formation of an electric field that accelerates the ions toward the sphere. This acceleration is apparent by the reduction in number density. Mass conservation dictates that density decreases inversely with velocity increase. The shape of the plasma potential also accelerates ions from the free stream to the wake behind the cylinder.

6 Eulerian Methods

This chapter continues the discussion of gas simulation methods by introducing numerical approaches based on the Eulerian (mesh-based) formulation. We describe solution approaches for the model advection-diffusion equation, but also introduce algorithms for other commonly encountered PDEs. We also cover the Finite Volume Method (FVM), vorticity - stream function method, and Vlasov solvers.

6.1 INTRODUCTION

In the prior chapter, we saw how to simulate gas flows through a Lagrangian approach in which we considered the motion of individual particles. As was also discussed in that chapter, such approaches are generally only applicable to rarefied gases and plasmas. At higher densities, it is more computationally efficient to consider only the macroscopic gas properties such as density, stream velocity, and temperature. Much of the undergraduate aerospace engineering curriculum involves deriving equations that establish a relationship between these properties of interest. As an example, from our understanding of mass conservation, we know that the total mass of a closed system cannot change (here we are taking the liberty to ignore nuclear reactions and matter-antimatter annihilation). We can thus write

$$\frac{dm}{dt} = 0 \tag{6.1}$$

We can next consider a small control volume through which gas of mass density ρ flows with some velocity $\vec{v}$. By balancing the mass inside the volume, $\int_V \rho dV$, with the normal net flux through the box boundaries, $\Gamma_n = \rho \vec{v} \cdot \hat{n}$, we arrive at the fluid form of the mass conservation equation (also known as continuity equation),

$$\frac{\partial \rho}{\partial t} - \nabla \cdot (\rho \vec{v}) = 0 \tag{6.2}$$

Note that ρ is now mass density instead of charge density as was the case in the prior chapter (unfortunately such a reuse of Greek letters is common). This equation can similarly be applied to just a single material species. It may also be preferred to consider number density, $n = \rho/m$. We then have

$$\frac{\partial n}{\partial t} - \nabla \cdot (n\vec{v}) = \dot{n}_{chem} \tag{6.3}$$

Here $\dot{n}_{chem}$ captures species mass creation (or destruction) through, for example, chemical reactions.

DOI: 10.1201/9781003132233-6   **214**

Similarly, the rate of change of momentum can also be shown to equal to the total force applied to the object,

$$\frac{d(m\vec{v})}{dt} = \vec{F} \tag{6.4}$$

This equation simplifies even further if the mass remains constant,

$$m\frac{d\vec{v}}{dt} = \vec{F} \tag{6.5}$$

When applied to a fluid parcel, this equation becomes

$$\rho\frac{D\vec{v}}{Dt} \equiv \rho\left(\frac{\partial\vec{v}}{\partial t} + \vec{v}\cdot\nabla\vec{v}\right) = -\nabla p + \mu\nabla^2\vec{v} + \rho\vec{F} + \vec{R} \tag{6.6}$$

The terms on the left side capture $md\vec{v}/dt$, while the terms on the right side are various forces acting on the fluid parcel. Here we include the influence of a pressure gradient, the "pull" of nearby molecules in viscous flows, and some arbitrary body forces (such as gravity). We also include momentum transfer $\vec{R}$ term to capture different fluid species interacting with each other. Due to the vector nature, Equation 6.6 actually provides relationships for u, v, and w. For example, for the x direction we have the following:

$$\rho\left(\frac{\partial u}{\partial t} + u\cdot\nabla u\right) = -\nabla p + \mu\nabla^2 u + \rho F_x + R_x \tag{6.7}$$

Finally, the energy conservation equation can be written in the following form:

$$\frac{\partial}{\partial t}\left(\frac{3}{2}nkT\right) + \nabla\cdot\left(\frac{5}{2}nkT\vec{u}\right) - \nabla\cdot(K\nabla T) = S \tag{6.8}$$

where K is the heat conduction coefficient and S is a source term accounting for energy sinks and losses.

6.2 ADVECTION-DIFFUSION EQUATION

The actual details of these equations are not important for the following discussion. Of importance is noting that they all provide a relationship for a time rate of change of some property (such as density, velocity, or temperature) in terms of other spatially-varying terms. Similar observations can be made about governing equations from all engineering and science disciplines. These terms can be grouped into four categories:

1. A *time derivative* term introduces the possibility of the property changing with time.
2. An *advective* term governs how the property is transported by the action of a flow field.

3. A *diffusive* term controls how sharp gradients get smoothed out.
4. A *source* term allows for the creation (or destruction) of the property within the domain.

Instead of focusing on the actual fluid dynamics conservation equations, we can construct a model equation that retains just one term from each category. This is known as the advection-diffusion (A-D) equation:

$$\frac{\partial \psi}{\partial t} = -\nabla \cdot (\psi \vec{v}) + \nabla \cdot (D\nabla \psi) + R \tag{6.9}$$

Here $\psi = \psi(\vec{x}, t)$ is some generic property of interest. This equation is also referred to as the Convection-Diffusion equation when specifically applied to fluid flows.

In order to visualize the meaning of the four terms, let's imagine that a bucket of coffee is dumped into a moving river. We can let $\psi = n$ be the number density (molecular concentration) of coffee molecules. At time $t < 0$, we have $n = 0$ everywhere. The insertion of coffee is represented by the source term, $R = R(\vec{x}, 0)$. In this example, R is a function that is non-zero only within the prescribed region Ω_{drop}. A simple example is the step function,

$$R = \begin{cases} 0 & \vec{x} \notin \Omega_{drop} \\ n_0 & \vec{x} \in \Omega_{drop} \end{cases} \tag{6.10}$$

This expression introduces a discontinuity along the boundary of Ω_{drop}. Obviously, such a sharp transition is not physical. Random motion arising from molecular collisions will lead to the coffee molecules finding their way out of the region of high concentration into the region of low concentration. This process of smoothing out discontinuities is captured by the diffusive term $\nabla \cdot (D\nabla \psi)$. Given an infinitely large time interval, these random scattering events lead to the molecular concentration equilibrating across the entire domain. However, in our river analogy, coffee molecules are also colliding with water molecules already flowing by with some net velocity. Momentum transfer collisions between these two types of molecules lead to the coffee molecules acquiring a net velocity component in the river flow direction. This "entrainment" of the coffee molecules in the ambient flow field is captured by the advective term, $\nabla \cdot (\psi \vec{v})$. Finally, the time derivative term $\partial \psi / \partial t$ represents the time variation of the concentration.

6.2.1 DIFFUSION EQUATION

Let's assume that there is no fluid flow $\vec{v} = 0$ (a lake instead of a river using our coffee drop analogy). The advection term vanishes, $\nabla \cdot (n\vec{v}) = 0$. For uniform medium, the diffusion coefficient can be assumed to be independent of position, $\nabla D = 0$. Then, $\nabla \cdot (D\nabla n)$ simplifies to $D\nabla^2 n$. Finally, if there are no sources, $R = 0$, the advection-diffusion equation reduces to

$$\frac{\partial n}{\partial t} = D\nabla^2 n \tag{6.11}$$

This is the unsteady diffusion equation. It tells us that the rate of change of density in respect to time is proportional to its Laplacian (divergence of the gradient). This relationship is also known as the heat equation since it governs the evolution of temperature due to heat conduction, $\partial^2 T/\partial t = \alpha \nabla^2 T$.

Diffusion arises from the molecules randomly scattering away from regions of high concentration. We can imagine that after a sufficient amount of time, the spatial variation in density equilibrates such that $\nabla n = 0$. Once this happens, the density field stops evolving with time, $\partial n/\partial t = 0$. This condition in which the property of interest no longer varies temporally is known as steady state. The steady-state diffusion equation reads

$$\nabla^2 n = 0 \tag{6.12}$$

This is also known known as the Laplace equation. The closely related Poisson's equation features a non-zero right hand side, $\nabla^2 \psi = b$. We area already familiar with its solution methods from Section 2.6.2.

6.2.2 PDE CLASSIFICATION

Although Equations 6.11 and 6.12 are similar, the presence of the time derivative in the first equation has a major impact on the behavior of the solution. This term governs the evolution of the density (or other property of interest) and as such, we need to specify the initial state from which the solution should start evolving from. This is known as the *initial condition*. The Laplacian term controls the evolution due to spatial gradients. It is not possible for a computer to resolve an infinitely large domain and hence we need to limit the computation to a bounded domain. This domain may also include some internal features (objects) that may impose certain restrictions on the property of interest, such as zero density inside solid objects or no flow through walls, as discussed in Section 5.2. These are examples of boundary conditions, specifically Dirichlet (condition on the dependent variable) and Neumann (condition specified for the derivative of the dependent variable) types.

A linear, second order partial differential equation for $u = u(x,y)$ can be written in the form

$$A u_{xx} + B u_{xy} + C u_{yy} + D u_x + E u_y + F = 0 \tag{6.13}$$

where $u_{xx} = \partial^2 u/\partial x^2$, $u_{xy} = \partial^2 u/(\partial x \partial y)$, and $u_x = \partial u/\partial x$. This equation is similar to the equation for conic sections. Just as is the case with conics, a PDE can be characterizing according to the discriminant

$$\mathcal{D} = B^2 - 4AC \tag{6.14}$$

as elliptic, parabolic, or hyperbolic. An elliptic equation is one for which $\mathcal{D} < 0$. The steady state diffusion (Laplace) equation is an example of this type. This can be seen by considering a 2D case $u_{xx} + u_{yy} = 0$. We have $A = 1$, $B = 0$, $C = 1$ and hence $B^2 - 4AC = -4$. Elliptic equations share an important feature that their solution is not dependent on the initial condition. Boundary conditions propagates through the domain instantaneously overwriting any initial state information.

The next category, a parabolic PDE, has $\mathcal{D} = 0$. An example of this type is the one-dimensional unsteady diffusion equation, $u_t - u_{xx} = 0$. Here t plays the role of y, and we have $E = 1$, $A = -1$, $B = 0$, and $C = 0$, thus leading to $\mathcal{D} = 0$. The initial conditions now play a role, however their importance vanishes with t. Specifically, a parabolic equation becomes elliptic as $t \to \infty$. This correlation between elliptic and parabolic equations is in fact exploited by some matrix solvers that solve an elliptic problem by recasting it as a time-dependent problem in pseudo-time. As will be seen shortly, some algorithms for parabolic PDEs require only matrix-vector multiplications.

Next, $\mathcal{D} > 0$ gives us a hyperbolic PDE. An example of this system is the second order wave equation $u_{tt} - c^2 u_{xx} = 0$. We have $A = 1$, $B = 0$, and $C = -c^2$ and hence $B^2 - 4AC > 0$. The initial conditions in a hyperbolic system never vanish. Instead, they propagate through the domain with some characteristic wave speed. Boundary conditions are needed primarily to account for the finite computational domain, with care needed to avoid numerical reflection at the domain edge. Finally, PDEs "found in the wild" typically contain more than just two dependent variables; this is in fact the case with the model Advection-Diffusion equation. Such PDEs are of mixed type and exhibit a combination of the above mentioned behaviors.

6.2.3   ADVECTIVE TERM

While it is customary to use the central difference (CDS) for the diffusive term, there are different approaches for the advective term $\nabla \cdot (\vec{v}n)$,

$$
\begin{aligned}
\nabla \cdot (\vec{v}n) &= \left(\frac{\partial}{\partial x}\hat{i} + \frac{\partial}{\partial y}\hat{j} + \frac{\partial}{\partial z}\hat{k} \right) \cdot \left(un\hat{i} + vn\hat{j} + wn\hat{k} \right) \\
&= \frac{\partial(un)}{\partial x} + \frac{\partial(vn)}{\partial y} + \frac{\partial(wn)}{\partial z}
\end{aligned}
$$

We consider two: CDS and Upwind Difference Scheme (UDS). From the Taylor Series, we know that the central difference for the first derivative is

$$
\left. \frac{\partial f}{\partial x} \right|_{i,j} = \frac{f_{i+1,j} - f_{i-1,j}}{2\Delta x} + O(2) \tag{6.15}
$$

which for $f = un$ becomes

$$
\left. \frac{\partial(un)}{\partial x} \right|_{i,j} = \frac{u_{i+1,j}n_{i+1,j} - u_{i-1,j}n_{i-1,j}}{2\Delta x} + O(2) \tag{6.16}
$$

This is the CDS scheme for the advective term.

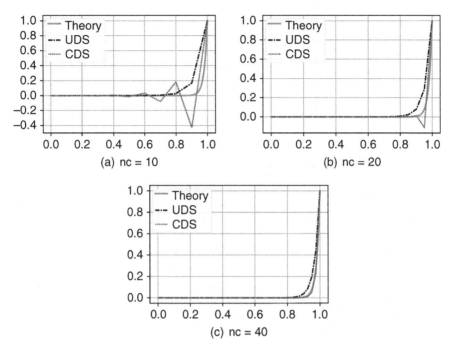

Figure 6.1 Comparison of UDS and CDS schemes for the advective term.

Alternatively, we can consider the velocity direction and use one-sided first-order differencing. If $u > 0$, mass is advected from left to right and hence we use

$$\left.\frac{\partial un}{\partial x}\right|_{i,j} = \frac{u_{i,j}n_{i,j} - u_{i-1,j}n_{i-1,j}}{\Delta x} + O(1) \qquad ;u > 0 \qquad (6.17)$$

Otherwise, we use one sided derivative on the right

$$\left.\frac{\partial un}{\partial x}\right|_{i,j} = \frac{u_{i+1,j}n_{i+1,j} - u_{i,j}n_{i,j}}{\Delta x} + O(1) \qquad ;u < 0 \qquad (6.18)$$

The use of these methods is demonstrated in `ad1-steady.py`. This code solves the steady-state form of the 1D advection-diffusion equation with boundary conditions discussed later in Section 6.3.2. Figure 6.1 compares the solution obtained with the CDS and UDS schemes to the theoretical solution given by the solid gray line. We can see that neither scheme is particularly good at resolving the analytical solution using 10 mesh cells. The upwind scheme is more diffusive at it spreads the sharp increase in the solution over a wider spatial distance. The central difference scheme on the other hand produces an oscillatory behavior with negative densities. The remaining figures demonstrate the evolution of the solution as the mesh resolution is increased to 20 and 40 cells, respectively. With 20 cells, CDS now appears to start capturing the sharp increase, but is still exhibiting the non-physical negative density kink. This kink disappears if 40 cells are used, in which case the CDS prediction agrees with the theoretical solution. UDS, on the other hand continues to demonstrate numerical diffusion. As can be seen, the UDS scheme avoids non-physical negative densities at the cost of artificial numerical diffusion. In production codes, higher-order flux-conserving schemes are used to capture sharp transitions (for instance in shocks) without artificially increasing the width of the discontinuity.

6.3 INTEGRATION SCHEMES

6.3.1 FORWARD TIME CENTRAL SPACE (FTCS)

Let's now consider the unsteady form of Equation 6.9 with $\nabla D = 0$ and $\vec{v} = 0$,

$$\frac{\partial \psi}{\partial t} = D\nabla^2 \psi + R \tag{6.19}$$

Here D is the diffusion coefficient and R is a source term. We limit the following discussion to the 2D form with $\nabla^2 \psi = \partial \psi^2 / \partial x^2 + \partial \psi^2 / \partial y^2$. We have already learned in Chapter 2 that the second derivative can be discretized using the central difference scheme,

$$\frac{\partial^2 \psi}{\partial x^2} \approx \frac{\psi_{i-1,j} - 2\psi_{i,j} + \psi_{i+1,j}}{\Delta^2 x} \tag{6.20}$$

The time derivative can be discretized using the Forward Euler method

$$\left(\frac{\partial \psi}{\partial t}\right)^k = \frac{\psi^{k+1} - \psi^k}{\Delta t} + O(1) \tag{6.21}$$

giving us (in 2D)

$$\frac{\psi_{i,j}^{k+1} - \psi_{i,j}^k}{\Delta t} = D\left[\frac{\psi_{i-1,j}^k - 2\psi_{i,j}^k + \psi_{i+1,j}^k}{\Delta^2 x} + \frac{\psi_{i,j-1}^k - 2\psi_{i,j}^k + \psi_{i,j+1}^k}{\Delta^2 y}\right] + R_{i,j}^k \tag{6.22}$$

This scheme is known as Forward Time Central Difference (FTCS). Its popularity stems from its simplicity. Because of the reliance on the Forward Euler scheme, it is only first-order accurate in the time direction. It may be tempting to write the time derivative using the second-order central difference, i.e.,

$$\left(\frac{\partial \psi}{\partial t}\right)^k = \frac{\psi^{k+1} - \psi^{k-1}}{2\Delta t} + O(2) \tag{6.23}$$

However, this approach would result in decoupling of consecutive solutions as ψ^{k+1} does not depend on ψ^k. Instead, as needed, the accuracy in the time integration can be improved by utilizing multistep methods such as Adams–Bashforth or Runge-Kutta, as discussed in Section 6.3.4.

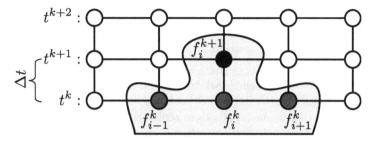

Figure 6.2 Computational stencil for 1D FTCS integration scheme.

The computational stencil for a 1D FTCS scheme is shown in Figure 6.2. It is an explicit scheme, since values at time $k+1$ depend solely on data from time k. In other words,

$$\psi^{k+1} = f(\psi^k) \tag{6.24}$$

Let's return to the unsteady diffusion equation,

$$\frac{\partial \psi}{\partial t} = D\nabla^2 \psi \tag{6.25}$$

We have

$$\frac{\psi_{i,j}^{k+1} - \psi_{i,j}^k}{\Delta t} = D\left[\frac{\psi_{i-1,j}^k - 2\psi_{i,j}^k + n_{i+1,j}^k}{\Delta^2 x} + \frac{\psi_{i,j-1}^k - 2\psi_{i,j}^k + \psi_{i,j+1}^k}{\Delta^2 y}\right] \tag{6.26}$$

or

$$\psi_{i,j}^{k+1} = \psi_{i,j}^k + D\Delta t\left[\frac{\psi_{i-1,j}^k - 2\psi_{i,j}^k + \psi_{i+1,j}^k}{\Delta^2 x} + \frac{\psi_{i,j-1}^k - 2\psi_{i,j}^k + \psi_{i,j+1}^k}{\Delta^2 y}\right] \tag{6.27}$$

The linear system in Equation 6.27 can be rewritten as

$$\vec{\psi}^{k+1} = \vec{\psi}^k + D\Delta t\mathbf{A}\vec{\psi}^k \tag{6.28}$$

or

$$\vec{\psi}^{k+1} = (\mathbf{I} + D\Delta t\mathbf{A})\,\vec{\psi}^k \tag{6.29}$$

where $\mathbf{A}$ is, for a 2D problem, a five-banded coefficient matrix. This form makes it apparent that the FTCS scheme does not require solving a $\mathbf{A}\vec{x} = \vec{b}$ type system for $\vec{x}$. Instead, each time integration step consists solely of a single matrix-vector multiplication!

Implementation of the FTCS method for a 2D unsteady heat equation with $\psi = T$ is demonstrated in `heat2d_unsteady.py`. This program is a derivative of the steady-state heat equation solver from Section 2.6.2. It begins by initializing the standard 5 point stencil on all internal nodes within a coefficient matrix $\mathbf{A}$. Subsequently, 20 points are randomly selected and reassigned to represent Dirichlet boundary conditions.The solver is implemented using Matplotlib's FuncAnimation to support the generation of an animation plot illustrating the solution evolution,

```
ani = FuncAnimation(fig, update, 201, repeat=False)
```

The update function is called 201 times, and on each call, it first calls the heat_solver function to integrate the solution using the FTCS scheme. Then, every 10 times, the displayed plots are replotted. Each plot is also saved to a file to support animation movie creation.

```
1  def update(i):
2     T2d = heat_solver()       # get new solution
3
4     #plotting
5     if i%10==0:
6        print("Time step: %d"%i)
7        lev = np.linspace(0,2,10)
8        ax.clear()
9        ax.contourf(T2d,cmap='hot_r',levels=lev)
10       ax.contour(T2d,colors='black',levels=lev)
11       plt.savefig('img/heat2d_%d.png'%i,dpi=300)
```

The solver marches the solution forward using Equation 6.29,

```
1  def heat_solver():
2     global T
3     T = (I+D*dt*A)*T
4     return np.reshape(T,(nj,ni))
```

Note that this method requires an additional correction for Neumann boundaries. This is not done here for simplicity and as such the simulation models a problem with Dirichlet walls. The evolution of the solution is visualized in Figure 6.3.

6.3.2 VON NEUMANN STABILITY ANALYSIS

These results were obtained using $D = 10^{-2}$, $\Delta t = 4 \times 10^{-3}$, $\Delta x = 1/4$, and $\Delta y = 1/3$. Experimenting with the code, you will observe that some changes to these parameters lead to the solution exhibiting a checkerboard pattern instead of the smoothly varying temperatures. This is an indication of solver divergence.

To study the solution stability in more detail, we can consider a one-dimensional form of the A-D equation. This example is implemented in ad1-ftcs.py. Following an example from Ferziger and Perić 2002, this code implements a FTCS solver for

$$\frac{\partial(\rho\phi)}{\partial t} + \frac{\partial(\rho u\phi)}{\partial x} - \frac{\partial}{\partial x}\left(D\frac{\partial\phi}{\partial x}\right) = 0 \tag{6.30}$$

The analytical solution for $t \to \infty$ is given by

$$\phi = \phi_0 + \frac{\exp(xP_e/L) - 1}{\exp(P_e) - 1}(\phi_L - \phi_0) \tag{6.31}$$

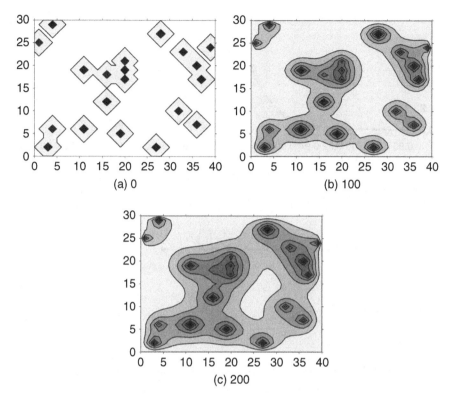

Figure 6.3 Solution to the unsteady heat equation after the given number of time steps.

where $P_e = \rho U L / D$ is called the Peclet number, and ϕ_0 and ϕ_L are the prescribed values for ϕ for $x = 0$ and $x = L$, respectively. The code gives us the chance to experiment with different values for the time step Δt, cell size Δx, as well as the diffusion coefficient D in order to observe how these changes affect the solution.

Figure 6.4(a) shows the solution converging toward the analytical solution. This plot was generated by running the code for 100 $\Delta t = 10^{-2}$ steps on a mesh discretized into 40 cells. Seeing this behavior, you may decide to reduce the run time by simulating 50 $\Delta t = 2 \times 10^{-2}$ steps. The solutions are expected to be identical, but instead, we now observe an oscillating pattern with negative densities as shown in Figure 6.4(b). Alternatively, you decide to refine the original solution using a finer mesh. Using the original $\Delta t = 10^{-2}$ but doubling the number of cells to 80 gives us the results plotted in Figure 6.4(c). Finally, we also notice that using the CDS scheme for the advective term, we are unable to get a converged solution regardless of how small the time step is made if there is no diffusion, $D = 0$.

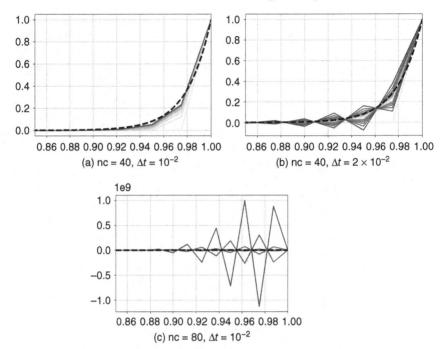

Figure 6.4 Effect of domain resolution and time step on FTCS solution.

This behavior can be investigated further by performing Von Neumann stability analysis. This method is named after John von Neumann, a prominent early 20th-century physicist, mathematician, and computer scientist associated with Los Alamos and the Manhattan project. Let's consider FTCS for a 1D problem,

$$\frac{\psi_i^{k-1} - \psi_i^k}{\Delta t} + \frac{(\psi u)_{i+1}^k - (\psi u)_{i-1}^k}{2\Delta x} - D\frac{(\psi u)_{i-1}^k - 2(\psi u)_i^k + (\psi u)_{i+1}^k}{\Delta^2 x} = 0 \quad (6.32)$$

We can rewrite this equation in terms of ψ_i coefficients,

$$\psi_i^{k+1} = (1 - 2d)\psi_i^k + \left(d - \frac{c}{2}\right)\psi_{i+1}^k + \left(d + \frac{c}{2}\right)\psi_{i-1}^k \quad (6.33)$$

with

$$d = \frac{D\Delta t}{\Delta^2 x} \quad \text{and} \quad c = \frac{u\Delta t}{\Delta x}$$

This allows us to write the method as

$$\vec{\psi}^{k+1} = \mathbf{A}\vec{\psi}^k \quad (6.34)$$

Von Neumann hypothesized that boundary conditions are rarely the source of instability and hence can be ignored. Then, utilizing the above relationship, it follows that

$$\vec{\psi}^{k+1} = \mathbf{A}\vec{\psi}^k = \mathbf{A}\left(\mathbf{A}\vec{\psi}^{k-1}\right) = \mathbf{A}^2\left(\mathbf{A}\vec{\psi}^{k-2}\right)\ldots = \mathbf{A}^{k+1}\psi^0 \qquad (6.35)$$

In other words, the solution after some number of time steps is achieved by a repetitive multiplication of the input vector by the matrix $\mathbf{A}$.

If we let the initial $\vec{\psi}^0 = \vec{\psi} + \vec{\epsilon}$ where $\vec{\epsilon}$ is some initial error, we clearly see that we need the coefficients of $\mathbf{A}$ be such that

$$|\mathbf{A}^{k+1}\vec{\epsilon}| < |\mathbf{A}^k\vec{\epsilon}| \qquad (6.36)$$

otherwise the initial error "blows up".

A general linear system of the form $\vec{f}'(t) + A\vec{f}(t) = 0$ will have a solution in the form $\vec{f} = \sum^l c_l e^{\lambda_l t}\vec{\eta}_l$, where the RHS consists of constant terms relating to the initial conditions, an exponential term containing an eigenvalue λ_l and the corresponding eigenvector $\vec{\eta}_l$. For the error to decay, we need all eigenvalues to be less than 1. By ignoring boundary conditions and using constant cell spacing, we obtain a tridiagonal matrix with constant terms on the three diagonals. This is known as Toeplitz matrix. This special matrix has the property that all eigenvectors have a form of sines and cosines. We can then write

$$\psi_j^k = \lambda^k e^{i\alpha j} \qquad (6.37)$$

where $i = \sqrt{-1}$ and α is some arbitrary wave number. By substituting this form into Equation 6.33 we obtain

$$\lambda = 1 + 2d(\cos\alpha - 1) + i2c\sin\alpha \qquad (6.38)$$

The magnitude of a complex number is the sum of the squares of the real and imaginary parts, hence

$$\lambda^2 = [1 + 2d(\cos\alpha - 1)]^2 + 4c^2\sin^2\alpha \qquad (6.39)$$

We can now consider some special cases. If $d = 0$ (no diffusion), $\lambda^2 > 1$ for any value of α and the system is *unconditionally unstable*. It will not converge, regardless of the chosen values of Δx and Δt. On the other hand, if $c = 0$ (no advection), then $\lambda^2 = [1 + 2d(\cos\alpha - 1)]^2$, which is < 1 for some values. This is known as *conditional stability*. For the full equation, we obtain $c < 2d$.

Summarizing, for FTCS we have

$$\frac{D\Delta t}{\Delta^2 x} < \frac{1}{2} \qquad (6.40)$$

and

$$\frac{u\Delta t}{\Delta x} < 2\frac{D\Delta t}{\Delta^2 x} \qquad (6.41)$$

The stability requirement for the FTCS method is thus

$$\Delta t < \frac{\Delta^2 x}{2D} \qquad (6.42)$$

We can see that decreasing the cell size Δx by 2 requires decreasing the integration time step Δt by 4. Often we are interested in long-term solutions so we would like Δt to be as large as possible. This stability requirements of FTCS make it impractical for simulations in which fine spatial details need to be resolved.

Let's now consider some alternatives. Consider the following ODE

$$\frac{d\psi(t)}{dt} = f(\psi(t)) \tag{6.43}$$

We can discretize this system using the forward Euler method as

$$\psi^{k+1} = \psi^k + f(\psi^k)\Delta t \tag{6.44}$$

However, there are other possibilities such as

$$\psi^{k+1} = \psi^k + f(\psi^{k+1})\Delta t \tag{6.45}$$

and

$$\psi^{k+1} = \psi^k + f(\psi^{k+0.5})\Delta t \tag{6.46}$$

These schemes are known as backward Euler and midpoint, respectively. These are examples of *implicit* methods in which the integration from time k to $k+1$ requires a solution from the future time step. The benefit of implicit methods is that they are *unconditionally stable*. Large time step can produce non-physical oscillations or otherwise non-physical results; however, the results will be bounded and will not grow to infinity. The downside of implicit methods is that they lead to a system of equations, which needs to be solved using a linear solver. For example, for the backward Euler scheme we have

$$\psi^{k+1} = \psi^k + \mathbf{C}\psi^{k+1}$$
$$(\mathbf{I} - \mathbf{C})\psi^{k+1} = \psi^k \tag{6.47}$$

or in general $\mathbf{A}\psi^{k+1} = f(\psi^k)$. We may find that the total computational time with FTCS using a large number of smaller Δt time steps is actually shorter than with an implicit scheme that utilizes fewer larger Δt steps, but now requires a matrix solver at each time step.

6.3.3 PREDICTOR-CORRECT METHOD

There are two classes of methods that attempt to recover the stability of implicit methods without requiring the time-consuming matrix solver: predictor-corrector and multipoint. The first method uses known values to generate an estimated future solution using forward Euler,

$$\psi^* = \psi^k + f(\psi^k)\Delta t \tag{6.48}$$

This new temporary solution is then used to evaluate the average forcing term used to march the solution forward "for real",

$$\psi^{k+1} = \psi^k + \frac{1}{2}\left[f(\psi^k) + f(\psi^*)\right]\Delta t \tag{6.49}$$

6.3.4   MULTIPOINT METHODS

Instead of using the predicted future value, we can perform the time integration using several "old" values from previous time steps. As an example, the second order Adams-Bashford method is given by

$$\psi^{k+1} = \psi^k + \frac{\Delta t}{2}\left[3f(\psi^k) - f(\psi^{k-1})\right] \tag{6.50}$$

Other schemes utilizing different past data points can also be found in literature. These methods require retaining the solution vector for one or more past time steps. They also require the initial use of some alternate scheme (such as the forward Euler method) until the sufficient number of past solutions is obtained. Multipoint methods can also be written in an implicit fashion. For example, the third-order Adams-Moulton implicit method is given by

$$\psi^{k+1} + \psi^k + \frac{\Delta t}{12}\left[5f(\psi^{k+1}) + 8f(\psi^k) - f(\psi^{k-1})\right] \tag{6.51}$$

6.3.5   RUNGE-KUTTA METHODS

Another class of integration schemes that you may already be familiar with are the Runge-Kutta methods. Instead of utilizing solutions from past time steps, these methods implement a multipoint scheme based on intermediate solutions between times k and $k + 1$. There are multiple variants, with the most popular being the fourth-order method, RK4. For a general problem

$$\frac{\partial \psi}{\partial t} = f(t, \psi) \tag{6.52}$$

we first compute four intermediate solutions using

$$\begin{aligned}
k_1 &= f(t, \psi^k) \\
k_2 &= f(t + 0.5\Delta t, \psi^k + 0.5\Delta t k_1) \\
k_3 &= f(t + 0.5\Delta t, \psi^k + 0.5\Delta t k_2) \\
k_4 &= f(t + \Delta t, \psi^k + \Delta t k_3)
\end{aligned} \tag{6.53}$$

The solution is then advanced per

$$\psi^{k+1} = \frac{\Delta t}{6}\left[k_1 + 2k_2 + 2k_3 + k_4\right] \tag{6.54}$$

For the more applicable case where the right hands side is only a function of ψ, this integration algorithm reduces to

$$\psi^a = \psi^k + \frac{\Delta t}{2} f(\psi^k)$$

$$\psi^b = \psi^k + \frac{\Delta t}{2} f(\psi^a)$$

$$\psi^* = \psi^k + \Delta t f(\psi^b)$$

$$\psi^{k+1} = \psi^k + \frac{\Delta t}{6} \left[f(\psi^k) + 2f(\psi^a) + 2f(\psi^b) + f(\psi^*) \right] \qquad (6.55)$$

Note that a wide range of variants exist with useful properties such as "low-storage" versions that minimize additional vectors required to perform higher-order integration. Readers are referred to Chapter 10 of Bewley 2018 and references therein for an introduction to a wide array of such algorithms.

6.3.6 LEAPFROG AND DUFORT-FRANKEL METHODS

Another integration scheme is

$$\psi^{k+1} = \psi^{k-1} + f(\psi^k) 2\Delta t \qquad (6.56)$$

This is an example of a Leapfrog method since the time at which f is evaluated "leapfrogs" over the times at which ψ is advanced. This method is unconditionally unstable for the advection-diffusion equation; however, it can be stabilized using some tricks. One such a trick involves replacing the ψ^k term with $(\psi^{k-1} + \psi^{k+1})/2$,

$$\psi^{k+1} = \psi^{k-1} + f\left(\frac{\psi^{k-1} + \psi^{k+1}}{2} \right) 2\Delta t \qquad (6.57)$$

This is known as DuFort-Frankel method.

6.3.7 EULER IMPLICIT METHOD

Implicit methods may sometimes be preferred, with backward Euler being the simplest. Applying it to a one dimensional A-D equation, we have

$$\psi_i^{k+1} = \psi_i^k + \left[-u \frac{\psi_{i+1}^{k+1} - \psi_{i-1}^{k+1}}{2\Delta x} + \frac{D}{\rho} \frac{\psi_{i-1}^{k+1} - 2\psi_i^{k+1} + \psi_{i+1}^{k+1}}{\Delta^2 x} \right] \Delta t \qquad (6.58)$$

which can be rearranged as

$$(1 + 2d)\psi_i^{k+1} + \left(\frac{c}{2} - d \right) \psi_{i+1}^{k+1} + \left(-\frac{c}{2} - d \right) \psi_{i-1}^{k+1} = \psi_i^k \qquad (6.59)$$

or

$$A_p \psi_i^{k+1} + A_e \psi_{i+1}^{k+1} + A_w \psi_{i-1}^{k+1} = Q_p \qquad (6.60)$$

where

$$A_p = (1 + 2d), \quad A_e = \left(\frac{c}{2} - d\right), \quad A_w = \left(-\frac{c}{2} - d\right), \quad Q_p = \psi_i^k \qquad (6.61)$$

which is $\mathbf{A}\psi^{k+1} = \psi^k$ linear system.

6.3.8   CRANK-NICOLSON METHOD

In our prior discussion on the Leapfrog method in Chapter 1, we observed that it is preferred to evaluate the forcing term at the midpoint,

$$\frac{\psi^{k+1} - \psi^k}{\Delta t} = f\left(\psi^{k+0.5}\right) \qquad (6.62)$$

Using linear interpolation, the above expression can be rewritten as

$$\frac{\psi^{k+1} - \psi^k}{\Delta t} = \frac{f\left(\psi^{k+1}\right) + f\left(\psi^k\right)}{2} \qquad (6.63)$$

This gives us the popular Crank-Nicolson scheme. This implicit method combines the stability of the backward Euler method with the second order accuracy of the trapezoidal method. For the A-D equation, we have

$$\frac{\psi^{k+1} - \psi^k}{\Delta t} = -\nabla \cdot \left(\vec{v}^{k+0.5}\psi^{k+0.5}\right) + D\nabla^2\psi^{k+0.5} + R^{k+0.5} \qquad (6.64)$$

or

$$\frac{\psi^{k+1} - \psi^k}{\Delta t} = -\frac{\nabla \cdot \vec{v}^{k+1}\psi^{k+1} + \nabla \cdot \vec{v}^k\psi^k}{2} + \frac{D\left[\nabla^2\psi^{k+1} + \nabla^2\psi^k\right]}{2} + R^{k+0.5} \qquad (6.65)$$

Collecting terms, we obtain

$$\psi^{k+1} + \frac{\Delta t}{2}\left[\nabla \cdot \vec{v}^{k+1}\psi^{k+1} - D\nabla^2\psi^{k+1}\right] = \psi^k + \frac{\Delta t}{2}\left[-\nabla \cdot \vec{v}^k\psi^k + D\nabla^2\psi^k\right] + R^{k+0.5} \qquad (6.66)$$

Here we are assuming that the source term R is not a function of ψ and can be evaluated independently at the mid-point step.

Pulling out the unknowns, we get

$$\left[I + (\Delta t/2)(\nabla \cdot \vec{v}^{k+1} - D\nabla^2)\right]\psi^{k+1} = \left[I + (\Delta t/2)(-\nabla \cdot \vec{v}^{k+1} + D\nabla^2)\right]\psi^k + R \qquad (6.67)$$

We can next introduce $\mathbf{C} = \nabla \cdot \vec{v}$ and $\mathbf{L} = D\nabla^2$ differencing operators, and rewrite the previous equation as

$$\left[\mathbf{I} - (\Delta t/2)(-\mathbf{C} + \mathbf{L})\right]\vec{\psi}^{k+1} = \left[\mathbf{I} + (\Delta t/2)(-\mathbf{C} + \mathbf{L})\right]\vec{\psi}^k + \vec{R}^{k+0.5} \qquad (6.68)$$

which is just a $\mathbf{A}\vec{\psi} = \vec{b}$ system.

6.4   VORTICITY-STREAM FUNCTION METHOD

We next demonstrate the use of Eulerian methods by implementing a solver for a two-dimensional (specifically axisymmetric) incompressible flow. Such flows can be simulated using a method called *vorticity - stream function*, or $\omega - \psi$. This formulation can be found in numerous computational fluid dynamic texts. For 2D flows we can define velocity components in terms of curl of a scalar potential ψ,

$$\text{XY:} \quad u = u_x = \frac{\partial \psi}{\partial y} \quad ; \quad v = u_y = -\frac{\partial \psi}{\partial x} \qquad (6.69)$$

$$\text{RZ:} \quad u = u_z = \frac{1}{r}\frac{\partial \psi}{\partial y} \quad ; \quad v = u_r = -\frac{1}{r}\frac{\partial \psi}{\partial x} \qquad (6.70)$$

These two expressions specify the formulation used for planar (XY) and axisymmetric (RZ) formulations.

Next, vorticity is defined as the curl of a vector field,

$$\omega = \nabla \times \vec{v}. \qquad (6.71)$$

For an axisymmetric flow, $\partial/\partial\theta = 0$, and u_θ, if present, is independent of v_r and v_z,

$$\omega = \omega_\theta = \frac{\partial v}{\partial z} - \frac{\partial u}{\partial r} \qquad (6.72)$$

6.4.1   STREAM FUNCTION

Substituting the definition of velocity in Equation 6.70 into Equation 6.72 leads to

$$\omega = -\frac{\partial}{\partial z}\left(\frac{1}{r}\frac{\partial \psi}{\partial z}\right) - \frac{\partial}{\partial r}\left(\frac{1}{r}\frac{\partial \psi}{\partial r}\right)$$

$$= -\frac{1}{r}\frac{\partial^2 \psi}{\partial z^2} - \frac{1}{r}\frac{\partial^2 \psi}{\partial r^2} + \frac{1}{r^2}\frac{\partial \psi}{\partial r} \qquad (6.73)$$

or

$$\frac{\partial^2 \psi}{\partial z^2} + \frac{\partial^2 \psi}{\partial r^2} - \frac{1}{r}\frac{\partial \psi}{\partial r} = -\omega r \qquad (6.74)$$

Note that this is almost $\nabla_r^2 \psi = -\omega r$ but with a negative sign on the third term on the left. The ψ term is known as *stream function*. Equation 6.74 is discretized using standard central difference as

$$\frac{1}{\Delta^2 z}(\psi_{i-1,j} - 2\psi_{i,j} + \psi_{i+1,j}) + \frac{1}{\Delta^2 r}(\psi_{i,j-1} - 2\psi_{i,j} + \psi_{i,j+1})$$

$$- \frac{1}{2r_{i,j}\Delta r}(\psi_{i,j+1} - \psi_{i,j-1}) = \omega_{i,j}r_{i,j} \quad (6.75)$$

The resulting system can then be solved using a matrix solver of choice. Our implementation uses a SOR-accelerated Jacobi solver. The algorithm iterates until the difference between consecutive solution drops below a user-defined tolerance, $||\psi^{k+1} - \psi^k|| < \epsilon_{tol}$.

6.4.2 VORTICITY TRANSPORT EQUATION

The temporal evolution of vorticity is given by the *vorticity transport equation*.
It is traditionally derived by taking a curl of the momentum equation, for
instance, see Salih 2013. It is given by

$$\frac{\partial \omega}{\partial t} + \vec{v} \cdot \nabla \omega = \nu \nabla^2 \omega$$

For the axisymmetric flow we have

$$\frac{\partial \omega}{\partial t} + u\frac{\partial \omega}{\partial z} + v\frac{\partial \omega}{\partial r} = \nu \left[\frac{\partial^2 \omega}{\partial z^2} + \frac{\partial^2 \omega}{\partial r^2} + \frac{1}{r}\frac{\partial \omega}{\partial r} \right] \tag{6.76}$$

Our implementation uses the fourth-order Runge-Kutta (RK4) method to
advance Equation 6.76 By collecting all non $\partial w/\partial t$ terms on the right hand
side, we have

$$\frac{\partial \omega}{\partial t} = R(\omega) \tag{6.77}$$

where

$$R(\omega) = \nu \left[\frac{\partial^2 \omega}{\partial z^2} + \frac{\partial^2 \omega}{\partial r^2} + \frac{1}{r}\frac{\partial \omega}{\partial r} \right] - u\frac{\partial w}{\partial z} - v\frac{\partial w}{\partial r} \tag{6.78}$$

we have

$$R^{(1)} = w^k + \frac{\Delta t}{2}R^k \quad ; \quad R^{(2)} = w^k + \frac{\Delta t}{2}R^{(1)} \quad ; \quad R^{(3)} = w^k + \Delta t R^{(2)} \tag{6.79}$$

$$w^{k+1} = w^k + \frac{\Delta t}{6}\left(R^k + 2R^{(1)} + 2R^{(2)} + R^{(3)} \right) \tag{6.80}$$

6.4.3 BOUNDARY CONDITIONS

Let's say we want to simulate gas flowing through a cylindrical cavity with
an inlet on one side and a small opening on the other end. This is visualized
in Figure 6.5. There are five types of boundary conditions to consider: 1) the
inlet, 2) the axis of revolution, 3) the wall, 4) the outlet on zmax, and 5) the
outlet on rmax.

Let's assume that the flow entering through the inlet is parallel to the
cylinder axis. Thus at the inlet we have $u_r = v = 0$ or

$$\left. \frac{\partial \psi}{\partial z} \right|_{inlet} = 0 \tag{6.81}$$

Substituting $v = 0$ into the vorticity equation, Equation 6.76, gives us

$$\left. \omega \right|_{inlet} = -\frac{\partial u}{\partial r} \tag{6.82}$$

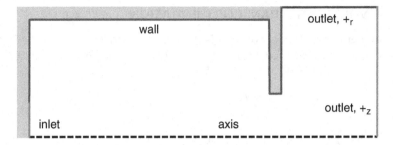

Figure 6.5 Boundaries for a vorticity-stream function example problem.

These two equations are next discretized using first order scheme as

$$\psi_{0,j} = \psi_{1,j} \tag{6.83}$$

and

$$\omega_{0,j} = (u_{0,j-1} - u_{0,j+1})/(2\Delta r) \tag{6.84}$$

Next let's consider the axis of revolution. We require $v = 0$ at $j = 0$ as there can be no flow across the axis of revolution. Therefore $\partial\psi/\partial z = 0$, and value of ψ is constant along the axis. We set this value to zero, giving us

$$\psi\big|_{axis} = 0 \tag{6.85}$$

A zero radial velocity also implies that along the axis $\omega = -\partial u/\partial r$. Axial symmetry implies $\partial()/\partial r = 0$ at $r = 0$, and thus

$$\omega\big|_{axis} = 0 \tag{6.86}$$

Next we move on to the wall boundary. The volumetric flow rate between two stream tubes is given by

$$Q = 2\pi(\psi_2 - \psi_1) \tag{6.87}$$

With $\psi_1 = 0$ being the axis of revolution, we have Dirichlet

$$\psi\big|_{wall} = \frac{1}{2}u_0 r^2_{inlet} \tag{6.88}$$

along the outer wall.

Vorticity boundaries along the wall are derived using similar approach to Salih 2013. Since the stream function is constant along a wall, derivatives of ψ vanish in the wall direction. Along the left wall, we have

$$\frac{\partial^2\psi}{\partial z^2}\bigg|_{wall} = -\omega r \tag{6.89}$$

Assuming the wall is at $i = L$, we can write the following Taylor series expansion

$$\psi_{L+1,j} = \psi_{L,j} + \left.\frac{\partial \psi}{\partial z}\right|_{L,j} \Delta z + \left.\frac{\partial^2 \psi}{\partial z^2}\right|_{L,j} \frac{\Delta z^2}{2}$$

Since $\partial \psi/\partial z = -vr$

$$\psi_{L+1,j} = \psi_{L,j} - v_{L,j} r \Delta z + \left.\frac{\partial^2 \psi}{\partial z^2}\right|_{L,j} \frac{\Delta z^2}{2}$$

or

$$\left.\frac{\partial^2 \psi}{\partial z^2}\right|_{L,j} = \frac{2(\psi_{L+1,j} - \psi_{L,j})}{\Delta z^2} + \frac{2v_{L,j} r}{\Delta z}$$

and finally

$$\omega\big|_{L,j} = \frac{2(\psi_{L,j} - \psi_{L+1,j})}{r\Delta z^2} - \frac{2v_{L,j}}{\Delta z} \tag{6.90}$$

Using similar approach, the boundary condition for a right wall at $i = R$ is found to be

$$\omega\big|_{R,j} = \frac{2(\psi_{R,j} - \psi_{R-1,j})}{r\Delta z^2} + \frac{2v_{R,j}}{\Delta z} \tag{6.91}$$

Along the top wall, $\partial \psi/\partial z = 0$ and Equation 6.74 reduces to

$$\left[\frac{\partial^2 \psi}{\partial r^2} - \frac{1}{r}\frac{\partial \psi}{\partial r}\right]_{wall} = -\omega r$$

Again we start by expanding the second derivative,

$$\psi_{i,T-1} = \psi_{i,T} - \left.\frac{\partial \psi}{\partial r}\right|_{i,T} \Delta r + \left.\frac{\partial^2 \psi}{\partial r^2}\right|_{i,T} \frac{\Delta r^2}{2}$$

Using $\partial \psi/\partial r = ur$, the above reduces to

$$\left.\frac{\partial^2 \psi}{\partial r^2}\right|_{i,T} = \frac{2(\psi_{i,T-1} - \psi_{i,T})}{\Delta r^2} + \frac{2u_{T,j} r}{\Delta r}$$

Substituting into the original equation,

$$\frac{2(\psi_{i,T-1} - \psi_{i,T})}{\Delta r^2} + \frac{2u_{i,T} r}{\Delta r} - u_{i,T} = -\omega r$$

or

$$\omega\big|_{i,T} = \frac{2(\psi_{i,T} - \psi_{i,T-1})}{r\Delta r^2} - \frac{2u_{i,T}}{\Delta r} + \frac{u_{i,T}}{r} \tag{6.92}$$

There is no bottom wall in this problem, but for generality, the matching boundary condition can be found to be

$$\omega\big|_{i,B} = \frac{2(\psi_{i,B} - \psi_{i,B+1})}{r\Delta r^2} + \frac{2u_{i,B}}{\Delta r} + \frac{u_{i,B}}{r} \tag{6.93}$$

Let's now consider the open boundary along the zmax edge. In general, the flow will be aligned with the z axis, however, there may be some non-zero v component due to jet expansion. As such, simply setting $\partial\psi/\partial z|_{zmax} = 0$ may not be valid, and would lead to an artificial compression of the exiting plume. Again following the approach in Salih 2013, we let

$$\frac{\partial\psi}{\partial z}\bigg|_{zmax} = -vr \tag{6.94}$$

which is differenced as $\psi_{ni-1,j} = \psi_{ni-2,j} - \Delta z v_{ni-1,j} r_j$. Vorticity boundary condition on the outlet is set as

$$\frac{\partial\omega}{\partial z}\bigg|_{zmax} = 0 \tag{6.95}$$

or $\omega_{ni-1,j} = \omega_{ni-2,j}$.

This finally brings us to r_{max}. The appropriate condition to apply here is not completely clear. Generally, we expect there to be very little flow along this wall. Setting no-flow boundary is analogous to making this boundary a wall, with $\psi = \psi_{wall}$ and ω set from Equation 6.92. However, a more appropriate boundary may involve requiring that the flow is solely in the radial direction, hence $u = 0$ and

$$\frac{\partial\psi}{\partial r}\bigg|_{rmax} = 0 \tag{6.96}$$

which is differenced as $\psi_{i,nj-1} = \psi_{i,nj-2}$. Vorticity boundary condition is set similarly to the zmax outlet

$$\frac{\partial\omega}{\partial r}\bigg|_{rmax} = 0 \tag{6.97}$$

or $\omega_{ni-1,j} = \omega_{ni-2,j}$. Given this artificially imposed normal flow direction, we can expect some boundary effect to develop. As such, the solution in the vicinity of the boundary may be incorrect and we may want to perform mesh-dependence analysis by running the simulation with the boundary farther removed from the exit orifice.

6.4.4   IMPLEMENTATION

This solver is demonstrated in `vorticity-rz.py`. The program begins by setting some user inputs such as Δt and the mesh spacing, as well as information on the inlet flow.

```
1   import numpy as np
2   import matplotlib.pyplot as pl
3
4   #main program
5   ni = 31
6   nj = 15
7
8   dt = 2.0e-3
```

```
 9   dz = 0.0025
10   dr = 0.0025
11
12   #parameters
13   u0 = 0.1            #inlet velocity
14   nu0 = 1.568e−5     #air kinematic viscosity at 300K
15   nu_k = 1            #artifical visocisity factor
16   nu = nu_k*nu0
17   inlet_nn = 3        #number of nodes making up inlet
18   outlet_nn = 3       #number of nodes in the outlet
19   outlet_dz = 6       #number of nodes outside the cavity
20   pos_z = np.arange(0,ni)*dz
21   pos_r = np.arange(0,nj)*dr
22
23   #generate geometry
24   node_type = makeGeometry()
```

The size of the inlet and the outlet are then set by specifying the number of nodes occupied by each, respectively. These values are subsequently used by a **makeGeometry** function to flag mesh nodes as open or as belonging to one of the boundary types,

```
 1   #flags nodes as follows:
 2   OPEN = 0
 3   WALL = 1
 4   INLET = −1
 5   OUTLET= −2
 6   def makeGeometry():
 7       node_type = np.zeros((ni,nj))
 8
 9       #left wall
10       node_type[0,0:ni] = WALL
11       node_type[0,0:inlet_nn] = INLET
12
13       #top wall
14       node_type[:ni−outlet_dz,nj−1] = WALL
15
16       #right wall
17       node_type[ni−3−outlet_dz:ni−outlet_dz, outlet_nn:] = WALL
18
19       return node_type
```

We next compute some additional parameters, such as the Reynold's number, mainly to indicate to the user whether the simulation parameters are within acceptable limits.

```
 1   # screen output
 2   d = nu*dt*(1/(dz*dz)+1/(dr*dr))
 3   print("d=%g, should be <= 0.5"%d)
 4   u_max = u0*(inlet_nn/outlet_nn)  # outlet mass conservation
 5   Re_x = u_max*dz/nu
 6   cx = u0*dt/dz
 7   Re_main = u0*pos_r[nj−1]*2/nu
 8   Re_outlet = u_max*pos_r[outlet_nn−1]*2/nu
 9   print("%g <= %g <= %g"%(2*cx,Re_x,2/cx))
10   print("u0 = %g, u_max = %g"%(u0,u_max))
11   print("Re_main = %g, Re_outlet = %g"%(Re_main,Re_outlet))
12
```

```
13   # set streamfunction boundary conditions
14   psi = initPsi()
```

We also call `initPsi` to initialize the stream function values on the nodes along the wall and the axis of rotation. This function is given by

```
1    # sets boundary conditions on psi
2    def initPsi():
3        psi = np.zeros((ni,nj))
4
5        psi[:0] = 0        # streamline along axis of rotations
6
7        # volumetric flow = 2*pi*psi
8        psi_wall = 0.5*u0*(pos_r[inlet_nn]**2)
9        print("Inlet flow rate: %g"%(psi_wall))
10
11       psi[node_type==WALL] = psi_wall
12
13       return psi
```

The rest of the main program involves setting up arrays for storing vorticity and velocity components. We also initialize figures for plotting. The program then jumps into the main integration loop which consists of calling `computePsi` to evaluate ψ per Equation 6.74. Next, the velocity components are evaluated by differencing ψ per Equation 6.70. Subsequently, ω is integrated by advancing Equation 6.76 using the RK4 method. New plots are generated every 100 time steps.

```
1    # set initial values to zeros
2    w = np.zeros_like(psi)
3    u = np.zeros_like(psi)
4    v = np.zeros_like(psi)
5
6    it = 0
7    fig1=pl.figure(1,figsize=(8,4))
8    fig2=pl.figure(2,figsize=(8,4))
9    fig3=pl.figure(3,figsize=(8,4))
10   fig4=pl.figure(4,figsize=(8,4))
11
12   # iterate
13   print("Starting main loop")
14   for it in range(1001):
15       # solve psi
16       psi = computePsi(w,psi,u,v)
17
18       # update u and v
19       u,v = computeVel(psi)
20
21       # advance w
22       w = advanceRK4(w,psi,u,v)
23
24       if (it%100==0):
25           flux,u_ave = make_plot(it)
26
27   print("Done!")
```

The code for computing ψ is given below. This function implements the Jacobi algorithm to iteratively solve the linear system. Convergence is checked once every 25 steps.

```
1    # solves  d^2psi/dz^2 + d^2psi/dr^2  = -w
2    def computePsi(w, psi, u, v):
3        psi2 = np.copy(psi)
4
5        idz2 = 1/(dz*dz)
6        idr2 = 1/(dr*dr)
7        for it in range(1001):
8            psi2[1:ni-1,1:nj-1] = (w[1:ni-1,1:nj-1]*pos_r[1:nj-1] +
9                                   idz2*(psi[0:ni-2,1:nj-1]+psi[2:ni,1:
                                       nj-1])+
10                                  idr2*(psi[1:ni-1,0:nj-2]+psi[1:ni
                                       -1,2:nj])-
11                                  1/(2*dr*pos_r[1:nj-1])*(psi[1:ni-1,2:
                                       nj]-psi[1:ni-1,0:nj-2])
12                                  )/(2*(idz2+idr2))
13
14            # replace values on boundary nodes with previous values
15            psi2[node_type>0] = psi[node_type>0]
16
17            # inlet, dpsi/dz = -v = 0
18            for j in range(nj):
19                if (node_type[0,j]==INLET):
20                    psi2[0,j] = psi2[1,j]
21
22            # dpsi/dz = -v on zmax
23            for j in range(nj):
24                if (node_type[ni-1,j]!=WALL):
25                    psi2[ni-1,j] = psi2[ni-2,j] - dz*(v[ni-1,j])*pos_r[j]
26
27            # dpsi/dr = -u = 0 on rmax
28            psi2[ni-outlet_dz:,nj-1] = psi2[ni-outlet_dz:,nj-2]
29
30            # copy-down solution
31            psi = np.copy(psi2)
32
33            #check for convergence
34            if (it%25==0):
35                R = np.zeros_like(psi)
36                R[1:ni-1,1:nj-1] = (w[1:ni-1,1:nj-1]*pos_r[1:nj-1] +
37                                    idz2*(psi[0:ni-2,1:nj-1]-2*psi[1:ni-1,1:nj-1]+
                                        psi[2:ni,1:nj-1])+
38                                    idr2*(psi[1:ni-1,0:nj-2]-2*psi[1:ni-1,1:nj-1]+
                                        psi[1:ni-1,2:nj])-
39                                    1/(2*dr*pos_r[1:nj-1])*(psi[1:ni-1,2:nj]-psi
                                        [1:ni-1,0:nj-2]))
40                R[node_type>0] = 0
41                norm = np.linalg.norm(R)
42                if (norm<1e-8):
43                    return psi2
44
45        print("computePsi failed to converge, norm = %g"%norm)
46        return psi
```

Velocity is computed by the following code

```
1    # differentiates psi to get velocity
```

```
2  def computeVel(psi):
3      u = np.zeros_like(psi)
4      v = np.zeros_like(psi)
5      for i in range (ni):
6          for j in range (1,nj-1):
7              # skip over walls, otherwise differencing on neighbors
                     will be off
8              if (node_type[i,j]==WALL):
9                  continue
10
11             # v = -dpsi/dz
12             if (i==0):
13                 v[i,j] = -(psi[i+1,j]-psi[i,j])/(dz*pos_r[j])
14             elif (i==ni-1):
15                 v[i,j] = -(psi[i,j]-psi[i-1,j])/(dz*pos_r[j])
16             else:
17                 v[i,j] = -(psi[i+1,j]-psi[i-1,j])/(2*dr*pos_r[j])
18
19             # u = dpsi/dr
20             u[i,j] = (psi[i,j+1] - psi[i,j-1])/(2*dr*pos_r[j])
21
22      # u velocity on the axis from q=2*pi*psi
23      u[:,0] = 2*psi[:,1]/((pos_r[1])**2)
24
25      # similar approach to get u velocity on nj-1
26      # first compute u[:,0.5]
27      u[:,nj-1] = (psi[:,nj-1] - psi[:,nj-2])/dr
28
29      u[node_type==WALL] = 0
30      # v=0 on axis
31      v[:,0] = 0
32
33      return (u,v)
```

Next, the function for integrating ω is given by

```
1  # advances vorticity equation using RK4
2  def advanceRK4(w,psi,u,v):
3      applyVorticityBoundaries(w,psi,u,v)
4
5      # compute the four terms of RK4
6      Rk = R(w)
7      w1 = w + 0.5*dt*Rk
8
9      R1 = R(w1)
10     w2 = w + 0.5*dt*R1
11
12     R2 = R(w2)
13     w3 = w + dt*R2
14
15     R3 = R(w3)
16     w_new = w + (dt/6.0)*(Rk + 2*R1 + 2*R2 +R3)
17
18     # return new value
19     return w_new
```

with boundaries set using

```
1  def applyVorticityBoundaries(w,psi,u,v):
```

```
2     dz2  =  dz*dz
3     dr2  =  dr*dr
4     # apply boundaries
5     for i in range(ni):
6         for j in range(nj):
7             count = 0
8             ww = 0
9
10            #left wall
11            if (i<ni-1 and node_type[i,j]==WALL and node_type[i+1,j
                  ]==OPEN):
12                ww += 2*(psi[i,j]-psi[i+1,j])/(pos_r[j]*dz2) - 2*v[i,
                      j]/dz
13                count += 1
14
15            # right wall
16            if (i>0 and node_type[i,j]==WALL and node_type[i-1,j]==
                  OPEN):
17                ww += 2*(psi[i,j]-psi[i-1,j])/(pos_r[j]*dz2) + 2*v[i,
                      j]/dz
18                count += 1
19
20            # top wall
21            if (j>0 and node_type[i,j]==WALL and node_type[i,j-1]==
                  OPEN):
22                ww += 2*(psi[i,j]-psi[i,j-1])/(pos_r[j]*dr2) - 2*u[i,
                      j]/dr + u[i,j]/pos_r[j]
23                count +=1
24
25            # set values
26            if (count>0):
27                w[i,j]  = ww/count
28
29    # outlet on right side, dw/dz = 0
30    w[ni-1,:] = w[ni-2,:]
31
32    # outlet on rmax, dw/dr = 0
33    for i in range(ni-outlet_dz, ni):
34        w[i,nj-1]=w[i,nj-2]
35
36    # inlet on left side
37    for j in range(1,inlet_nn):
38        if (j<nj-1):
39            du_dr = (u[0,j+1]-u[0,j-1])/(2*dr)
40        else:
41            du_dr = (u[0,j]-u[0,j-1])/dr
42            w[0,j] = -du_dr
43
44    # this should already be set, w=0 on axis
45    w[:,0] = 0
```

The right hand side of $\partial\omega/\partial t = R(\omega)$ is computed using the following function:

```
1    # computes RHS for vorticity equation
2    def R(w):
3        dz2 = dz*dz
```

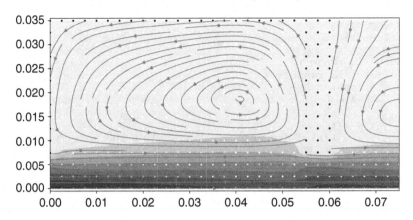

Figure 6.6 Velocity magnitude and streamlines computed with the $\omega - \psi$ method.

```
4        dr2 = dr*dr
5        # make copy so we use consistent data
6        r = np.zeros_like(w)
7        for i in range(1,ni-1):
8            for j in range(1,nj-1):
9                if (node_type[i,j]>0): continue
10
11               # viscous term, d^2w/dz^2+d^2w/dr^2+(1/r)dw/dr
12               A = nu*(
13                   (w[i-1][j]-2*w[i][j]+w[i+1][j])/dz2 +
14                   (w[i][j-1]-2*w[i][j]+w[i][j+1])/dr2 +
15                   (w[i][j+1]-w[i][j-1])/(2*dr*pos_r[j])))
16
17               # advective term u*dw/dz
18               B = u[i][j]*(w[i+1][j]-w[i-1][j])/(2*dz)
19
20               # advective term v*dw/dr
21               C = v[i][j]*(w[i][j+1]-w[i][j-1])/(2*dr)
22
23               r[i][j] = A - B - C
24       return r
```

Finally, the plotting function `makePlots` is used to generate plots. For brevity, the listing for this function is not included, but you will find it in the companion code repository. Figure 6.6 visualizes the flow field at the end of the simulation with the scatter markers used to indicate node types.

6.5 VLASOV METHODS

We now describe yet another modeling approach for gas flows that combines aspects of Eulerian methods with the kinetic treatment of the gas velocity distribution function. Newton's second law tells us that for a rigid body of constant mass, the rate of change of velocity is directly proportional to the total force,

$$\frac{d\vec{v}}{dt} = \frac{\vec{F}}{m} \qquad (6.98)$$

$\vec{F}$ is a function of position and velocity. These terms are identical for all molecules of the same material within a single VDF bin. The only way for a velocity bin to splinter is for some molecules to undergo collisions. We can write a conservation law stating that the total derivative of the velocity distribution function can change only through collisions,

$$\frac{df}{dt} = \left(\frac{\partial f}{\partial t}\right)_{col} \qquad (6.99)$$

Applying the chain rule leads to

$$\frac{\partial f}{\partial t} + \frac{\partial f}{\partial x}\frac{\partial x}{\partial t} + \frac{\partial f}{\partial y}\frac{\partial y}{\partial t} + \frac{\partial f}{\partial z}\frac{\partial z}{\partial t} + \frac{\partial f}{\partial u}\frac{\partial u}{\partial t} + \frac{\partial f}{\partial v}\frac{\partial v}{\partial t} + \frac{\partial f}{\partial w}\frac{\partial w}{\partial t} = \left(\frac{\partial f}{\partial t}\right)_{col} \qquad (6.100)$$

Realizing that $\partial x/\partial t = u$ and $\partial u/\partial t = a_x$ reduces the equation to

$$\frac{\partial f}{\partial t} + u\frac{\partial f}{\partial x} + v\frac{\partial f}{\partial y} + w\frac{\partial f}{\partial z} + a_x\frac{\partial f}{\partial u} + a_y\frac{\partial f}{\partial v} + a_z\frac{\partial f}{\partial w} = \left(\frac{\partial f}{\partial t}\right)_{col} \qquad (6.101)$$

Finally, utilizing $\vec{v} = u\hat{i} + v\hat{j} + w\hat{k}$ along with $\nabla = (\partial/\partial x)\hat{i} + (\partial/\partial y)\hat{j} + (\partial/\partial z)\hat{k}$ and $\nabla_u = (\partial/\partial u)\hat{i} + (\partial/\partial v)\hat{j} + (\partial/\partial w)\hat{k}$, the above equation can be further simplified into

$$\frac{\partial f}{\partial t} + \vec{v}\cdot\nabla f + \frac{\vec{F}}{m}\cdot\nabla_u f = \left(\frac{\partial f}{\partial t}\right)_{col} \qquad (6.102)$$

The partial differential equation in 6.102 is the fundamental relationship in gas dynamics called the *Boltzmann equation*. The kinetic methods described in the prior chapter approximate its solution using the simulation particles as velocity distribution samples. A direct integration would give us, conceptually, a way to resolve the evolution of some gas species without making the thermalization assumption inherent in CFD methods, while at the same time avoiding the noise of particle-based schemes. There is however one difficulty. Numerically resolving an arbitrary distribution function requires defining a grid in the *velocity space* in each spatial cell. This grid allows us to keep track of the values of $f(u, v, w)$ at particular (x, y, z) positions. This grid leads to enormous computational requirements. As an example, an extremely coarse gridding with just 20 unique velocity bins in each x, y, z direction leads to a $20 \times 20 \times 20 = 8000$ cell grid in each volume mesh cell. Typically a much finer resolution will be required. Since the VDF is spatially varying, this velocity grid needs to be defined for each mesh cell. Considering a relatively coarse $100 \times 100 \times 100$ computational grid, almost 30 Gb of RAM are needed to store just a single distribution. Numerical integration schemes generally require a secondary allocated memory block to temporarily store the f^{k+1} values, thus doubling the memory requirements. Of course, each material species requires its own array. Furthermore, assuming we are interested in resolving velocities in $\pm 10^6$ m/s (as needed for electrons in plasma simulations), this coarse binning provides only a 100 km/s resolution. A much finer velocity space mesh is likely to be needed, leading to an even more prohibitive memory requirement. Therefore outside the realm of massive parallel simulations, a direct solution of the Boltzmann equation for multi-dimensional flow does not appear feasible.

However, one area where Boltzmann solvers can be applied efficiently is in 1D-1V problems. Let's consider the collisonless form $(\partial f/\partial t)_{col} = 0$ of Equation 6.102 called the Vlasov equation. For the 1D-1V problem, we have the following:

$$\frac{\partial}{\partial t}f(x,u,t) + u\frac{\partial}{\partial x}f(x,u,t) - E(x,t)\frac{\partial}{\partial u}f(x,u,t) = 0 \tag{6.103}$$

Following Cheng and Knorr 1976, we further simplify the governing equation through normalization and by assuming $q = -1$ and $m = 1$. The Vlasov equation is then solved by splitting the integration into two parts

$$\frac{\partial f}{\partial t} + u\frac{\partial f}{\partial x} = 0 \tag{6.104}$$

$$\frac{\partial f}{\partial t} - E(x,t)\frac{\partial f}{\partial u} = 0 \tag{6.105}$$

Equation 6.104 above can be rewritten using forward difference as

$$\frac{f^{k+1}(x,u) - f^k(x,u)}{\Delta t} = -u\frac{\partial f^k}{\partial x} \tag{6.106}$$

which leads to

$$f^{k+1}(x,u) = f^k(x,u) - v\frac{\partial f^k}{\partial x}(x,u)\frac{\Delta t}{2} \tag{6.107}$$

It may not be obvious at first, but the term on the right hand side is actually a shift of the distribution function in the spatial direction. This arises because

$$f - \frac{\partial f}{\partial x}u\Delta t \approx f - \frac{\partial f}{\partial x}\frac{\Delta x}{\Delta t}\Delta t = f - \frac{\partial f}{\partial x}\Delta x \tag{6.108}$$

and we can write

$$f^{k+1}(x,u) = f^k(x - u\Delta t/2, u) \tag{6.109}$$

In the paper, the authors show that the electric field in Equation 6.105 needs to be evaluated at the half-time step for stability. We end up with the following three-step algorithm:

1. Integrate Equation 6.104 through a half-time step, $\Delta t/2$. Using PIC analogy, this is similar to using velocity from the previous time step to compute positions at the time (k+0.5).

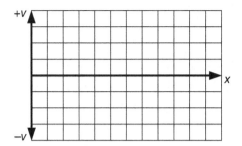

Figure 6.7 Computational grid for a 1D-1V Vlasov solver.

2. Compute the electric field using charge density at time (k+0.5) and use it to advance Equation 6.105 through a whole time step. This produces the new velocity at time (k+1).
3. Perform the second half-step integration of Equation 6.104. Using the particle analogy, this gives us the new position at the time (k+1) using the velocity at (k+1).

This integration is performed with the help of temporary arrays:

$$f^*(x, u) = f^k(x - u\Delta t/2, v) \tag{6.110}$$

$$f^{**}(x, u) = f^*(x, u + E(x, t)\Delta t) \tag{6.111}$$

$$f^{k+1}(x, u) = f^{**}(x - u\Delta t/2, v) \tag{6.112}$$

These equations are written for the normalized electrons and the force arising solely from the electric field. This field is computed using the distribution f^*, $E = E(f^*)$.

6.5.1 IMPLEMENTATION

The computational domain is a two-dimensional mesh as shown in Figure 6.7. The j component corresponds to velocity, and not a y position. Therefore, the data stored at a (i, j) grid node corresponds to the value of the distribution function at the i-th x position and the j-th v velocity. We let x go from zero to some length L. The velocity discretization needs to capture particles moving in both positive and negative directions. The limits for the j direction thus correspond to $(-v_{max})$ to $(+v_{max})$, where v_{max} is a user parameter.

The code utilizes two-dimensional arrays to store f, f^*, and f^{**}. While the integrations could proceed with just a single temporary buffer to advance the "old" state to the "new" one, we utilize all three for clarity. We also need arrays to store charge density and electric field. These are one-dimensional quantities since they vary solely with x. The allocation of these buffers is given by the code below. Here we utilize a helper function `newAndClear` that also sets the newly dynamically allocated buffer to zero.

```
1  double **f = newAndClear(ni,nj);     // f
2  double **fs = newAndClear(ni,nj);    // fs
3  double **fss = newAndClear(ni,nj);   // fss
4  double *ne = newAndClear(ni);        // number density
5  double *b = newAndClear(ni);         // Poisson solver RHS
6  double *E = newAndClear(ni);         // electric field
7  double *phi = newAndClear(ni);       // potential
```

6.5.2 INITIAL CONDITIONS

The next thing we need to do after allocating the memory is to load some initial distribution function. We assume that electrons are Maxwellian, and load a normalized variant,

$$f(u) = \frac{1}{\sqrt{\pi u_{th}^2}} \exp\left[-\frac{(u - u_d)^2}{u_{th}^2} \right] \tag{6.113}$$

As an example, let's consider the two-stream instability problem commonly studied in plasma physics. It models a periodic system consisting of two electron populations with identical uniform densities streaming through each other with similar, but opposite, speeds. Collisions are ignored, and hence interactions between the electrons arise solely from the electrostatic effects. We can load this initial state using

```
1   // set initial distribution
2   for (int i=0;i<ni;i++)
3       for (int j=0;j<nj;j++) {
4           double x = world.getX(i);
5           double u = world.getU(j);
6
7           double uth2 = 0.001;  // thermal speed, squared
8           double ud1 = 1.6;     // stream 1 drift speed
9           double ud2 = -1.3;    // stream 2 drift speed
10          double A = (1+0.02*cos(3*pi*x/world.L));
11          f[i][j] = 0.5/sqrt(uth2*pi)*exp(-(u-ud1)*(u-ud1)/uth2);
12          f[i][j] += 0.5/sqrt(uth2*pi)*exp(-(u-ud2)*(u-ud2)/uth2)*A;
13      }
```

This code iterates over all velocities j in each spatial coordinate i and for each, uses Equation 6.113 to evaluate the value of f. Two beams are loaded with $u_{d,1} = 1.6$ and $u_{d,2} = -1.3$ drift velocities, respectively, and $u_{th}^2 = 0.001$. We also introduce a small harmonic undulation to the second beam to model a tiny misbalance in number densities that would occur naturally. As will be seen shortly, this small difference in densities leads to the formation of an electric field that is responsible for providing coupling between the two populations.

6.5.3 SIMULATION MAIN LOOP

All that is left next is implementing the three-step method from Equations 6.110–6.112 in the simulation main loop:

```
1    for ( it =0; it <=2000; it ++) {
2        // compute f*
3        for (int i=0;i<ni; i++)
4            for (int j=0;j<nj; j++) {
5                double u = world.getU(j);
6                double x = world.getX(i);
7                fs[i][j] = world.interp(f,x-u*0.5*dt,u);
8            }
9
10       /* TODO: update E */
11
12       // compute f**
13       for (int i=0;i<ni; i++)
14           for(int j=0;j<nj; j++) {
15               double u = world.getU(j);
16               double x = world.getX(i);
17               fss[i][j] = world.interp(fs,x,u+E[i]*dt);
18           }
19
20       // compute f(n+1)
21       for (int i=0;i<ni; i++)
22           for(int j=0;j<nj; j++) {
23               double u = world.getU(j);
24               double x = world.getX(i);
25               f[i][j] = world.interp(fss,x-u*0.5*dt,u);
26           }
27
28       if (it%5==0)
29           saveVTK(it,world,scalars2D,scalars1D);
30   }
```

6.5.4 ELECTRIC FIELD

The above code is not complete, since it does not include the self-consistent electric field computation. As we know from the discussion in the section on the PIC method, in electrostatic plasma simulations, we obtain the electric field from $E_x = -\partial\phi/\partial x$. Plasma potential is computed from Poisson's equation, which for our normalized system reads

$$\frac{\partial^2 \phi}{\partial x^2} = n_e - 1 \qquad (6.114)$$

Clearly, we need to compute the electron number density at each spatial grid node. Number density is simply the number of molecules divided by the cell volume. Since the VDF gives us the number of molecules with some particular velocity, we can get the total particle count by integrating the VDF over all velocities,

$$N = \int_{-\infty}^{\infty} f(u)du \qquad (6.115)$$

We assume that the cell volume is unity and hence this integral gives us the number density. This integration is performed in the code with the trapezoid scheme,

```
1   //number density from integration of f with the trapezoidal rule
2   for (int i=0;i<ni;i++) {
3       ne[i] = 0;
4       for (int j=0;j<nj-1;j++)
5           ne[i]+=0.5*(fs[i][j+1]+fs[i][j])*du;
6   }
```

We then use any 1D Poisson solver. This example utilizes a solver based on the Gauss-Seidel scheme. Periodic systems such as this can also be solved using the Fourier transform.

6.5.5 INTERPOLATION

The above code used an `interp` function to evaluate $f(x,u)$ at arbitrary points that do not necessarily correspond to grid nodes. The example code uses the linear area-weighted scheme. While this method is simple to implement, it leads to unacceptable numerical diffusion to be used in production Vlasov codes. The VDF typically decays rapidly from the peak, and linear weighing over-estimates the number of moving particles. Production Vlasov codes instead employ higher-order interpolation methods. This interpolation function is defined as a method in a `World` object. It supports both periodic and open boundaries in x,

```
1   // linear interpolation - higher order method needed
2   double interp(double **f, double x, double u) {
3       double fi = (x-0)/dx;
4       double fj = (u-(-u_max))/dv;
5
6       // periodic boundaries in i
7       if (periodic) {
8           if (fi<0) fi+=ni-1;     //-0.5 becomes ni-1.5
9           if (fi>ni-1) fi-=ni-1;
10      }
11      else if (fi<0 || fi>=ni-1) return 0;
12
13      // return zero if velocity less or more than limits
14      if (fj<0 || fj>=nj-1) return 0;
15
16      int i = (int)fi;
17      int j = (int)fj;
18      double di = fi-i;
19      double dj = fj-j;
20
21      double val = (1-di)*(1-dj)*f[i][j];
22      if (i<ni-1) val+=(di)*(1-dj)*f[i+1][j];
23      if (j<nj-1) val+=(1-di)*(dj)*f[i][j+1];
24      if (i<ni-1 && j<nj-1) val+=(di)*(dj)*f[i+1][j+1];
25      return val;
26  }
```

6.5.6 VISUALIZATION

We also make our world periodic by setting `world.periodic=true;` and we also add a call to `world.applyBC()` after each VDF field is updated. This

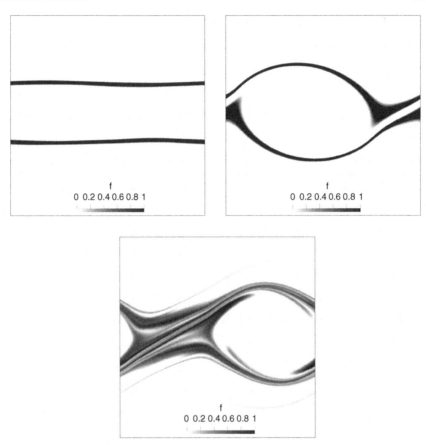

Figure 6.8 Evolution of velocity distribution function in a two-stream instability example.

function simply makes the values on the left and right edges identical. The code was run with time step $\Delta t = 1/8$ and increased 401×201 mesh resolution. Example results are shown in Figure 6.8. This demo simulates two electron beams interacting with each other in a domain containing neutralizing ion background. The numerical diffusion introduced by the linear interpolation operator is quite apparent if you run the code longer. The vortex eventually closes up due to numerical diffusion.

6.6 OTHER MODEL EQUATIONS

We now briefly introduce solution methods to several other PDEs commonly encountered in engineering applications. For a more detailed discussion, the reader is encouraged to review references such as Anderson et al. 2021, Ferziger and Perić 2002, and Jackson 1992.

6.6.1   FIRST-ORDER WAVE EQUATION

Consider the following first order equation

$$\frac{\partial u}{\partial t} + c\frac{\partial u}{\partial x} = 0 \tag{6.116}$$

which can alternatively be written in the index notation as

$$u_t + cu_x = 0 \tag{6.117}$$

where c is constant. A commonly used method to solve Equation 6.116 involves using finite-difference combined with the Lax method. The Lax scheme is explicit, first order in time.

$$U_i^{n+1} = \frac{U_{i+1}^n + U_{i-1}^n}{2} - c\frac{\Delta t}{\Delta x}\frac{U_{i+1}^n - U_{i-1}^n}{2} \tag{6.118}$$

Applying Taylor expansion around U_j^n

$$U_j^{n+1} = U_j^n + U_t\Delta t + U_{tt}\frac{\Delta t^2}{2!} + U_{ttt}\frac{\Delta t^3}{3!} + \dots$$

$$U_{j+1}^{n+1} = U_j^n + U_x\Delta x + U_{xx}\frac{\Delta x^2}{2!} + U_{xxx}\frac{\Delta x^3}{3!} + \dots$$

$$U_{j-1}^{n+1} = U_j^n - U_x\Delta x + U_{xx}\frac{\Delta x^2}{2!} - U_{xxx}\frac{\Delta x^3}{3!} + \dots$$

and substituting into Equation 6.118, we obtain the modified equation

$$U_t + cU_x = \frac{1}{2}U_{xx}(\frac{\Delta x^2}{\Delta t} - c^2\Delta t) + \frac{1}{3}U_{xx}(\Delta x^2 - c^2\Delta t^2) + \dots = O\left(\Delta t, \frac{\Delta x^2}{\Delta t}\right) \tag{6.119}$$

These equations utilize subscripts to indicate derivatives, $U_{xx} \equiv \partial^2 U/\partial x^2$.

The Lax scheme is not uniformly consistent. The truncation error of the Lax scheme is $O(\Delta t, \Delta x^2/\Delta t)$, and the stability condition is given by

$$\cos^2\beta + \gamma^2\sin^2\beta \le 1 \to \gamma \le 1 \tag{6.120}$$

where $\gamma = c\Delta x/\Delta t$, $\beta = k_m\Delta x$, $k_m = m\pi/L$, $m = 0, 1, 2\dots$. This leads to $\gamma \le 1$.

6.6.2   DIFFUSION EQUATION

Consider the diffusion equation of form

$$U_t = \nu U_{xx} \tag{6.121}$$

Common methods to solve this method utilize finite-difference using the Dufort-Frankel algorithm or the Crank-Nicolson algorithm. The Dufort-Frankel scheme is given by

$$\frac{U_i^{n+1} - U_i^{n-1}}{2\Delta t} = \alpha \frac{U_{i+1}^n - U_i^{n+1} - U_i^{n-1} + U_{i-1}^n}{\Delta x^2} \tag{6.122}$$

$$U_i^{n+1}(1 + 2r) = U_i^{n-1} + 2r(U_{i+1} - U_i^{n-1} + U_{i-1}^n) \tag{6.123}$$

where $r = \alpha \Delta t / \Delta x^2$. The Dufort-Frankel method is explicit, the truncation error is $O(\Delta t^2, \Delta x^2, (\Delta t/\Delta x)^2)$, and the scheme is stable for $r \geq 0$.

6.6.3 BURGER'S EQUATION

The inviscid Burger's equation is given by

$$U_t + F_x = 0 \tag{6.124}$$

while the viscous Burger's equation is given by

$$U_t + F_x = \nu U_{xx} \tag{6.125}$$

The Burger's equation may be solved using finite-difference with the Upwind algorithm or the Lax-Wendroff algorithm. For the inviscid Burger's equation, Equation 6.124, the Upwind method is given by

$$U_j^{n+1} - U_j^n = -\frac{\Delta t}{\Delta x} U_j^n (U_{j+1}^n - U_j^n), \quad U_j^n \leq 0 \tag{6.126}$$

$$= -\frac{\Delta t}{\Delta x} U_j^n (U_j^n - U_{j-1}^n), \quad U_j^n > 0$$

For the viscous Burger's equation, Equation 6.124, one may use the central difference approximation for the viscous term. Thus, the Upwind method for the viscous Burger's equation with $U_j^n \leq 0$ is given by

$$U_j^{n+1} - U_j^n = -\frac{\Delta t}{\Delta x} U_j^n (U_{j+1}^n - U_j^n) + \frac{\nu \Delta t^2}{\Delta x^2}(U_{j+1}^n - 2U_j^n + U_{j-1}^n) \tag{6.127}$$

For $U_j^n > 0$ we have

$$U_j^{n+1} - U_j^n = -\frac{\Delta t}{\Delta x} U_j^n (U_j^n - U_{j-1}^n) + \frac{\nu \Delta t^2}{\Delta x^2}(U_{j+1}^n - 2U_j^n + U_{j-1}^n) \tag{6.128}$$

The Lax-Wendroff method for the inviscid Burger's equation is given by

$$U_j^{n+1} - U_j^n = -\frac{\Delta t}{2\Delta x}(F_{j+1}^n - F_{j-1}^n) \tag{6.129}$$

$$+ \frac{\Delta t^2}{2\Delta x^2}\left[\frac{U_j^n + U_{j+1}^n}{2}(F_{j+1}^n - F_j^n) - \frac{U_{j-1}^n + U_j^n}{2}(F_j^n - F_{j-1}^n)\right]$$

6.6.4 MAXWELL'S EQUATIONS

The governing equations of electromagnetics are known as the Maxwell's equations, see Section 5.6.2. The propagation of electromagnetic wave can be solved from Faraday's law, Equations 5.66 and Ampere's law, Equation 5.67, once the two Gauss' laws, Equations 5.64 and 5.65, are imposed as the initial conditions. Equation 5.64 remains true by virtue of the charge conservation condition

$$\frac{\partial \rho}{\partial t} + \nabla \cdot \vec{J} = 0 \tag{6.130}$$

Many methods have been developed to solve the Maxwell's equation in the field of computational electromagnetics. The numerical schemes generally fall into the following two categories: the spectral method and the finite-different or finite-element time domain methods. In this section, we present the finite-difference time-domain (FDTD) of Yee 1966. Yee's FDTD method updates the E field and B field from Faraday's and Ampere's laws:

$$\frac{\partial E_x}{\partial t} = \left(\frac{\partial B_z}{\partial y} - \frac{\partial B_y}{\partial z} \right) c - J_x \tag{6.131}$$

$$\frac{\partial E_y}{\partial t} = \left(\frac{\partial B_x}{\partial z} - \frac{\partial B_z}{\partial x} \right) c - J_y$$

$$\frac{\partial E_z}{\partial t} = \left(\frac{\partial B_y}{\partial x} - \frac{\partial B_x}{\partial y} \right) c - J_z$$

$$\frac{\partial B_x}{\partial t} = - \left(\frac{\partial E_z}{\partial y} - \frac{\partial E_y}{\partial z} \right) c \tag{6.132}$$

$$\frac{\partial B_y}{\partial t} = - \left(\frac{\partial E_x}{\partial z} - \frac{\partial E_z}{\partial x} \right) c$$

$$\frac{\partial B_z}{\partial t} = - \left(\frac{\partial E_y}{\partial x} - \frac{\partial E_x}{\partial y} \right) c$$

In order to achieve second-order accuracy in space, the Yee method uses a staggered grid system, the Yee lattice. The B field components are defined at the center of the cell face, while the E field components are defined at the center of the cell edge. This staggered arrangement ensures central differencing in space for Equations 6.131 and 6.132. In order to achieve second-order accuracy in time, Equations 6.131 and 6.132 are updated using the leap-frog scheme, where the E field components are defined at the full time step, and the B field and current components are defined at half-time step. This leap-frog scheme ensures center differencing in time for Equations 6.131 and 6.132.

We use i, j, k to denote the location of the full grid point in the x, y, and z direction, respectively, and $i + 1/2$, $j + 1/2$, and $k + 1/2$ to denote the location at the half grid point in the x, y, and z direction, respectively. We use superscript n to denote the full time step, and $n - 1/2$ and $n + 1/2$ to denote the half-time step. Thus, the finite difference representations of Faraday and Ampere's laws using the Yee scheme are given by

$$
\frac{B_x^{n+1/2}(i,j+1/2,k+1/2) - B_x^{n-1/2}(i,j+1/2,k+1/2)}{c\Delta t} = \tag{6.133}
$$
$$
\frac{E_y^n(i,j+1/2,k+1) - E_y^n(i,j+1/2,k)}{\Delta z}
$$
$$
-\frac{E_z^n(i,j+1,k+1/2) - E_z^n(i,j,k+1/2)}{\Delta y}
$$

$$
\frac{E_x^n(i+1/2,j,k) - E_x^{n-1}(i+1/2,j,k)}{c\Delta t} = \tag{6.134}
$$
$$
-\frac{B_y^{n-1/2}(i+1/2,j,k+1/2) - B_y^{n-1/2}(i+1/2,j,k-1/2)}{\Delta z}
$$
$$
+\frac{B_z^{n-1/2}(i+1/2,j+1/2,k) - B_z^{n-1/2}(i+1/2,j-1/2,k)}{\Delta y}
$$
$$
-\frac{J_x^{n-1/2}(i+1/2,j,k)}{c}
$$

The stability requirement for the FDTD scheme is given by the Courant condition

$$
c\Delta t \leq \frac{1}{\sqrt{1/\Delta x^2 + 1/\Delta y^2 + 1/\Delta z^2}} \tag{6.135}
$$

To ensure the accuracy, the mesh size also needs to be much less than the wavelength.

6.7 FINITE VOLUME METHOD

The integration methods discussed so far have utilized the Finite Difference Method. FDM is just one of several discretization schemes employed in numerical analysis. The already familiar FDM is relatively simple but is applicable only to orthogonal meshes with cell edges aligned with the coordinate directions. You may have heard of the Finite Element Method (FEM). FEM is commonly used in mechanical analysis due to its ability to resolved complex geometries represented by arbitrarily shaped elements. The downside of FEM method is its complex mathematical formulation, see Hughes 2000 for implementation details.

Yet another method, the Finite Volume Method or FVM, offers a middle ground between Finite Difference and Finite Element. It can be applied to non-Cartesian, non-orthogonal meshes, akin to FEM, but with complexity not greatly exceeding that of FDM. As will be apparent shortly, this method is computed by considering fluxes across cell boundaries. This inherent conservation principle makes it naturally attractive to fluid dynamics simulations:

material flux leaving through one cell's face matches the flux entering the neighbor cell.

Let's return to our model advection-diffusion equation, 6.9. Instead of using Taylor Series to approximate the derivatives, we integrate the governing equation over some small control volume V,

$$\int_V \frac{\partial \psi}{\partial t} dV + \int_V \nabla \cdot (\vec{v}\psi) \, dV = \int_V \nabla \cdot (D\nabla \psi) \, dV + \int_V R dV \qquad (6.136)$$

The size of the control volume is given by the mesh resolution. We need to decide whether the solver should be cell- or node-centered. In the first case, the mesh cell becomes the control volume. For the second case, we construct a control volume centered on the node, with boundary vertices defined by the centroids of the neighboring cells. In both cases we assume that the property associated with the control volume is defined at its centroid. For cell-centered codes, this means the cell centroid, while for node-centered codes, the value is assumed to be known at the mesh nodes.

The time derivative can be pulled out of the first integral. We also assume that ψ is constant in the entire control volume, allowing us to write

$$\int_V \frac{\partial \psi}{\partial t} dV = \frac{\partial}{\partial t} \int_V \psi dV = \frac{\partial}{\partial t}(\psi \Delta V) \qquad (6.137)$$

where ΔV is the volume of the control region. We can similarly assume that the source term is constant over the control volume,

$$\int_V R dV = R\Delta V \qquad (6.138)$$

The Divergence Theorem is one of the several generalizations of the fundamental theorem of calculus to vector calculus. It tells us that the integral of divergence of a vector field over some volume is equal to the integral of flux through the enclosing surface,

$$\int_V \nabla \cdot \vec{f} dV = \oint_S \vec{f} \cdot \hat{n} dA \qquad (6.139)$$

Equation 6.139 allows us to rewrite the 2nd and 3rd in Equation 6.136 as surface integrals,

$$\int_V \nabla \cdot (\vec{v}\psi) \, dV = \oint_S (\vec{v}\psi) \cdot \hat{n} dA \qquad (6.140)$$

and

$$\int_V \nabla \cdot (D\nabla \psi) \, dV = \oint_S D\nabla \psi \cdot \hat{n} dA \qquad (6.141)$$

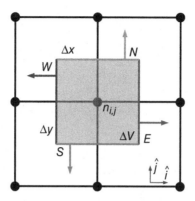

Figure 6.9 Control volume boundary with normal vector directions indicated.

6.7.1 SURFACE INTEGRAL: ADVECTION

With these substitutions, Equation 6.136 becomes

$$\frac{\partial}{\partial t}(\psi \Delta V) + \oint_s (\vec{v}\psi) \cdot \hat{n} dA = \oint_s (D\nabla\psi) \cdot \hat{n} dA + R\Delta V \qquad (6.142)$$

We need some way to evaluate the surface integral terms. We do this by considering all faces of the enclosing surface. This is where the actual geometry of the control volume makes appearance. For a two-dimensional problem, we assume unit depth and threat the four edges as the boundary. The surface integral is the rewritten as a sum over these four faces,

$$\oint_S (\vec{v}\psi) \cdot \hat{n} dA = \sum_{m=1}^{4} [(\vec{v}\psi) \cdot (\hat{n}\Delta A)]_m \qquad (6.143)$$

These faces are visualized in Figure 6.9. To simplify notation, we label the "faces" (edges) as E, N, W, and S for East, North, West, and South, respectively. Note the orientation of normal vectors $\hat{n}$. The vectors are oriented to point away from the control volume centroid. Using this orientation, we can now write the expressions for each of the four terms.

On the East face, we have $\hat{n} = \hat{i}$ and $\Delta A = \Delta y$. Thus

$$\left[(\vec{v}\psi) \cdot (\hat{n}\Delta A)\right]_E = (u\hat{i} + v\hat{j})_E \cdot \hat{i}(\psi \Delta A)_E$$
$$= (un)_E \Delta y \qquad (6.144)$$

Here $(un)_E$ is the value of normal flux across the East face. We similarly obtain the following expressions for the 3 remaining faces:

$$\left[(\vec{v}\psi) \cdot (\hat{n}\Delta A)\right]_N = (v\psi)_N \Delta x \tag{6.145}$$

$$\left[(\vec{v}\psi) \cdot (\hat{n}\Delta A)\right]_W = -(u\psi)_W \Delta y \tag{6.146}$$

$$\left[(\vec{v}\psi) \cdot (\hat{n}\Delta A)\right]_S = -(v\psi)_S \Delta x \tag{6.147}$$

Combining the terms leads to

$$\oint_S (\vec{v}\psi) \cdot \hat{n} dA = \left[(u\psi)_E - (u\psi)_W\right] \Delta y + \left[(v\psi)_N - (v\psi)_S\right] \Delta x \tag{6.148}$$

All that remains is evaluating the fluxes. We again assume that this expression is constant along the edge. Consider a control volume centered on node (i, j). Using the edge centroid position,

$$(u\psi)_E = u_{i+0.5,j}\psi_{i+0.5,j} \tag{6.149}$$

As we do not have a node here, we we approximate this value using linear averaging,

$$(u\psi)_E = \frac{u_{i+1,j}\psi_{i+1,j} + u_{i,j}\psi_{i,j}}{2} \tag{6.150}$$

With similar substitutions for the remaining terms, we obtain

$$\left[\oint_S (\vec{v}\psi) \cdot \hat{n} dA\right]_{i,j} = \left[(u\psi)_E - (u\psi)_W\right] \Delta y + \left[(v\psi)_N - (v\psi)_S\right] \Delta x$$

$$= \frac{u_{i+1,j}\psi_{i+1,j} - u_{i-1,j}\psi_{i-1,j}}{2}\Delta y + \frac{u_{i,j+1}\psi_{i,j+1} - u_{i,j-1}\psi_{i,j-1}}{2}\Delta x \tag{6.151}$$

6.7.2 SURFACE INTEGRAL: DIFFUSION

We can similarly rewrite the diffusion term,

$$D \oint_S \nabla\psi \cdot \hat{n} dA = D\left[(\nabla\psi)_E - (\nabla\psi)_W\right] \Delta y + D\left[(\nabla\psi)_N - (\nabla\psi)_S\right] \Delta x \tag{6.152}$$

While functionally similar to Equation 6.151, this expression contains gradient terms. In general, gradients can be evaluated for arbitrary surface shapes using the Stokes theorem. However, in the case of rectangular cells aligned with the coordinate directions, we can just utilize the already familiar Finite Difference,

$$(\nabla\psi)_{i,j,E} = \frac{\psi_{i+1,j} - \psi_{i,j}}{\Delta x} \tag{6.153}$$

or

$$D \oint_S \nabla\psi \cdot \hat{n} dA = D\left[\frac{\psi_{i+1,j} - 2\psi_{i,j} + \psi_{i-1,j}}{\Delta x}\right] \Delta y$$

$$+ D\left[\frac{\psi_{i,j+1} - 2\psi_{i,j} + \psi_{i,j-1}}{\Delta y}\right] \Delta x \tag{6.154}$$

6.7.3 COMBINING TERMS

Substituting these expressions into Equation 6.142 and using $\Delta V = \Delta x \Delta y$, we obtain

$$\frac{\partial \psi}{\partial t} + \frac{u_{i+1,j}\psi_{i+1,j} - u_{i-1,j}\psi_{i-1,j}}{2\Delta x} + \frac{u_{i,j+1}\psi_{i,j+1} - u_{i,j-1}\psi_{i,j-1}}{2\Delta y} =$$
$$D\left[\frac{\psi_{i+1,j} - 2\psi_{i,j} + \psi_{i-1,j}}{\Delta^2 x} + \frac{\psi_{i,j+1} - 2\psi_{i,j} + \psi_{i,j-1}}{\Delta^2 y}\right] + R \quad (6.155)$$

This is the same discretization as obtained with the Finite Difference. This is the advantage of the FVM method. It is applicable to cells of arbitrary shapes, but in the case of rectangular cells aligned with the coordinates, it reduces to the familiar Finite Difference equations.

6.7.4 AXISYMMETRIC FORMULATION

To demonstrate the use of FVM with non-Cartesian meshes, let's consider the Poisson's equation (steady-state diffusion equation with a non-zero source) in axisymmetric coordinates,

$$\nabla^2 \psi = S \quad (6.156)$$

Utilizing the definition of a Laplacian in cylindrical coordinate,

$$\nabla_r^2 \equiv \frac{1}{r}\frac{\partial}{\partial r} + \frac{\partial^2}{\partial r^2} + \frac{\partial^2}{\partial \theta^2} + \frac{\partial^2}{\partial z^2} \quad (6.157)$$

and using the axisymmetric approximation, $(\partial \psi / \partial \theta = 0)$, Equation 6.156 becomes

$$\frac{\partial^2 \psi}{\partial \theta^2} + \frac{1}{r}\frac{\partial \psi}{\partial r} + \frac{\partial^2 \psi}{\partial z^2} = S \quad (6.158)$$

where S is some source term. This equation can then be discretized using FDM as

$$\frac{\psi_{i,j-1} - 2\psi_{i,j} + \psi_{i,j+1}}{\Delta^2 r} + \frac{1}{r_{i,j}}\frac{\psi_{i,j+1} - \psi_{i,j-1}}{\Delta r} + \frac{\psi_{i-1,j} - 2\psi_{i,j} + \psi_{i+1,j}}{\Delta^2 z} = S_{i,j}$$
$$(6.159)$$

Let's now consider the Finite Volume approach. Just as was done in the prior section, we integrate the governing equation over a control volume and apply the divergence theorem to obtain

$$\oint_S (\nabla \psi) \cdot \hat{n} dA = \int_V S dV \quad (6.160)$$

which is again rewritten as a sum over the faces

$$\sum_{m=1}^{4} [(\nabla \psi) \cdot \hat{n} dA]_m = S \Delta V \quad (6.161)$$

Figure 6.10 Cell volume for an axisymmetric formulation.

In the axisymmetric model, each cell represents a rectangular toroid centered on the axis of rotation as visualized in Figure 6.10. This leads to two important observations. First, the cell volume increases with the radial distance r. Secondly, while the east and west faces have the same area, the area of the north face exceeds the southern one. Let the radial distance at the cell centroid be given by $r_{i,j}$. The cell size in the radial and axial dimensions is given by Δr and Δz, respectively. We then have

$$\Delta V = 2\pi r \Delta r \Delta z \tag{6.162}$$

and

$$(\Delta A \hat{n})_E = 2\pi r \Delta r \hat{e}_z \tag{6.163}$$
$$(\Delta A)\hat{n}_N = 2\pi (r + 0.5\Delta r)\Delta z \hat{e}_r \tag{6.164}$$
$$(\Delta A \hat{n})_W = 2\pi r \Delta r (-\hat{e}_z) \tag{6.165}$$
$$(\Delta A \hat{n})_S = 2\pi (r - 0.5\Delta r)\Delta z (-\hat{e}_r) \tag{6.166}$$

Substituting these definitions into Equation 6.161, we obtain

$$(\nabla\psi)_E \cdot \hat{e}_z 2\pi r \Delta r + (\nabla\psi)_N \cdot \hat{e}_r 2\pi (r + 0.5\Delta r)\Delta z -$$
$$(\nabla\psi)_W \cdot \hat{e}_z 2\pi r \Delta r - (\nabla\psi)_S \cdot \hat{e}_r 2\pi (r - 0.5\Delta r)\Delta z = S 2\pi \Delta z \Delta r \tag{6.167}$$

The gradient has the same form in cylindrical coordinates as in Cartesian and hence we have $(\nabla\psi)_E = \partial\psi/\partial z$. This term is approximated on the east face as

$$(\nabla\psi)_E \approx \frac{\psi_{i+1,j} - \psi_{i,j}}{\Delta z} \tag{6.168}$$

On the north face we similarly obtain

$$(\nabla\psi)_N \approx \frac{\psi_{i,j+1} - \psi_{i,j}}{\Delta r} \tag{6.169}$$

With these definitions, Equation 6.167 becomes

$$\frac{\psi_{i,j-1} - 2\psi_{i,j} + \psi_{i,j+1}}{\Delta^2 r} + \frac{1}{r_{i,j}} \frac{\psi_{i,j+1} - \psi_{i,j-1}}{\Delta r} + \frac{\psi_{i-1,j} - 2\psi_{i,j} + \psi_{i+1,j}}{\Delta^2 z} = S_{i,j}$$

(6.170)

which is the same relationship obtained using the Finite Difference Method in Equation 6.159, but without needing to explicitly define the Laplacian in cylindrical coordinates. For further reading on finite volume methods, particularly with regards to stability and accurately handling hyperbolic systems, readers are referred to Leveque 2002 and the associated implementations in the CLAWPACK software package.

7 Interactive Applications

This chapter introduces approaches for developing interactive simulation codes that run in web browsers. Such codes can be of use on systems that lack a dedicated compiler and for prototyping solvers that benefit from user interaction. We cover HTML, style sheets, Javascript, and 3D rendering with WebGL.

7.1 HTML

It may seem curious to find a chapter on web development in a book focusing on scientific computing. As you will see shortly, HTML and Javascript can be extremely useful tools for numerical analysis. First, the very nature of websites makes it simple to add interactivity and graphical output to your application. Second, just about every computer has a web browser and a text editor installed. These two tools are all that is needed to develop code even on systems that do not have any programming language installed. Such may be the case with lab computers used for data acquisition. Since web pages are just text files, they also simplify code sharing with colleagues who may not be proficient (or have access to) compilers.

If you have never developed one, you may be surprised to learn that websites are essentially just a collection of text documents written in a format known as *Hypertext Markup Language* or *HTML*. HTML, and its generalization *XML*, uses *tags* to describe a nested tree structure. The tree starts off with a single root *element* which branches off into multiple children. Each element is indicated by an opening `<tag>`. It is closed with a corresponding `</tag>`. The element can contain multiple nested child *nodes* which can be additional elements or can be just textual string. Each element can also be assigned attributes, such as

```
1  <!-- some element with attributes -->
2  <tag1 attr1="attribute value" attr2="another attribute">
3     <arg>child element</arg>
4  </tag1>
5  <tag2>another element</tag2>
```

Two special attributes `id` and `class` are used to specify the element's identification name and a list of classes it belongs to. We will learn more about classes once we discuss styling. Comments are included within `<!- ->`.

Elements without any child nodes can be closed as part of the opening tag as in `<tag ... />`. The need to close tags is one area where XML differs from HTML. XML has a strict requirement that elements be properly nested. HTML, likely due to an abundance of non-conforming legacy web pages, is less demanding. While the code below will cause an error if saved as an `.xml` document, it will load successfully if interpreted as `.html`

```
1  <tag1>
```

DOI: 10.1201/9781003132233-7

Figure 7.1 Visual representation of our first HTML code from page 259.

```
2  <tag2 param1="abc">  <!-- tag2 is not closed -->
3  </tag1>
```

We need to add **</tag2>** or replace line 2 with **<tag2 param1="abc"/>** to make the file XML compliant.

In the examples above, we utilized tags such as **<tag1>**, **<tag2>**, and **<arg>**. These are not real tags you would find in an HTML document, and are used just to demonstrate the tree structure. The HTML standard defines a collection of tags, and their desired purpose, that are understood by all compliant web browsers. Consider the following simple web page with leading indentation added for emphasis:

Listing 7.1 A very basic HTML document
```
1   <!DOCTYPE html>
2   <html>
3       <head>
4           <meta charset="UTF-8"></meta>
5           <title>Getting Started</title>
6       </head>
7       <body>
8           <h1>Hello from HTML</h1>
9       </body>
10  </html>
```

When you visit any website, the *web server* returns a document like this, although with more content, to the web browser. The browser then uses predefined style rules to display the elements. Using Mozilla Firefox, the listing above may display as shown in Figure 7.1. You can confirm this by saving the code into a file with an .html extension and double clicking on it. As you can see, we don't need a web server, or even Internet connectivity, to display a web page from the local file system.

The code starts with a **<!DOCTYPE html>** declaration. It signals that the document utilizes newer tags introduced in the early 2000s revision to the HTML standard known as *HTML5*. XML documents instead start with

```
<?xml version="1.0" encoding="UTF-8"?>
```

The next tag **<html>** (or **<xml>** for XML) acts as the root element, from which all other tags inherit. It contains two child elements: **<head>** and **<body>**. The former serves as a container for any "peripheral" envelope information. Properties such as the document character encoding given by `<meta charset />`, or the page title `<title>`...`</title>` go here. The character encoding controls how characters are displayed. Any file saved on a disc is essentially just a sequence of bytes, which themselves are just numbers. The legacy ASCII encoding used a single byte to represent the different letters of the alphabet. Since a byte can hold only values from 0 to 255, this convention allowed the computer to represent only 256 unique characters. This was sufficient for letters of the standard Roman alphabet, along with special characters for encoding line breaks, tabs, or spaces. But it is not capable of supporting additional international alphabets, special characters, and more recently, emojis. This is where Unicode came in. It is an extension of ASCII that allows for a much wider number of characters in order to support internalization. UTF-8 is one of the most common ways to encode these additional characters. It stores the common Latin characters using just a single 8-bit byte, just like ASCII, while allowing for inclusion of addition multi-byte characters. It is ubiquitous in web development. The `<head>` is also where we place social-media specific information that are used to display "cards" on social media sites. A brief description that may be shown in search engine results also goes here.

The `<body>` section contains the actual content to be displayed in the browser window. Here we use the `<h1>` tag to display a top "level 1" heading. Headings help search engine page crawlers to understand the page layout. By default, headings are printed using a larger font size, and we use it here primarily for this effect. Headings can be defined down to the `<h6>` level.

7.1.1 COMMON ELEMENTS

There are about a hundred different HTML elements, with websites such as `w3schools.org` and `developer.mozilla.org` providing a good overview of their usage. You don't need to learn them all. Some of the most common tags include:

- `<p>` delineates a text paragraph. This tag is generally used to add white space between blocks of text. A line break can also be added using `<br>`.
- `<b>`, `<i>`, and `<u>` add bold, italics, and underline formatting. Note that this is just the default behavior, and it is possible to redefine the applied style.
- `<a>` adds a clickable link. The destination is specified in an `href` attribute, while the displayed text is included as a child text node. Thus `Examples available <a href="https://example.com">here</a>` displays "Examples available here", and clicking on "here" will load the specified web page.
- `<h1>` through `<h6>` add headings, as already discussed.
- `<img>` inserts an image from the path given by `src` attribute. Image width and height can be set using `width` and `height` attributes, with

size given in pixels. As an example,

`<img src="img/case1.png" width="400px" height="300px" />`
loads an image file `case1.png` found in a `results` sub-directory of
the location where the currently displayed `.html` document is located.
This is known as a relative path. We can also specify the absolute
path by starting the source URL with `http://`, `https://` or `file://`
prefix. While this allows us to load images from other people's websites, doing so is a bad practice and should be avoided. First, each
request results in data having to be sent by the remote webserver
which may impose data transfer costs on the host owner. Secondly,
we have no control over the image. If the image is changed on the
host computer, the modified image will be shown on our page. It is
possible to configure the web server to prevent this direct linking,
but this topic is beyond the scope of this text. The displayed image
is scaled to 400 × 300 pixels.

- `<ol>` and `<ul>` add an ordered (numbered) and unordered (bulleted)
  list. The actual list items are specified using `<li>`. For example:

```
1  Common variable types:
2  <ol>
3  <li><b>int</b> for whole numbers</li>
4  <li><b>float</b> for real numbers </li>
5  <li><b>string</b> for storing text</li>
6  </ol>
```

displays the following:

```
Common variable types:
1. int for whole numbers
2. float for real numbers
3. string for storing text
```

- `<div>` and `<span>` are used to group elements into a single block with
  some specified visual style and page position. The primary difference
  between the two is that by default a `<div>` is displayed as a *block*
  with a line break inserted after the closing `</div>` tag, A `<span>` is
  displayed as an *inline* element with additional elements continuing
  on the same line.

- `<form>` groups various `<input>` elements such as text fields, buttons,
  check boxes, radio buttons, and drop-down menus. When a button
  with attribute `type="submit"` is clicked, all data is transmitted to
  an URL specified in the form's `action` attribute. For example,

```
1  <form action="next.php">
2  Your name: <input name="name" /> <br>
3  Affiliation: <br>
4  <input type="radio" name="affi" value="st">Student<br>
5  <input type="radio" name="affi" value="pd">Post-Doc<br>
6  <input type="radio" name="affi" value="prof">Professor<br>
7  <input type="submit" value="Next">
8  </form>
```

produces the output shown in Figure 7.2. Clicking "Next" sends the

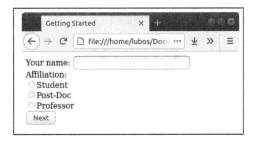

Figure 7.2 Example of using forms to collect user input.

collected input to a web page file `next.php` located in the same directory as the document currently being displayed. This file is given a `.php` extension. PHP is a common programming language used on servers to process data and generate dynamic web pages. It is discussed briefly in Section 7.1.4. In Section 7.3.1 we will see how to process form input locally using callbacks.

7.1.2 STYLES

Right now, all elements are display using their default *style*, but these can be modified. HTML is similar to LaTeX (discussed in Section 8.9) in that there is a clear separation between the *content* and the *visual representation*. Styling can be specified directly for each element through the `style` attribute. As this would quickly lead to a lack of consistency, it is recommended to specify common visual style to be applied to all elements of a given type (such as `<p>`), or to elements sharing the same *class*. Properties specific to a single element can be specified using that element's *id*. These styling rules can be specified in the `<head>` of the document within a `<style>` element, or they can be imported through an external *cascading style sheet* (or .css) files. The name is supposed to indicate that style applied to a top-level element "cascades" down, or is inherited, by all child elements.

Styles listed in a .css document or in the `<style>` tag are listed with the following syntax:

```
selector { property1:value; property2:values; ....}
```

Here is an example:

```
1  <!DOCTYPE html>
2  <html>
3  <head>
4      <meta charset="UTF-8">
5      <style>
6      h1 { color: white;  text-align: center;
7           font-size: 5em; text-shadow: 0px 0px 10px #422;
8           border: 5px dashed black;  border-radius: 50px;
9           margin: 10px 20px; padding: 30px;}
10     </style>
```

Figure 7.3 The visual representation is changed completely by specifying element styles.

```
11  </head>
12  <body>
13     <h1>Hello from HTML</h1>
14  </body>
15  </html>
```

The produced output is shown in Figure 7.3. Note that there is no change to the `<body>` section from Listing 7.1 yet the output looks much different from Figure 7.1.

Below is a summary of the most common style properties. A more expanded list can be found online at sites such as `www.w3schools.com/cssref`.

- `color` specifies the color of the element. We can use predefined names such as "white", "red" or "black", hexadecimal RGB (red-green-blue) color codes as in `#RRGGBB` where RR, GG, and BB is a value from 00 to FF (00 to 255), or `rgba(R, G, B,A)`. For example, `#220000` is a dark red color, while `#FF00FF` is a brilliant purple. It is also possible to specify the color using a shorter `#RGB` syntax, in which case the values are duplicated. `#422` is identical to `#442222` and produces a gray color with a reddish tone that we used to create the glow around the lettering in Figure 7.3. `rgba(255, 180, 0, 0.8)` syntax specifies an orange color with an 80% opacity. `background-color` specifies the background color.

- `border` specifies border width, type (solid, dashed, etc.), and color, with parameters following `border-width border-style border-color`, such `border: 5px dashed green`. Curved borders can be specified using `border-radius`, with parameters consisting of one to four values, indicating the radius to be used on all corners (if a single value) up to specifying each corner individually.

- `text-shadow` and `box-shadow` allow us to add a shadow to text or the element border. The syntax is `text-shadow: horizontal-shift vertical-shift [blur-radius] color`. In the above example, we create a "glow" around the letters by specifying zero shift, but a non-zero blur radius.

- `margin` and `padding` specify the amount of white space on the outside and inside the element. We can specify a single value which is then applied to all four faces, two values for the horizontal and vertical offset, or four values to set each side.

- **text-align** and **vertical-align** are used to control horizontal and vertical alignment. Common options include **left**, **right**, **center**, **top**, **bottom**, and **middle**.
- **font-size** changes the font size. Size can be specified in pixels (px), percent of normal font size (em), or cm or in. Additional styling is possible using **font-family** to change the font type , and **font-shadow** to include shadows. Web fonts that would typically not be installed on the user's system can be included via the **@import** tag

```
1  @import url('https://fonts.googleapis.com/css?family=
2            Sigmar+One&display=swap');
3  h1 { font-family: 'Sigmar One', cursive;}
```

- **display** controls how the element is displayed. Options include **inline**, **block**, **inline-block**, and **flex**. The inline-block option is particularly popular as it can be used to stack fixed-width **<div>**s next to each other. We specify block dimensions using **width** and **height**. The more recently introduced **flex** type simplifies development of *responsive websites* that display satisfactorily both on a large monitor and a small cell phone screen and can grow and wrap according to given rules.

7.1.3 CLASSES AND IDS

In the above example, we specified properties for the **h1** element. This styles all such elements. We may however be interested in styling some specific headings, or items in general, differently. HTML elements can be assigned one or more *classes* and a unique *id*. As an example, the div below is assigned two classes "banner"and "overlay" and has an id "ad01":

<div class="banner overlay" **id=**"ad01"> ... **</div>**

We use . and # to limit style selectors only to that class (dot) or id (hash). Dependence can also be specified, so that we can target an element belonging to a parent element of a given type, class name, or id. For example:

```
1  div, img { /* applies to any div or img */ }
2  div.banner { /* only applies to <div>s of class "banner" */ }
3  .banner { /* applies to any element of class "banner" */ }
4  #ad01 { /* applies only to the element with id=ad01 */ }
5  div.banner p { /* only <p>s within a <div> of class "banner" */ }
```

Note that in CSS, we utilize C-style comments.

7.1.4 DYNAMIC WEB PAGES

The listings introduced so far were all examples of *static* web pages. The same content is delivered to every visitor. We can include *dynamic* behavior through *server-side* and *client-side* scripting. Server-side scripting involves the use of code running on the web server to dynamically modify the HTML document delivered to the web browser. The web visitor never sees the server code. One popular server-side scripting language is PHP. With PHP, we use the `echo` command to output information, with the output destination being the HTML file. Below is an example of PHP being used to dynamically modify screen output

```
1  <h1><?php
2  if (date("w")==0) echo "We are closed"; else
3     echo "Welcome";
4  ?></h1>
```

Note the addition of PHP code within `<?php` and `?>` delimiters. If the page is visited on a Sunday (day of the week with code 0), the page outputs a notification that the business is closed. The client web browser receives either

```
<h1>We are closed</h1>
```

or

```
<h1>Welcome</h1>
```

with no indication that the heading text was created dynamically. Another common use for PHP is to check for authentication and to display information specific to the logged-in user, or to update a database with user-specified information, as in the `<form>` example from the prior section. This works only if you have a PHP interpreter installed and properly configured.

Given this brief crash course in PHP, you may be relieved to hear that we are actually not going to be discussing PHP anymore. This chapter focuses only on *client-side scripting*. This term refers to code that is included in the HTML file available to the client. The code is compiled on the fly and executed by the web browser (as opposed to the web server). The de facto standard scripting language is *JavaScript*, often abbreviated to JS. What makes client-side scripting so useful is that all processing happens on the client side (meaning within the computer or a smartphone). The web server is not involved at all, and as such, we can use it to write simulation codes that run without any online connectivity.

7.2 JAVASCRIPT

Javascript code is included within a `<script>` tag:

```
1  <html>
2  <head><meta charset="UTF-8"></head>
3  <body>
4  <p>I am a paragraph</p>
5  <script>
6  alert("Hello from Javascript!");
7  </script>
8  </body>
9  </html>
```

Upon loading, the page will display a pop up message box. Such alerts used be common in the early days of the Internet, but are now rarely used.

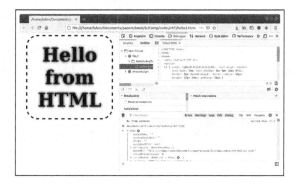

Figure 7.4 The console, along with a source code debugger, are accessible through the Web Developer Tools.

7.2.1 DEVELOPER TOOLS

We can also write to the *console* using

```
console.log("Hello console!");
```
The console is visible through the browser Web Developer Tools. This is a side of a web browser that you may not even be aware to exists. It can be accessed by pressing F12 in Firefox, or Ctrl+Shift+I in Chrome. Alternatively, you can just right-click in the browser window, and select an "Inspect" option. Besides displaying the console, the Developer Tools toolbar also contains a style editor that you can use to dynamically adjust element styles and a source code debugger that can be used to step through a JS code line by line. Besides being useful for outputting messages, the console is also a JavaScript interpreter. It is particularly useful for exploring element properties, as demonstrated in Figure 7.4.

7.2.2 SYNTAX

Javascript utilizes syntax that will be familiar from C. However, it is an *untyped language* and variables can hold data of arbitrary types including arrays of dictionary-like objects. Variables should be declared using **var**, **let**, or **const** keywords, but this is not strictly enforced. Variables scope is the current block and its sub-blocks. The interpreter searches for a variable declaration anywhere in the given block and such the two codes produce a different result:

```
1  var a = 1;
2  function f(){
3      a = 2;
4  }
5  f(); console.log(a);    // outputs 2
```
and

```
1  var a = 1;
```

```
2   function f(){
3       a = 2;
4       var a;
5   }
6   f(); console.log(a);      // outputs 1
```

In the second example, the interpreter treats a = 2 as an assignment to a local variable despite the var a; statement appearing after the assignment. In practice, this can lead to unexpected code behavior changes if a function that is expected to be modifying a global variable is accidentally made to operate on a local one by introducing a var in a later part of the function code.

Due to weak typing, we can get unexpected results. For example, it is possible to add strings and numbers,

```
1   var x = 2;          // declare variable
2   y = "hello" + x;    // y = hello2
```

Javascript introduces a === (triple-equals) comparison that ensures that both the values and types agree. For instance,

```
1   "1"==true    // true
2   "1"===true   // false
```

The first statement is true since the string "1" value can be converted to an integer 1, and any non-zero integer is considered to be true when cast to a Boolean. The second statement evaluates to false since a string is not of a Boolean type. JS also introduces a **typeof** operator that can be used to check the variable type. It is often used to check for undefined or non-existent data, such as

```
1   if (typeof(el)=="undefined") {
2     /* do something */
3   }
```

The example below summarizes the use of functions, for loops, and conditional statements. This code outputs integers 0 through 9, but skips over the specified item.

```
1   <body>
2   <script>
3   function foo(skip) {      // declare function
4       var msg = "";         // empty string
5       for (var i=0;i<10;i++) { if (i!=skip) msg += i; }
6       console.log(msg);
7       return 1;    // example of a return value
8   }
9
10  foo(7);    // call the function
11  </script>
12  </body>
```

7.2.3 MATHEMATICS

Javascript contains several built-in libraries. One of these, Math implements standard mathematical functions such as cos and exp. It also gives us an ability to sample random numbers using `Math.random()` and to truncate real values to the preceding integer using `Math.floor()`). This is also where constants such π are defined. To see the full listing, simply open the console and type in

```
>> Math
```

and then click the "down arrow" to expand the listing.

Another curiosity of Javascript is that all entries, even numerals, are essentially objects containing member functions. For instance, we can use `toFixed` or `toExponential` to convert a number to a string with an appropriate formatting:

```
>> 123.876.toFixed(1)
"123.9"
>>123.876.toExponential(2)
"1.24e+2"
```

7.2.4 ARRAYS AND OBJECTS

Besides strings, numbers, or Booleans, Javascript variables can hold *arrays* and *objects*. Arrays are collectors accessed using a numeric index, while JS objects are analogous to Python's dictionaries, and provide access using named keys. An empty array and an empty object are declared as

```
1  var arr = [];
2  var obj = {};
```

We subsequently add items into the array using

```
arr[4] = 2;
```

This command automatically resizes the array to hold 5 elements (JS indexing starts at 0), with the first 4 elements being set to `undefined`. Data can also be appended to the end of the array using `push`, as in

```
arr.push("ok");
```

As this example illustrates, it is perfectly acceptable for a JS array to hold a mix of numbers, strings, or any other data types. The number of entries is obtained from

```
1  arr.length;
```

Object data is assigned using the dot notation:

```
1  obj.x = 1.2;
2  obj.y = 2.0;
```

The object keys can be iterated using a "for-in" loop,

```
1  for (key in obj) {
2      console.log(key+" = "+obj[key]);
3  }
```

The above code will print

```
x = 1.2
y = 2
```

As illustrated object data can be accessed using the array selector brackets when the key name is stored in a variable.

7.2.5 OBJECT-ORIENTED PROGRAMMING

Object-oriented programming has also made its way into Javascript. Here is an example of a simple class used to encapsulate data:

```
1   class Point {
2       x = 0;                    // member variables
3       y = 0;
4       constructor(x,y) { // constructor
5           this.x = x;
6           this.y = y;
7       }
8       dist() {                  // class method
9           return Math.sqrt(this.x*this.x+this.y*this.y);
10      }
11  }
12
13  var p1 = new Point(1.0,3.0);     // instantiate
14  console.log("Distance of P1 from origin is: "+p1.dist());
```

Object Oriented Programming is a relatively new addition to JS, and as such, you may encounter legacy code that accomplishes encapsulation using functions:

```
1   var Point = function(x,y) {
2       Point.prototype.init = function(x,y) { // constructor
3           this.x = x;
4           this.y = y;
5           console.log(x,y);
6       }
7       Point.prototype.dist = function() {          // class method
8           return Math.sqrt(this.x*this.x+this.y*this.y);
9       }
10      this.init(x,y);  // code
11  }
12  var p1 = new Point(1.0,3.0);     // instantiate
```

7.2.6 ACCESSING HTML ELEMENTS

The most common use of Javascript is to dynamically modify HTML elements. This is done using a global pre-defined variable called **document**. This object stores all information about the current web page. It contains many functions, including **getElementById**, which, as expected, returns the reference to the element of the given id

```
1   <body>
2   <h1 id="heading">Hello World</h1>
3   <script>
4   var e = document.getElementById("heading");  // e is a reference
        to the element
5   e.innerHTML = "*** "+e.innerHTML+" ***";
6   </script>
```

Figure 7.5 Dynamically modified text with a dynamically added element.

7   </body>

Here we dynamically add asterisks to the header message. The `innerHTML` property corresponds to the child text node of the requested element. In this example, it is initially set to `Hello World`, which we modify to read ***** Hello World *****. Another option for accessing elements is using `document.getElementsByTagName`. This function returns an array of all elements sharing the specified tag, such as "h1".

Using Javascript, it is also possible to dynamically create elements. Consider the following code. It dynamically adds an image below the heading:

```
1  // create a new element
2  var img = document.createElement("img");
3  img.src = "strawberry.jpg"
4  img.width = 200;
5  e.parentElement.insertBefore(img,e.nextElementSibling);
```

For this to work, an image file called **strawberry.jpg** needs to reside in the same folder as the displayed html document. New elements are created using `document.createElement`. After we finish setting the appropriate attribute, we add it using `parent.insertBefore(new_element, reference)` where `reference` indicates the insertion point. The `parent` object is the element from which `reference` descends, which in this case is <body>. Now, note that we want to add the image *after* the heading, but are not using an "insertAfter" function. This is because there actually is no such function. Instead, to insert after a reference element, we need to insert before that element's next sibling, as demonstrated in the code. In the end, we obtain the result in Figure 7.5.

7.2.7   ANIMATION

Animation can be included using `setInterval`, `setTimeout` or `request-AnimationFrame` functions. The first version calls a specified *callback* function continuously at a fixed interval. It is useful when we need to accomplish some action at a given frequency. The `setTimeout` function is similar, but the callback is called just once. In the listing below, the `<h1>` element is initially empty. We use Javascript to add text to it, one letter at a time to produce a typewriter-like effect.

```
1   <h1 id=" ele "></h1>
2   <script>
3   var e = document.getElementById(" ele ");
4   var i = 0;
5   function draw() {
6       var string = "Hello World";
7       if (i<string.length) {  // if i not yet at the end
8           e.innerHTML += string[i];  // add i-th letter
9           i++;                       // increment i
10      }
11  };
12  var interval = setInterval(draw, 200); // call draw every 200 ms
13  </script>
```

In order to actually draw an animation, we would like to refresh the screen as quickly as possible. The actual frequency will depend on the available computational resources, and the overall workload on the operating system. This is where `requestAnimationFrame` comes in. It is a member function of a predefined `window` object, that, similarly to `document`, stores information about the browser window. This function executes the callback the next time the window is ready to be repainted. The call happens only once, and therefore, the drawing function needs to add another request in order to continue animating. To illustrate the concept, we use the `style` property of the element reference variable `e` to randomly change the text color:

```
1   e = document.getElementById(" ele ");
2   function changeColor() {
3       //the right side creates a string such as "rgb(123,220,102)"
4       e.style.color="rgb("+Math.floor(Math.random()*255)+","+
5                            Math.floor(Math.random()*255)+","+
6                            Math.floor(Math.random()*255)+")";
7       window.requestAnimationFrame(changeColor);  // keep animating
8   }
9   changeColor();  // initial call
```

Assigning

```
e.style.color = "red";
```
is identical to

```
e.style.color = "rgb(255,0,0)";
```
which is identical to

```
<style> #ele {color:rgb(255,0,0);} </style>
```
. In the example code, we use `Math.random` and `Math.floor` to construct a random color tuple. The colors change on every window redraw. Notice that we initiate the animation with a direct call to the `changeColor` function. Just before the function ends, it makes another call to `requestAnimationFrame` requesting to be executed again. This sequence will continue forever - or until the window is closed.

7.3   CANVAS

Probably the most exciting element introduced in the HTML5 revision was
<canvas>. It provides a rectangular area into which we can paint 2D or 3D
scenes. Let's take a look at the following example:

```
1  <!DOCTYPE html>
2  <html>
3  <head>
4  <meta charset="UTF-8">
5  <style>
6  canvas {border: 1px solid black;}
7  </style>
8  </head>
9  <body>
10 <canvas width=600 height=500 id="canv">Oops!</canvas>
11
12 <script>
13 var c = document.getElementById("canv");
14 var ctx = c.getContext("2d");
15 ctx.beginPath();
16 ctx.moveTo(20,20);
17 ctx.lineTo(400,300);
18 ctx.lineTo(300,500);
19 ctx.closePath();    // return to starting point
20 ctx.fillStyle = "gray";
21 ctx.fill();          // fill the path
22
23 ctx.beginPath();
24 ctx.arc(150,240,50,50*Math.PI/180, 1.8*Math.PI);
25 ctx.fillStyle = "white";
26 ctx.strokeStyle = "#000000";   // black
27 ctx.lineWidth = 5;
28 ctx.setLineDash([15, 10]);   // 15 pixels on, 10 off
29 ctx.fill();
30 ctx.stroke();
31 </script>
32 </body>
33 </html>
```

We start by using getElementById to get the canvas element and store
its handle in a variable called c. We then obtain canvas' 2D *context* using
c.getContext("2d"). The context is the object that actually does the drawing. Later in this chapter, we will learn about another context type that
does the drawing using a graphics card (GPU) that is often used to produce
real-time 3D scenes. But for now, we stay focused on 2D drawing. It is generally performed by defining *paths*. Each path is started using beginPath.
We then use moveTo to specify the starting point. The origin is located in
the top-left corner and hence positive y moves the position toward the bottom edge. It is possible to use scale(1,-1) to flip the scene vertically, although we don't do so in these examples. From the current point, we create a
straight line to another point using lineTo. If we wanted a curved path, we
could use bezierCurveTo. The closePath command returns to the starting
point. The entire path is then filled usind fill() using the color specified in

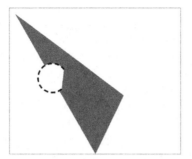

Figure 7.6 Javascript-based drawing using the `<canvas>` element.

fillStyle. We could also use `stroke()` to show the outline. The stroke and fill style can also be controlled using `fillStyle`, `strokeStyle`, `lineWidth`, and `setLineDash`. This is demonstrated in the second object which adds a fractional circle. The produced output is visualized in Figure 7.6.

Canvas drawing can be combined with `requestAnimationFrame` to generate animated output. Let's take a look at another example continuously drawing randomly-positioned, randomly-sized circles, as shown in Figure 7.7. Instead of being filled with a solid color, we use `createRadialGradient` to vary the color of each from a bright red in the center to pure white slightly beyond the circle boundary. This extension of the gradient limit ensures that the edge of the each circle can still be seen on the white background. We also wrap the `<canvas>` in a `<div>` and set the background color on a `<body>` to create a "card".

```
1  <head>
2  <style>
3      body {background-color:  white;}
4      h1 {margin:0.2em}
5      div.cdiv {
6          display:inline-block;
7          text-align:center;
8          border-radius:20px;
9          box-shadow: 0px 0px 10px #464;
10         padding: 1em 2em;
11         margin: 10px 50px;
12         background-color:#eee;
13     }
14     canvas {
15         border: 2px solid black;
16         background-color:white;
17     }
18 </style>
19 </head>
```

The document body is shown below.

```
1  <body>
2  <div class="cdiv">
3  <h1>Canvas Example</h1>
```

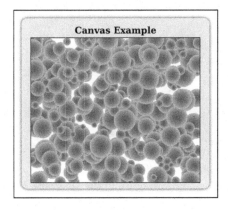

Figure 7.7 Output from the example code on page 273.

```
4   <canvas width=600 height=500 id="canv"> </canvas>
5   </div>
6
7   <script>
8   var c = document.getElementById("canv");
9   var ctx = c.getContext("2d");
10
11  function draw() {  // draws a randomly positioned circle
12      var x = Math.random()*c.width;
13      var y = Math.random()*c.height;
14      var r = 10+Math.random()*30;
15
16      // create gradient
17      var grd = ctx.createRadialGradient(x, y, 0, x, y, 1.3*r);
18      grd.addColorStop(0, "red");
19      grd.addColorStop(1, "white");
20
21      // draw a random circle
22      ctx.beginPath();
23      ctx.arc(x,y,r, 0, 2*Math.PI);
24      ctx.fillStyle = grd;
25      ctx.fill();
26
27      // keep animating
28      window.requestAnimationFrame(draw);
29  }
30
31  draw();      // initial call
32  </script>
33  </body>
```

7.3.1 USER INTERACTION

The above animation starts as soon as the page loads. It may be desired to wait for a signal from the user such as the press of a button. A button is added using HTML `<input type="button">` element.

```
1   <div class="cdiv">
2   <h1>Canvas Example</h1>
3   <canvas width=600 height=500 id="canv"> </canvas>
4   <br>
5   <input type="button" id="toggle" value="Start" onclick=toggle(
        this) />
6   </div>
```

We style it using CSS:

```
1   #toggle { font−size:2em; color:blue; }
2   #toggle.running {color:red}
```

The second declaration turns the text red if the button is assigned class "running". In the case of conflict, selectors specific to an element id have precedence over those specified for class or tag types. Therefore, we specify the red color using both the element id and the class to force the color change.

Of importance is the `onclick` attribute. This code adds an *event listener* that fires whenever the button is clicked. There are many such event handlers, allowing us to detect events such as a value being entered into a text field, the user tabbing out to the next element, the mouse hovering over an element, or the window being resized. The callback function body is specified in the `<script>` section

```
1   var running = false;
2   function toggle() {
3       running = !running;  //toggle boolean
4       var button = document.getElementById("toggle");
5       if (running) {
6           button.value="Stop";
7           draw();
8       }
9       else { button.value="Start"; }
10      button.classList.toggle("running"); // add or remove class
11  }
```

The primary purpose of this function is to toggle a Boolean global variable called **running**. Depending on the current state, we also modify the text shown by the button. When the animation is running, the button reads "Stop" to make it obvious that clicking it again will stop the animation. This can be seen in Figure 7.8. In the first *true* clause of the *if* statement, we also make a call to the **draw** function to start off the animation. This call replaces the initial call on line 31 from the prior version given on page 7.3.

We also modify the code in **draw()** to request a new animation frame only if the **running** flag is true,

```
1   function draw() {
2       ...
3
4       // keep animating
5       if (running)
6           window.requestAnimationFrame(draw);
7   }
```

This single modification is all that is needed to stop the animation once the "Stop" button is pressed.

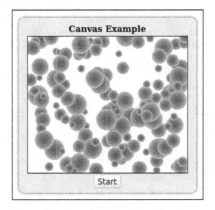

Figure 7.8 Second version adds a button to toggle the animation.

7.3.2   KEYBOARD AND MOUSE INTERACTION

Event listeners can also be added directly through Javascript. The syntax is

```
1  document.addEventListener(event_name,callback);
```

A particularly useful event is "keydown" which fires whenever a key is pressed on the keyboard. There are also **keyup** and **keypress** events, with the last one returning the ASCII character code for the pressed key instead of a button code returned by the others. Event callback functions get called with a single parameter containing information about the event. We simply ignored it in the previous example, since we had only one input element connected to a **onClick** event. However, in this case, we need to make use of this argument to determine exactly which key was pressed. Here is an example

```
1  function handleKeyPress(event) {
2      console.log("You pressed a key with code "+event.keyCode);
3  }
4  document.addEventListener("keydown",handleKeyPress);
```

Opening the console under Developer Tools lets us see output such as

```
You pressed a key with code 13
You pressed a key with code 32
You pressed a key with code 16
```

In our code, we would like to use the 'c' button to clear the canvas. We will also use the left and right arrow keys to change the color of the spheres. Utilizing the console output, we can determine that the code for 'c' is 67, while the left and right arrows are indicated by 37 and 39.

Callbacks are often added via anonymous lambda functions. Whether to use regular functions or lambdas is a personal choice; however, many programmers prefer the anonymous route as it directly ties the code with the callback without introducing any possibility of the function being called from somewhere else. Lambda functions also help us avoid needing to come up with arbitrary names such as `handleClick1`, `handleClick2`, and `handleClick3` if multiple handlers for the same event type are needed. They are defined using

```
var f = function(args) { /* body */ };
```

Here `f` is a reference to the function, which could be used to execute the function code `f(arg)`. Instead of storing the function reference in a variable, we can just instantiate as part of the call to `addEventListener`,

```
1  document.addEventListener("keydown", function(event) {
2      ...
3  });
```

You will often also encounter syntax as

```
1  var f = (function A(e) {
2          var my_data = e;
3          return function B() {
4              ...
5              console.log(my_data);
6          }
7      })(e);
8  f();    // calls B();
```

This parentheses-enclosed expression is called a *closure*. It is a way to store private variables, such as `my_data` with a function. Data encapsulation can be accomplished in a more elegant fashion using `classes`, and is often not even needed, and as such, we won't be using closures in our examples. Below is an example implementation of the keyboard listener:

```
1   var color_index = 0;      // global variable
2   document.addEventListener("keydown", function(event) {
3       if (event.keyCode == 67) {  // C key
4           ctx.fillStyle = "white"; // clear canvas by painting a
                    white rectangle
5           ctx.fillRect(0,0,c.width,c.height);
6       }
7       else if(event.keyCode==39)  { // right arrow key
8           color_index+=1;if(color_index>2) color_index=0;}
9       else if(event.keyCode==37)  { // left arrow key
10          color_index-=1;if(color_index<0) color_index=2;}
11  });
```

The listener responds to a press of the 'c' key by painting a white rectangle over the entire canvas to clear the previously drawn circles. It also listens to the left and right arrow keys for which it cycles a `color_index` variable through values [0,1,2]. The color index can be used in conjunction with an array of color code names:

```
1   var sphere_colors = ["red","green","blue"];
2
3   function draw() {
4       ...
5       grd.addColorStop(0, sphere_colors[color_index]);
6       grd.addColorStop(1, "white");
```

The arrow keys thus change the color of painted spheres.

We can also listen to mouse clicks by adding a `mousedown` event to the element of interest, canvas in this case. The basic syntax is:

```
1  //add mouse button listener
2  c.addEventListener("mousedown",function(event) {
3     const rect = c.getBoundingClientRect();
4     const x = event.clientX - rect.left;
5     const y = event.clientY - rect.top;
6     console.log(x,y);
7  });
```

The **event** object passed to the listener includes properties `clientX` and `clientY` containing the coordinates (in pixels) of the mouse click. Another property of interest is `button` with numeric code 0, 1, or 2 indicating which button is pressed. A simple way to explore the available attributes is to output the object to the console using `console.log(event)` and then use the drop-down view as illustrated in Figure 7.4.

Despite adding the listener to the canvas element, the click coordinates are given respective to the top left corner of the browser window. More useful are the coordinates with respect to the canvas element. These offsets can be computed by subtracting the left and top positions of the element bounding rectangle, which is retrieved using `getBoundingClientRect`. These coordinates can then be passed to the `draw` function to use in lieu of sampling random position by replacing the `console.log` line with a `draw(x,y)`. Since these values are passed in, the computation of circle origin in the `draw` function is commented out:

```
1  function draw(x,y) {  // draws a randomly positioned circle
2     // var x = Math.random()*c.width;
3     // var y = Math.random()*c.height;
4     ...
5
6     // if (running) window.requestAnimationFrame(draw);
7  }
```

The call to `requestAnimationFrame` is also commented out since we only want to draw a circle in response to a mouse click. We also add a note to indicate the new features using:

```
1  <style>
2     ...
3     div.note {color:gray;font-size:0.8em;}
4  </style>
5  ...
6  <input type="button" id="toggle" value="Start" onclick=toggle()/>
7  <div class="note">Press 'C' to clear, left and right arrows to
         change color</div>
```

Figure 7.9 shows the possible output we get with the new code.

7.3.3 ADDING DYNAMICS

The code currently simply paints a new circle at the position of the mouse click. But what if we wanted the "balls" to move so that we could develop a simple molecular dynamics simulation? We can add this dynamism by first integrating equations of motion through a small time step Δt. The canvas is then cleared, and the balls are painted at their new positions. In order to do

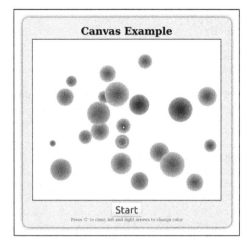

Figure 7.9 Canvas example with keyboard and mouse click listener.

this, we need to keep track of the added balls. Properties needed to define a single ball can be encapsulated into a custom `Ball` data type,

```
1   class Ball {
2       constructor(x,y,u,v,r) {
3           this.x = x;  this.y = y;
4           this.u = u;  this.v = v;
5           this.r = r;
6       }
7   }
```

We also modify the mouse click listener to append a new Ball object to an initially empty array:

```
1    var balls = [];  // empty array
2    c.addEventListener("mousedown",function(event) { // click
         listener
3        const rect = c.getBoundingClientRect()
4        const x = event.clientX - rect.left
5        const y = event.clientY - rect.top
6        const u = -20+Math.random()*40;      // random x velocity
7        const v = -50+Math.random()*100;     // random y velocity
8        const r = 10+Math.random()*30;       // random radius
9        balls.push(new Ball(x,y,u,v,r));     // append new object
10   });
```

The ball is initialized at the mouse click location and is given random initial $u \in [-20 : 20)$, $v \in [-50 : 50)$ velocity. We also sample random radius, $r \in [10 : 40)$. Next, we add a function to advance positions,

```
1    var gx = 0;
2    var gy = 9.81;
3
4    function advance() {
5        const dt = 0.1;
6        for (var i = 0;i<balls.length;i++) {
```

```
7          var ball = balls[i];
8          ball.u += gx*dt;
9          ball.v += gy*dt;
10         ball.x += ball.u*dt;
11         ball.y += ball.v*dt;
12
13         if (ball.y>c.height) {  // bottom wall impact
14             ball.y = 2*c.height-ball.y;
15             ball.v = -ball.v;
16         }
17
18         //remove out of domain objects
19         if (ball.x<-ball.r || ball.x>c.width+ball.r) {
20             balls.splice(i,1);   // remove 1 element at position i
21         }
22      }
23  }
```

Since the initial velocity was not rewound, this integration code implements the Forward Euler method. It is sufficient to demonstrate the JS programming concepts, but in a real numerical tool, you would want to implement the Leapfrog or another higher-order method. Of interest may be the if-block on line 13. Since vertical positions increase in the top to bottom direction, the y coordinate of the bottom edge is canvas.height. We let this boundary act as a perfectly elastic specular wall. As we have seen in Chapter 1, this wall impact can be simulated by mirroring the normal component of position and reversing the normal component of velocity. The left and right walls are assumed to be open and we remove any objects leaving through them. We wait for the entirety of the ball to be out of domain, implying that $x \geq x_{right} + r$, before removing it. The removal is accomplished by calling array.splice(index,number) function which removes the specified number of items starting at the given index. balls.splice(i,1) removes the single object at the current position i.

This function is called from draw on each window repaint. The draw function is updated as follows:

```
1   function draw() {
2       advance();  // update positions
3
4       // clear canvas;
5       ctx.fillStyle = "white";
6       ctx.fillRect(0,0,600,500);
7
8       for (var i = 0;i<balls.length;i++) {
9           var ball = balls[i];
10          var grd = ctx.createRadialGradient(ball.x, ball.y, 0, ball.x, ball.y, 1.3*ball.r);
11
12          grd.addColorStop(0, "red");
13          grd.addColorStop(1, "white");
14          ctx.beginPath();
15          ctx.fillStyle = grd;
16          ctx.arc(ball.x,ball.y,ball.r, 0, 2*Math.PI);
```

```
17        ctx . fill ();
18      }
19      if (running) window.requestAnimationFrame(draw);
20 }
```

The main difference from the prior version is that we begin by clearing the canvas. We then loop through all balls. Similarly to what was done in **advance**, we start by grabbing a reference to the *i*-th entry to eliminate having to write **balls[i]** every time we need to access object properties. The rest of the code is identical to the prior version. Finally, we update the keyboard listener to clear the object array on the 'C' keypress, and let the arrow keys update the *x* and *y* components of acceleration:

```
1 document.addEventListener("keydown", function(event) {
2      if (event.keyCode == 67) {   // C key
3          balls = [];      // clear array
4      }
5      else if (event.keyCode==38) gy -= 0.2;   // up
6      else if (event.keyCode==40) gy += 0.2;   // down
7      else if (event.keyCode==37) gx -= 0.2;   // left
8      else if (event.keyCode==39) gx += 0.2;   // right
9 });
```

Finally we dynamically "click" the Start button to start simulation on page load. While it is possible to dispatch a click event using **button.dispatchEvent(new Event("click"))**, we can just call the **toggle** function directly:

```
toggle(document.getElementById("toggle"));
```

7.3.4   COLORMAP

As the final step, we can use a *colormap* to map a scalar value (such as speed, $\sqrt{u^2 + v^2}$) to colors. A basic *blue-to-red rainbow* mapping for $s \in [0,1]$ can be achieved using the following code:

```
1 function valToRGB(val) {
2      if (val<0) val = 0;   // clamp to [0,1]
3      if (val>1) val = 1;
4      var a = (1-val)*4.0;      // invert and scale to [0,4]
5      var i = Math.floor(a);   //integer part
6      var f = Math.floor(255*(a-i));   //fractional part as [0,255]
            integer
7      switch(i) {
8          case 0:  r=255; g=f; b=0; break;
9          case 1:  r=255-f; g=255; b=0; break;
10         case 2:  r=0; g=255; b=f; break;
11         case 3:  r=0; g=255-f; b=255; break;
12         case 4:  r=0; g=0; b=255; break;
13     }
14     return "rgb("+r+","+g+","+b+")";
15 }
```

The function returns an **rgb** string of the appropriate color for the given scalar value. For instance, for **valToRGB(0.6)**, we obtain **rgb(102,255,0)**. The color sequence, given by the 5 cases, actually maps from red to blue. The

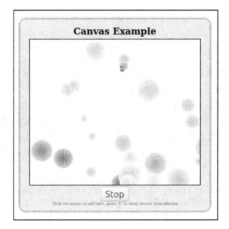

Figure 7.10 Final version of the "bouncy balls" example implementing event listeners, animations, and visualization using a colormap.

provided scalar value is inverted on line 4 to achieve the desired blue-to-red variation. We add the hook to this function by replacing the fixed color stop value in **draw** with

```
1  speed = Math.sqrt(ball.u*ball.u+ball.v*ball.v);
2  grd.addColorStop(0, valToRGB(speed/100));
```

And that's it. You will find the complete code in **canvas4.html**. A screenshot is also plotted in Figure 7.10.

7.4 FILE ACCESS

Let's say that you are working on a lab computer running a data acquisition system that continuously writes out collected telemetry to a text file. The computer does not have any compilers or interpreters installed, but you would like to periodically visualize the collected data and export a processed output to a file. Using some of the techniques discussed already, you may decide to put together a simple Javascript code to perform this data reduction. There is only one problem: the code needs access to the local file system to read the input data and subsequently save the reduced output.

While browser-based development can greatly simplify developing interactive codes, it has a major shortcoming when it comes to file I/O. Due to security reasons, the access to the local file system is greatly restricted. After all, you don't want some arbitrary websites to snoop around the files on your computer. But not all is lost. There are actually three ways to access files in modern HTML5 documents. Unfortunately, all but one requires user intervention and thus are not well suited for automated processes.

7.4.1 XMLHTTPREQUEST

You surely have experienced a website featuring a "Load More" button that, after clicking, shows additional articles or blog comments without reloading the rest of the page. This is accomplished using an XMLHttpRequest object, which implements a mechanism for Javascript to load data from a specified path. This communication paradigm is dubbed "Asynchronous JavaScript and XML", or AJAX for short. In a typical website, the path would specify a server-side script which would parse provided arguments, and return appropriate data. Let's say that the folder containing your .html file contains a signal.csv file containing data such as

```
t,signal
0, 3.19808
0.0030015, 2.62349
0.006003, 3.27875
0.0090045, 3.13399
...
```

It used to be possible to "hijack" this mechanism to open local files, as long as they were located in the same, or descendant, directory as the source .html document. However, this no longer seems to be the case. Running the following code

```
1  var xhttp = new XMLHttpRequest();
2  xhttp.onreadystatechange = function() { // call on status change
3      if (this.readyState == 4 && this.status == 200) { // if data
           received and OK
4          // do something with the data
5          document.getElementById("log").innerHTML = xhttp.
               responseText;
6      }
7  };
8  xhttp.open("GET", "./signal.csv", true);  // request file
9  xhttp.send();
```

will most likely result in an error message such as

```
Cross-Origin Request Blocked: The Same Origin Policy disallows
reading the remote resource at file:///codes/ch7/signal.csv.
(Reason: CORS request not http).
```

7.4.2 EXTERNAL JAVASCRIPT FILES

Therefore, we need to utilize other alternatives. One possibility is to modify the input file to convert it to a Javascript multiline string.

```
1  const data = 't,signal
2  0, 3.19808
3  0.0030015, 2.62349
4  0.006003, 3.27875';
```

Note that the string is delimited by single back ticks, '. The file can the be include as an external Javascript code:

```
<script src=" inputs . js "> </script>
```
For instance, we can write the following

```
1  <!DOCTYPE html>
2  <html>
3  <head>
4  <meta http−equiv='Content−Type' content='text/html;charset=utf−8'
       ></meta>
5  <script src=" signal . js "></script>
6  </head>
7  <body>
8  <textarea id=" log "></textarea>
9
10 <script>
11 document . getElementById (" log ") . innerHTML = data ;
12 </script>
13 </body>
14 </html>
```

which produces output similar to Figure 7.11. Here we also demonstrate the use of `<textarea>` to include an editable and resizable text container.

Clearly, this method is best used for including input databases and other inputs that can be pre-formatted into the multiline string. But in that case, we may as well just utilize native Javascript syntax and specify an options object

```
const options = {
param1: 123,
param2: 'xyz',
param3: true
};
```

7.4.3 FILE INPUT BUTTON

The second option is to utilize the *FileReader* interface introduced in HTML5 that operates on `File` objects. While this method allows us to open any file from the local file system, the file must be interactively selected by the user. This is in contrast with the above scheme, which loads the specified file without any user intervention. There are two ways to obtain the `File` container: either by using a new `<input type="file">`, or by implementing a drop zone into which the file can be dragged from file explorer.

The first of these variants adds a "Browse" button that the opens a File Select dialog box that the user can use to select a file. We can optionally specify valid file types, and toggle the ability to select multiple files. On file selection, the button to trigger a "input" event that we need to listen to:

```
1  <body>
2  <input type=" file " id=" fs " accept=" text/csv " oninput=" read ( this ) "
       /><br>
3  <textarea id=" log "></textarea>
```

The `type="file"` element contains an attribute called `files` which stores an array of the selected files represented by `File` objects. We grab a

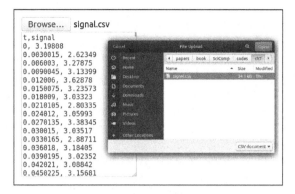

Figure 7.11 Data loaded through a Javascript multiline string.

reference to the first, and the only since multiselect is not enabled, file and store in a variable called `file`. Next, we create a new `FileReader` object. This object contains several functions including `readAsText`, `readAsDataURL`, and `readAsDataArray` for loading files as text, base-64 encoded binary, and raw binary data. The first option loading text documents is the appropriate one to use. On completion, the reader dispatches a "load" event. The event listener receives a single argument, as is customary, which contains all relevant information about the event. The actual loaded data is stored in `target.result`. The code for the event listener should be specified prior to launching the read function, as otherwise the file may get loaded before the listener is fully integrated. The entire code is listed below. It produces the output shown in Figure 7.11.

```
1  <script>
2
3  var input_file_name;
4  function read(el) {
5      var file = el.files[0];  // the first selected file
6      var reader = new FileReader();
7      reader.onload = function(event) {
8          const data = event.target.result;
9          document.getElementById("log").innerHTML = data;
10         console.log("Done reading "+file.name);
11         input_file_name = file.name;    // save for later
12     };
13     reader.readAsText(file);
14 }
15 </script>
16 </body>
```

7.4.4 DRAG AND DROP

Another option is to implement a *drag and drop* functionality. We first need to define a "drop zone". A `<div>` is typically used for this purpose,

```
1  <style>
2  div#flexbox {display:flex; flex-wrap:wrap; margin-bottom:1em; max-
       width:600px;}
3  div#drop_zone {display:flex;  height:80px;  width:200px;
4                 border:  2px dashed gray;  border-radius:10px;
5                 color:  gray;  align-items:center;  justify-content:
                      center;}
6  </style>
7  </head>
8  <body>
9  <div id="flexbox">
10 <div id="drop_zone">Drop a .csv file here</div>
11 </div>
```

The <div> called **drop_zone** is placed inside another <div> which is styled
to **display:flex**. This relatively new display style makes it easier to vertically (**align-items**) and horizontally (**justify-content**) center text in child
elements, which is surprisingly not trivial do otherwise.

Next, we extend this drop zone element with listeners for three events:
drop, **dragover**, and **dragleave**. Of these only the first is absolutely necessary,
however, the other two improve the user interface by allowing to style the
element when a file is hovered above it.

```
1  <script>
2  var dropZone = document.getElementById('drop_zone');
3  dropZone.addEventListener('drop', handleDrop);
4  dropZone.addEventListener('dragover', handleDragOver);
5  dropZone.addEventListener('dragleave', handleDragLeave);
```

The **drop** event listener receives an event object that contains an attribute
dataTransfer which in turn contains an array of **File** objects called **files**.
Since this matches the expected input of our already existing reader, we simply
need to pass the **evt.dataTransfer** data object to the read function:

```
1  function handleDrop(evt) {
2      evt.stopPropagation();  // prevent default file drop response
3      evt.preventDefault();
4      evt.target.style.background="white";
5      read(evt.dataTransfer);  // contains files array
6  }
```

Here you can also notice the presence of **stopPropagation** and **preventDefault**
functions. Event listeners are chained, with multiple code blocks possibly lis-
tening to the same event. The default web browser response to file drop is to
open the file, replacing the currently active page. These two functions prevent
this additional undesirable behavior from occurring. We also reset the background color, which as you will see shortly, is dynamically changed while the
file hovers above the drop zone.

The two remaining listeners are listed below:

```
1  function handleDragOver(evt) {
2      evt.dataTransfer.dropEffect = 'copy';
3      evt.currentTarget.style.background="#eee";
4  }
5
```

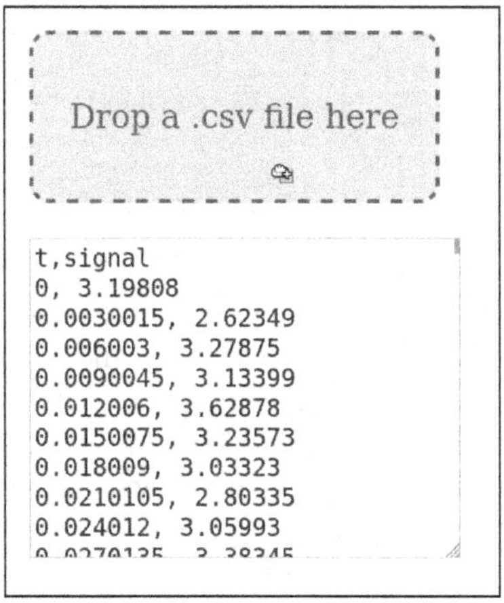

Figure 7.12 Data loading using drag and drop functionality.

```
6   function handleDragLeave(evt) {
7       // currentTarget is the original element
8       evt.currentTarget.style.background="white";
9   }
```

As was alluded to earlier, these two functions simply add a sensory feedback by changing the cursor style and the background color while the file hovers over the drop zone. One leave, we reset the view. This visual change could alternatively be implemented by toggling a class, as was demonstrated previously. The drop zone can be seen in the sample output in Figure 7.12.

7.4.5   FILE OUTPUT

When it comes to outputting the files, there is really only one option: a *data hyperlink*. The URL protocol of this link is of type **data:**, followed by the contents. A download link can be added using the function below, which is called from the **oninput** event handler in **read**:

```
1   function createOutput(data) {
2       var save_link = document.createElement("a");
3       save_link.id = "save_link";
4       var out_name = input_file_name.substr(0,input_file_name.length
            -4)+"_filtered.csv"
5       save_link.innerHTML="Download "+out_name;
6       save_link.setAttribute('download',out_name);
7       var text_area = document.getElementById("log");
8       text_area.parentNode.insertBefore(save_link,text_area.
            nextElementSibling);
9
10      // build output string
11      var lines = data.split("\n");     // split into lines
12      var text="t,signal,squared\n";
13      for (var i=1;i<lines.length;i++) { // skip header
14          var pieces = lines[i].split(","); // split by comma
```

```
15        var y = parseFloat(pieces[1]);  // convert second piece to
          a number
16        text += pieces[0]+","+y.toFixed(5)+","+y*y+"\n";
17     }
18
19     save_link.setAttribute('href',"data:text/csv;charset=utf-8,"+
          encodeURIComponent(text));
20   }
```

Lines 2 through 8 create a new `<a>` element, with the link text set to "Download signal_filtered.csv", if the input file is called `signal.csv`. We also set the link's `download` attribute to this file name. The link is inserted after the `log` text area. Lines 11 through 17 build the file content. We are saving a text `.csv` file, we just start with a string initialized to the appropriate headers. The input buffer is split into individual lines at the new line character `\n`. Since the first line is the header, we skip over it. For the subsequent lines, we split each line at a comma to separate the individual values, and do some demonstration data filtering. The `split` command returns a string variable, which is converted to a numeric value using `parseFloat`. Finally, once the output buffer gets constructed, we use the command on line 20 to include it as part of the address specified by the link. This is possible due to the `data:` protocol. As shown here, we also specify a .csv file type, which will most likely result in the file being opened with a spreadsheet program such as Microsoft Excel or LibreOffice Calc. The `encodeURIComponent` replaces non-alphabetic characters with URL-friendly UTF-8 escape sequences. For instance, the text `Hello world?` becomes `"Hello%20world%3F"`.

The concepts from this section are further demonstrated in `filter.html` example code available in the code demos directory. Figure 7.13 shows an example output. This script allows us to load a .csv file, visualize the data with a given point skip and optional markers, and to compute a cleaned-up signal by applying one of four possible filter types. The processed data can be downloaded as a .csv file. The $x - y$ plot is generated using code just as this:

```
1  ctx.beginPath();
2  for(i = 0; i < points.length; i+=skip) {
3      x = x0[0] + points[i][0]*scale_x;
4      y = x0[1] - points[i][1]*scale_y;
5      if (i==0) ctx.moveTo(x, y);
6      else ctx.lineTo(x,y);
7  }
8  ctx.stroke();
```

If enabled, markers plotted using a similar loop in which we fill and stroke a complete 2π arc.

7.5 WEBGL

All our graphics to this point utilized the 2D context. The canvas element also implements a `webgl` (and version 2.0 `webgl2`) context. The primary difference between the 2D and the `WebGL` contexts is that the latter performs the drawing

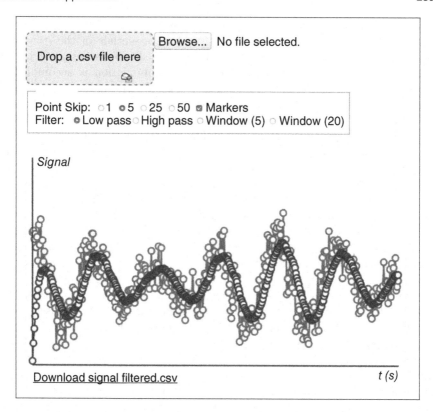

Figure 7.13 Example data processing web-based program.

using the graphics cards. As is discussed in more detail in Chapter 9, graphics cards are special devices optimized for performing the same operation on multiple data in parallel. This is the case when it comes to *shading* shapes on the screen. In order to draw a filled circle using the CPU, we need to iterate through every screen pixel within the bounding box and determine if it is inside or outside the perimeter. If inside, we set the color of the pixel. There is no particular ordering in which the pixels need to be assigned; we just need every pixel to get processed. A graphics card can process a large number, usually over a thousand, pixels concurrently. As you can imagine, this results in a significant reduction in draw time, and hence giving us the ability to render more complex geometries. You can find amazing demos of what is possible with webGL at `https://webglsamples.org/`, `https://www.khronos.org/webgl/wiki/Demo_Repository`, and `https://threejs.org/examples` (`three.js` is a popular Javascript library for working with webGL).

Due to the WebGL code running on the GPU, it is also no longer written in Javascript. Instead, the code is written in *OpenGL ES Shading Language* (or GLSL ES), which is similar to C but with additional data types and standard library functions. The ES stands for Embedded Subset, and it indicates that the language retains just the bare essentials of the otherwise rather large OpenGL library. The code first needs to be *compiled* into a shader *program*. As you will see shortly, this compilation is accomplished using Javascript. Each program consists of two algorithms: a *vertex shader* and a *fragment shader*. While the nomenclature may be new, the idea behind these functions is straightforward. WebGL and OpenGL operate on surface meshes defined, primarily, of triangles. The triangle vertices have some arbitrary three dimensional (x, y, z) coordinates. The purpose of the *vertex shader* is to transform these vertex coordinates to a 3D *clip space* extending from -1 to 1 in all directions. The first two (x, y) dimensions correspond to the screen coordinates. The z position indicates the depth used when painting overlapping elements. In 3D applications, this typically involves multiplying the position through a *transformation matrix* to take into account camera translation and rotation, and also through a *projection matrix* to apply perspective to make objects farther away appear smaller. The vertex shader operates only on the vertices, and the GPU launches as many copies as there are vertices.

Once the triangle vertices have been transformed to the appropriate 2D coordinates, the pixels making up each triangle face need to be assigned a color. This is where the *fragment shader* comes in. Fragment is the term openGL uses for a geometry primitive, such as a triangle. Each pixel can know its relative position in respect to the fragment vertices and can then use this information to, for instance, linearly interpolate vertex data to obtain nonuniform shading. Fragment shaders also often utilize *textures* to transfer a 2D image (the texture) onto the object using defined texture "UV" coordinates to produce realistic renderings. The fragment shader is launched for each pixel that needs to be drawn, with the GPU utilizing some built in processing to determine how to map a triangle into pixels.

7.5.1 SHADER PROGRAM

Let's now review an elementary webGL program. As always, we start by defining a <canvas> element into which we do the drawing.

```
1  <!DOCTYPE html>
2  <html>
3  <head>
4  <meta charset="UTF-8">
5  <style> canvas {border:1px solid black;} </style>
6  </head>
7  <body>
8  <canvas width="600" height="400" id="canv">Canvas not supported!
       </canvas>
```

We next create the context, and also add calls to three functions to be defined shortly:

```
1  var c = document.getElementById("canv");
2  var gl = c.getContext('webgl2');
3  var program = initWebGL();
```

```
4   initDrawing () ;
5   drawScene () ;
```

The **initWebGL** function specifies the shaders and compiles them into the shader program,

```
1   function initWebGL () {
2       if (!gl) return null;
3
4       // Vertex shader program
5       const vsSource = '
6           attribute vec4 aVertexPosition;
7
8           void main () {
9               gl_Position = aVertexPosition;
10          }
11      ';
12
13      // Fragment shader program
14      const fsSource = '
15          void main (void) {
16              gl_FragColor = vec4 (0.8 ,0.8 ,0.8 ,1.0) ;
17          }
18      ';
19
20      const vertexShader = loadShader (gl , gl.VERTEX_SHADER, vsSource) ;
21      const fragmentShader = loadShader (gl , gl.FRAGMENT_SHADER,
            fsSource ) ;
22
23      const program = gl.createProgram () ;
24      gl.attachShader (program , vertexShader ) ;
25      gl.attachShader (program , fragmentShader ) ;
26      gl.linkProgram (program) ;
27
28      if (!gl.getProgramParameter (program , gl.LINK_STATUS)) {
29          alert ('Error:  ' + gl.getProgramInfoLog (shaderProgram)) ;
30          return null;
31      }
32
33      gl.useProgram (program) ;    // activate program
34
35      return program;
36  }
```

The two shaders are listed starting on lines 5 and 14. We will discuss their details later, but for now, take a note that they are defined as Javascript multiline strings. We next use another helper function, **loadShader** to compile each shader source code into an object. Subsequently, we create a new shader program using **createProgram**, to which we attach the two shaders. Once complete, we call **linkProgram**. In case something went wrong, we show a pop up alert box with the error message.

The **loadShader** function is listed below. It starts by creating a shader of the specified type (**VERTEX_SHADER** or **FRAGMENT_SHADER**, assigning the source code, and attempting to compile it. If everything goes well, the shader object is returned to the calling function.

```
 1  function loadShader(gl, type, source) {
 2     const shader = gl.createShader(type);
 3     gl.shaderSource(shader, source);
 4     gl.compileShader(shader);
 5
 6     if (!gl.getShaderParameter(shader, gl.COMPILE_STATUS)) {
 7        alert('Compilation error: ' + gl.getShaderInfoLog(shader));
 8        gl.deleteShader(shader);
 9        return null;
10     }
11     return shader;
12  }
```

7.5.2 SHADERS

We can note that both shaders contain a

```
 1  void main(void) { }
```

function that assigns value to predefined variables. Specifically, the vertex shader sets `gl_Position`, while the fragment shader sets `gl_FragColor`. Let's start with the latter one. `gl_FragColor` is a 4-component floating point array of a predefined type `vec4`. The four components specify the red, green, and blue intensity as a value in 0 to 1, with the last item being the opacity. The selected values produce a light gray shading. The individual components can be accessed as `r,g,b,a` or `x,y,z,w` members. Furthermore, multiple components can be assigned at once, and their order can be arbitrarily specified to "swizzle" data around:

```
 1  mediump vec4 A,B;
 2  A.xy = vec2(0.2,0.0);
 3  B = A.xyxy;
```

Note the presence of `mediump` in front of the `vec4` variable declaration. This keyword, along with `lowp` and `highp`, specifies the precision of floating point numbers. The actual details, including the number of bits used, are not mandated by the standard and as such may be implementation specific. Also note that we are always including the decimal point, even when working with whole numbers such as 0.0 or 1.0. OpenGL is highly type sensitive, and does not support operations with operands of mixed types. For instance, the following

```
mediump float f = 2.0+5;
```

will generate a compilation error due to a lack of a '+' operation between a const 'float and' a 'const int'. To fix it, we simply replace the 5 with a 5.0;

7.5.3 DRAWING ELEMENTS

In our vertex shader, we simply copy the vertex positions in from a `vec4` attribute (more on these in 7.5.4) `aVertexPosition` to the output `gl_Position` variable. The positions are set by the Javascript code. Transferring data to a GPU is a somewhat convoluted process, as it follows the legacy OpenGL implementation. It contains of four major steps:

T We first create a new buffer using `createBuffer`.

T While you would think that we can next directly transfer data to the buffer, this is not how OpenGL works. Instead, we first bind the buffer to a named target using `bindBuffer`. There are two primary targets available: `ARRAY_BUFFER` used for vertex attributes and `ELEMENT_ARRAY_BUFFER` used to specify element connectivity.

T We then use `bufferData` to actually transfer the data. The values are stored in the buffer currently bound to the specified target.

T Finally, for vertex data, we associate the attribute variable with the buffer currently bound to `ARRAY_BUFFER` using `vertexAttribPointer`. We use `getAttribLocation` to grab the handle to the attribute of interest. We also activate an attribute variable using `enableVertexAttribArray`.

Here is what it looks like in practice:

```
1   function initDrawing () {
2       const positionBuffer = gl.createBuffer ();
3       gl.bindBuffer (gl.ARRAY_BUFFER, positionBuffer);
4       const positions = [
5           -1.0,-1.0,
6           -0.8,0.7,
7            0.0,0.2,
8            0.9,0.8 ];
9       gl.bufferData (gl.ARRAY_BUFFER, new Float32Array (positions), gl
            .STATIC_DRAW);
10
11      const numComponents = 2;
12      const type = gl.FLOAT;
13      const normalize = false;
14      const stride = 0;
15      const offset = 0;
16      var aVertexPosition = gl.getAttribLocation (program, '
            aVertexPosition ');
17      gl.vertexAttribPointer (aVertexPosition, numComponents, type,
            normalize, stride, offset);
18      gl.enableVertexAttribArray (aVertexPosition);
19  }
```

In this example, we will be drawing two triangles as a *triangle fan*. A triangle fan is a way to plot a continuous surface strip with the least number of unique vertices. The two triangles are given by vertices (0,1,2) and (2,1,3). The vertex positions are specified on lines 7 through 10. Unlike when painting into the 2D context, the coordinates now vary from -1 to 1, and the (-1,-1) origin is in the bottom left corner. We use `Float32Array` Javascript function to convert the generic JS array into a C-like single-precision float buffer. We grab a handle to the attribute variable on line 19. Then on line 20, we use `vertexAttribPointer` to map the buffer data to the attribute. Here we need to specify various information about how the received data was packed. Specifically, we tell the GPU that we are sending just two components per vertex. Since the vertex position is actually a `vec4`, only the first two `.xy` components will be assigned, with the rest set to zero. This is fine. We also tell the GPU that the sent data was of type float, per `Float32Array`, and that we don't want to renormalize the data to [0,1], and that there is no skip between vertices, and that the data begins at index 0. Values in `buffer[0:1]` will be assigned to `aVertexPosition[0].xy`, and `buffer[2:3]` will go to `aVertexPosition[1].xy`, and so on. Since it can become hard to remember to order of arguments, it is customary to use block-local temporary variables to specify these properties, as is the case here. Finally, the attribute is enabled.

With the vertex positions set, we are now ready to draw the triangles. But before doing that, we first clear the canvas. It is accomplished using the following code

```
1   function drawScene() {
2       gl.clearColor(1.0, 1.0, 1.0, 1.0);   // fully opaque white
3       gl.clearDepth(1.0);                  // also clear depth info,
        value in [0,1]
4       gl.enable(gl.DEPTH_TEST);            // enable depth testing
5       gl.depthFunc(gl.LEQUAL);             // make closer things
        obscure farther ones
6       gl.clear(gl.COLOR_BUFFER_BIT | gl.DEPTH_BUFFER_BIT);  // clear
        canvas
```

In this example, we also illustrate how depth testing, which uses the pixel's z coordinate, can be enabled, as it is disabled by default. It is not relevant to our example, but is important to specify for plotting 3D data in which the triangles may drawn in an arbitrary order.

The actual code to draw the triangles is simple:

```
1       const offset = 0;
2       const vertexCount = 4;
3       gl.drawArrays(gl.TRIANGLE_STRIP, offset, vertexCount);
4   }   // end of draw()
5   </script>
6   </body>
7   </html>
```

There are two functions at our disposal: `drawElements` and `drawArrays`. The first function draws the primitives using connectivity specified by the buffer bound to `ELEMENT_ARRAY_BUFFER`. The second command does not need the connectivity array and assumes that the vertices are stored in the correct order. This method is simpler to use but is not appropriate for complex geometries with vertices shared among non-contiguous elements. That is not the case here, however. We tell the GPU to paint a triangle strip, in which the connectivity is (0,1,2) and (2,1,3) for the first 2 triangles. Other primitive types include `POINTS`, `LINE_STRIP`, `LINE_LOOP`, `LINES`, `TRIANGLE_FAN`, and `TRIANGLES`. In a triangle fan, vertex 0 is shared by all elements. The first 2 triangles are given by vertices (0,1,2), and (0,2,3). The triangles mode uses vertices (0,1,2) and (3,4,5) for the first 2 triangles. The output from this code is shown in Figure 7.14.

Figure 7.14 Output from the initial WebGL example code.

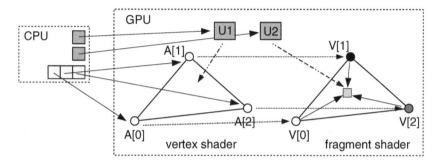

Figure 7.15 Visualization of attributes, uniforms, and varying data types.

7.5.4 ATTRIBUTES, VARYINGS, AND UNIFORMS

. Let's now return to the vertex shader. We can see that it declares a global vec4 variable called aVertexPosition that is prefixed with the keyword attribute. Vertex shader globals come in one of three "flavors": attribute, varying, or uniform. When discussing them, it is important to remember that the shader runs on the GPU, which has its own memory. Furthermore, a GPU contains hundreds, if not thousand, computational cores, leading to multiple shader codes executed in parallel. As such, each shader is its own independent unit that exists only for the brief span of its computation. These keywords specify the purpose of the variable, and how they are mapped in the GPU memory.

An *attribute* is a unique value stored for each fragment vertex. It is set by the CPU (in our case Javascript) code and essentially acts as an input to the vertex shader. A *varying* is also a value stored per vertex but is computed and set by the vertex shader. The CPU code cannot access varying data directly. A varying variable is an output of a vertex shader and an input to the fragment shader, with the GPU automatically linearly interpolating the value from the fragment vertices (three for a triangle) onto each pixel position. Finally, a uniform is a constant value that doesn't change with the vertex or pixel coordinate. These concepts are visualized in Figure 7.15.

Let's now see how these data objects can be used in practice. Instead of painting the two triangles with uniform color, we will use a new attribute aValue to assign a scalar value to each vertex. The fragment shader will then use a user-selected colormap to transform the value into a color. We also demonstrate vertex transformations by applying a rotation and a projection matrix to the vertex shader.

First, let's modify the vertex shader to the following:

```
1  const vsSource = '
2      attribute vec3 aVertexPosition;
3      attribute mediump float aValue;
4      varying mediump float vValue;
5      uniform mat3 uRotMat;
6
7      void main() {
8          mediump vec3 rotPos = uRotMat*aVertexPosition;
9          rotPos.z += 1.5;      // move farther away
10         gl_Position = vec4(rotPos, rotPos.z);
11         vValue = aValue;
12     } ';
```

The type of vertex position was modified from vec4 to vec3. The primary reason for doing this was to demonstrate how a vec4 could be constructed from a vec3 on line 9. Next, notice the addition of a new floating point attribute aValue, a similar varying vValue, and a uniform uRotMat 3×3 matrix. and a There is no need to start the variable names with the type prefix, as is done here; however, it is a common practice to further identify the variable type.

OpenGL natively supports matrix-vector multiplication, as is the case on line 8. Here we transform the position per $\vec{x}_{rot} = \mathbf{R}\vec{x}$. Rotating a two-dimensional coordinate $(x, y, 0)$ about x or y by a non-zero angle will result in a non-zero z component. The 3D position needs to be *projected* to the 2D screen. In order for the rotation to look realistic, we need to capture *perspective*: objects closer to us appear larger. A simple way to accomplish this perspective projection is to scale the x and y coordinates by z, $x_{persp} = x/z$. The larger the z, the smaller the coordinate value. This operation is so common place that OpenGL does it natively. However, instead of scaling by the z coordinate, the scaling uses the fourth w component. We perform this assignment during the construction of the 4-component gl_Position from the 3-component position on line 9.

The shader also copies the provided vertex attribute aValue to the corresponding varying float vValue. This varying is used subsequently by the fragment shader. We have the following new code:

```
1  const fsSource = '
2      precision mediump float;        // set default float precision
3      uniform int uColormap;
4      varying mediump float vValue;
5      void main(void) {
6          float f = clamp(vValue,0.0,1.0); // make sure in limits
7          if (uColormap==0)               // reds
```

```
8              gl_FragColor = vec4(f,0,0,1);
9         else if (uColormap==1)                   // grayscale
10             gl_FragColor = vec4(f,f,f,1);
11        else if (uColormap==2) {                  // red-orange-white
12            float a = f*3.0;
13            float r, g, b;
14            int i = int(a);
15            float fi = a-floor(a);   // fractional part
16            if (i==0) {r = fi; g=0.0; b=0.0;}
17            else if (i==1) {r=1.0;g=fi;b=0.0;}
18            else if (i==2) {r=1.0;g=1.0;b=fi;}
19            else if (i==3) {r=1.0;g=1.0;b=1.0;}
20            gl_FragColor = vec4(r,g,b,1.0);
21        }
22     } ';
```

Just as before, the objective of this shader is to assign a value to the vec4
gl_FragColor variable. Instead of using a hard-coded color, we utilize the
value from the varying vValue. Remember that this is simply a copy of the
aValue attribute set for each vertex node. Unlike the vertex shader, the fragment shader runs for each triangle pixel. The GPU code automatically inter-
polates the varying value from the triangle vertices onto the pixel. Therefore,
even though they share the same three vertices, each pixel in a triangle will receive a unique value for vValue. At this point, we translate the scalar through
a colormap to obtain a corresponding RGB tuple. We already saw on page
281 how to generate the classic blue-to-red rainbow. In the example above,
we implement a few additional options, including "reds", in which the color
varies from black to red by assigning vec4(f,0,0,1), "grays" in which the
shading varies from black to white using vec4(f,f,f,1), and a red-orange-
white heatmap, where f=vValue. Since these mappings assume that $f \in [0,1]$,
we use a built-in clamp function to specify lower and upper limits. Finally,
instead of hardcoding the colormap, we chose it dynamically using a uniform
integer uColormap.

We also need to update the Javascript code to populate the newly specified
attributes and uniforms. We first set the aValue attribute using a function
called setData:

```
1   function setData() {
2       const buffer = gl.createBuffer();
3       gl.bindBuffer(gl.ARRAY_BUFFER, buffer);
4
5       const values = [0.2, 1.0, 0.4, 0.0];
6       gl.bufferData(gl.ARRAY_BUFFER, new Float32Array(values), gl.
            STATIC_DRAW);
7
8       const numComponents = 1;
9       const type = gl.FLOAT;
10      const normalize = false;
11      const stride = 0;
12      const offset = 0;
13      var loc = gl.getAttribLocation(program, 'aValue');
14      gl.vertexAttribPointer(loc, numComponents, type, normalize,
            stride, offset);
```

```
15     gl.enableVertexAttribArray(loc);
16 }
```

The syntax should look quite familiar from `initDrawing`. The primary difference is that we are sending just 4 floats since we have only 4 vertices. These numbers, specified on line 5, are the vertex values that get mapped to colors. We also add a call to this function just after `initDrawing`:

```
1 initDrawing();
2 setData();
```

All that remains at this point is to set the two uniforms. To make the visualization more interesting, we will add animation using `requestAnimationFrame` and on each pass, will increment the angle θ through which the triangles get rotates. Rotation about the y axis is achieved using

$$\mathbf{R}_y = \begin{bmatrix} \cos(\theta) & 0 & \sin(\theta) \\ 0 & 1 & 0 \\ -\sin(\theta) & 0 & \cos(\theta) \end{bmatrix} \tag{7.1}$$

We add the following code to the start of `drawScene`.

```
1 function drawScene() {
2     theta += 1*Math.PI/180.0;        // increment theta by 1 degree
3     var loc = gl.getUniformLocation(program, 'uRotMat');
4     var rotMatrix = [Math.cos(theta),0,Math.sin(theta),
5                      0, 1, 0,
6                     -Math.sin(theta),0,Math.cos(theta)];
7     gl.uniformMatrix3fv(loc, false, rotMatrix);   // transfer data
```

Uniforms are simpler to set than attributes as they need just one item to be uploaded instead of a buffer of unique per-vertex values. On line 2, we use `getUniformLocation` to get the pointer to the specified variable. Subsequently, we use `uniformMatrix3fv` to set a `mat3` object using a one-dimensional vector of floats. This function assumes that the vector data is organized following the row-major order ($[a_{0,0}, a_{0,1}, a_{0,2}, a_{1,0} \ldots]$). The second argument, `false` in our case, could be used to optionally transpose the data if ordered by columns.

We also need to set the integer `uColormap`. To enable interactive selection, we add a drop-down list with id `#cmap` to the `<body>` section of the web page,

```
1 <fieldset>
2 Colormap: <select id="cmap">
3 <option value="2">Heat</option>
4 <option value="1">Gray</option>
5 <option value="0">Reds</option>
6 </select>
7 </fieldset>
```

The `<fieldset>` is used align the drop-down list item with the top of the canvas:

```
1 <style>
2 fieldset {display:inline-block;border:0;vertical-align:top}
3 </style>
```

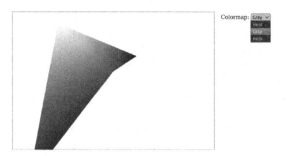

Figure 7.16 Modified version implementing rotation and a colormap.

Then, returning to the `drawScene` function, we again use `getUniformLocation` to grab a pointer to the `uColormap` variable. We then use `uniform1i` to set the value to the currently selected item in the drop down:

```
1  loc = gl.getUniformLocation(program, 'uColormap');
2  gl.uniform1i(loc, document.getElementById("cmap").value);
```

Finally, we add a hook to animate the scene:

```
1      window.requestAnimationFrame(drawScene);
2  }  // end of drawScene()
```

Because of the 1° increment of θ added to the beginning of `drawScene`, we see the two triangles spinning about the vertical axis. As expected, the sections farther away from the camera shrink in size. This output can be seen in Figure 7.16.

7.5.5 TEXTURES

Given that this is a book on scientific computing, and not game development, it may not be immediately obvious why the preceding discussion should be relevant. It is not unreasonable to imagine needing to develop a solver for some Eulerian system and wanting to visualize the mesh-based results with a contour plot. In the earlier discussion of the 2D context, we saw how to draw simple shapes, gradient filled spheres, and even $x - y$ plots (Section 7.4.5). A contour plot is more tricky, as it requires assigning a value to each pixel. While direct pixel manipulation is possible with the 2D context using `getImageData` and `putImageData`, doing the calculation in Javascript will not be efficient. Furthermore, this is exactly the type of an operation GPUs excel in.

Let's assume that our solution is known on a 2D Cartesian grid with $n_i \times n_j$ nodes. Since each cell is a rectangle, while WebGL operates on triangles, one possible solution is to create a surface mesh composed of $n_i n_j$ nodes and $2(n_i - 1)(n_j - 1)$ triangles. We would need to specify element connectivity and use `gl.drawElements` to draw the surface. The computed solution values, let's say temperatures, would be stored as vertex `attribute float` data.

We would then use a fragment shader essentially identical to the one given in the previous section to color each triangle with the local **varying float** transformed by a colormap.

But turns out there is a simpler method. Instead of using a value interpolated from the three vertices, we can use an interpolated *coordinate* to sample color from a two-dimensional image called a *texture*. This concept is visualized in Figure 7.17.

We begin by modifying the vertex shader. Instead of passing an attribute value to the fragment shader, we now pass a **vec2** attribute containing the (u, v) texture coordinates. These are floats in the range of $(0, 0)$ to $(1, 1)$, with the first value corresponding to the bottom left corner of the texture image. The shader code now reads as follows:

```
1  const vsSource = '
2      attribute vec4 aVertexPosition;
3      attribute vec2 aTextureCoord;
4      varying highp vec2 vTextureCoord;
5
6      void main() {
7          gl_Position = aVertexPosition;
8          vTextureCoord = aTextureCoord;
9      } ';
```

The top part of the fragment shader is modified to the following:

```
1  const fsSource = '
2      precision mediumImp float;
3      uniform int uColormap;
4      varying highp vec2 vTextureCoord;
5      uniform sampler2D uSampler;
6      void main(void) {
7          float f = texture2D(uSampler, vTextureCoord).r;
8          f = floor(f*16.0)/16.0;     // show only 16 unique colors
9          f = clamp(f,0.0,1.0);       // make sure value is in limits
10         if (uColormap==1)           // grayscale
11             gl_FragColor = vec4(f,f,f,1);
12         ...
```

Notice the absence of the **varying vValue**. We instead obtain the value at this pixel location using **texture2D(uSampler,vTextureCoord).r**. The **uSampler** uniform indicates which texture should be used, as multiple textures can be defined. Even though its type is **sampler2D**, this variables is essentially an integer holding the value 0. This value indicates that the default first texture unit should be used. **texture2D** function retrieves data from a 2D texture. OpenGL also defines samplers for 1D and 3D textures, as well as samplers from specialty areas such as shadow maps. On line 8, we demonstrate how to visualize only a reduced number of color values, 16 in this case, instead of the continuous coloring used before. This approach makes it easier to see individual contour levels, as plotted in Figure 7.18.

You may have noticed that we assign to **f** the red value returned by the **texture2D** command. Textures were developed as a way to add realism to rendered images. Therefore, the original use was limited to storing 4-byte

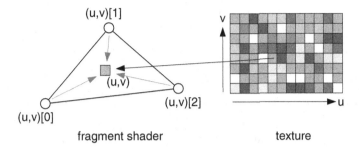

Figure 7.17 Instead of using vertex attributes, the fragment shader can sample data from a texture.

(r, g, b, a) tuples. The 4 values in range [0,255] would mapped to the 4 floating point values making up the vec4 gl_Fragcolor, with precision limited to $1/256 \approx 0.004$. However, version 2.0, WebGL introduced the ability to store arbitrary floating point values. One of the "texture filterable" options that allows us to interpolate floating point value is GL_R16F, which maps the provided data to the red channel. Enabling this support requires creating a webgl2 context, i.e.,

```
var gl = c.getContext('webgl2');
```
While this context is natively supported by desktop browsers, the support on mobile devices is a bit more limited. For instance, as of this writing, on an iPhone, we need to navigate to Settings for Safari, then selecting Advanced and Experimental Features to arrive at a page containing a huge list of optional toggles. One of these controls the support for WebGL 2.0.

We would like to use the entire canvas to draw the contour plot, and hence we modify the coordinates of the two triangles accordingly:

```
1  const positions = [
2      -1.0, -1.0,
3      -1.0, 1.0,
4       1.0, -1.0,
5       1.0, 1.0   ];
```
Triangle one goes from the bottom left corner to the top left corner and finishes at the bottom right corner. The second triangle of the strip starts at this point, then returns to the first triangle's second vertex, and finishes at the top right corner. As such, each triangle covers half of the canvas. We also need to assign the texture coordinates. This is done in initDrawing:

```
1  const textureCoordBuffer = gl.createBuffer();
2  gl.bindBuffer(gl.ARRAY_BUFFER, textureCoordBuffer);
3
4  const textureCoordinates = [
5      0.0,  0.0,
6      0.0,  1.0,
7      1.0,  0.0,
8      1.0,  1.0  ];
9
```

```
10      gl.bufferData(gl.ARRAY_BUFFER, new Float32Array(
            textureCoordinates), gl.STATIC_DRAW);
11
12      const num = 2;          // 2 values per coordinate
13      const type = gl.FLOAT;  // data type
14      const normalize = false; // don't normalize
15      const stride = 0;       // bytes between sets
16      const offset = 0;       // bytes to offset start by
17      var textureCoord = gl.getAttribLocation(program, '
            aTextureCoord');
18      gl.bindBuffer(gl.ARRAY_BUFFER, textureCoordBuffer);
19      gl.vertexAttribPointer(textureCoord, num, type, normalize,
            stride, offset);
20      gl.enableVertexAttribArray(textureCoord);
```

The steps are identical to setting vertex positions. Two coordinates are set per vertex. Following the just described node positions, the first vertex maps to $(u, v) = (0, 0)$, the second one to $(u, v) = (0, 1)$, and so on. We also make texture 0 the active one (although it is active by default), and then assign the value of 0 to the uSampler uniform to indicate we want to be sampling from this texture.

```
1       gl.activeTexture(gl.TEXTURE0);
2       gl.uniform1i(gl.getUniformLocation(program, 'uSampler'), 0);
```

We next need some data to visualize. In the example heat-webgl.html code, this data comes from the solution of the $\nabla^2 T = 0$ heat equation. Here for simplicity create a one dimension buffer holding data for a 40×40 grid. Data is packed using the familiar $j \cdot n_i + i$ indexing. The original setData is replace with the following

```
1    function setData() {
2       // pack some example data
3       const ni = 40;
4       const nj = 40;
5
6       var data = Array(ni*nj).fill(0);
7       for (var j=0;j<nj;j++)
8           for (var i=0;i<ni;i++) {
9               var u = j*ni+i;
10              data[u] = u/(ni*nj);
11          }
```

Next, we create a new texture and bind it to the TEXTURE_2D handle. Just as with attribute data, data transfer is accomplished by first binding the target to a named transfer point, and then working with that transfer point.

```
1    const texture = gl.createTexture();
2    gl.bindTexture(gl.TEXTURE_2D, texture);
```

The next step is to actually transfer the data. This is done using the following code:

```
1    const level = 0;
2    const internalFormat = gl.R16F  // return as a 16-bit red channel
3    const width = ni;
4    const height = nj;
5    const border = 0;
```

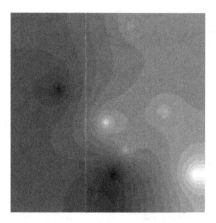

Figure 7.18 Contour plot visualized using WebGL textures.

```
6  const srcFormat = gl.RED;
7  const srcType = gl.FLOAT;
8  gl.texImage2D(gl.TEXTURE_2D, level, internalFormat, width, height
     , border, srcFormat, srcType, new Float32Array(data));
```

Textures support multiple resolution levels. We want to transfer to the default level 0. We also specify the texture dimensions, and note there is no border around the data of interest. The **internalFormat** property specifies how the data should be interpreted by the shader. As described above, we map the float to the red channel. We also specify the source data. **gl.RED** indicates that we are providing just a single component. **gl.FLOAT** specifies that the data will be in the form of a **Float32Array**.

Finally, we specify some texture parameters:

```
1  gl.texParameteri(gl.TEXTURE_2D, gl.TEXTURE_WRAP_S, gl.
     CLAMP_TO_EDGE);
2  gl.texParameteri(gl.TEXTURE_2D, gl.TEXTURE_WRAP_T, gl.
     CLAMP_TO_EDGE);
3  gl.texParameteri(gl.TEXTURE_2D, gl.TEXTURE_MIN_FILTER, gl.LINEAR);
4  gl.bindTexture(gl.TEXTURE_2D, texture); // bind to texture unit 0
```

The first setting imply that texture coordinates outside the valid [0,1] range should be clamped to these values. Alternatively, we may specify that values should wrap around. WebGL uses S and T, instead of the more common U and V notation for the coordinates. The third line specifies how data should be interpolated for fractional coordinates. We specify linear interpolation. And that's it! Now using just two triangles, we are able to rapidly plot detail colormaps as shown in Figure 7.18. The example code **heat-webgl.html** plots iterative solver convergence by using **requestAnimationFrame** to perform a single pass of the Gauss-Seidel solver, followed by a screen update accomplished by uploading new texture data.

8 Software Engineering

This chapter discusses various important code development topics, such as debugging, use of build systems, software libraries, version control, code testing, and documentation. We also provide a brief introduction to LATEX.

8.1 DEBUGGING

It is quite unusual for a simulation code to work correctly on the first try. More likely, the result will contain non-physical values, or the program will terminate prematurely due to a crash. The term "crash" indicates a state in which the program attempted to do something illegal, which resulted in the operating system shutting it down. The most common type encountered in C and C++ code is "segmentation fault". It is sent by the memory controller when a program attempts to access data outside its protected space. The operating system listens to these faults (or exceptions) being raised, and sends the program an appropriate *signal* (at least this is the terminology used by the Unix/Linux environment). Memory access violation leads to a SIGSEGV. Another signal, SIGFPE, indicates a floating point error. The default behavior for most exceptions is for the program to terminate, although it is possible to define a custom exception handler. However, it is not always feasible to recover from the source of the error. Your custom exception handler may generally at best print out additional information and dump data to a restart file.

Debugging is the process, as you have guessed, of looking for software errors, or bugs. The commonly told story is that early computers, which were the size of an office room and used large electric components, were afflicted by actual bugs getting zapped in the circuitry. Whether true or not, the term "bug" has stuck. Being skilled at debugging is just as important as being able to code. Otherwise, you will easily find yourself stuck for days or weeks "chasing a bug". This time could be better spent developing new features or producing analysis results. The following paragraphs summarize our experience in code debugging.

8.1.1 COMMON ERRORS

Debugging involves breaking up the problem into small tractable tasks and confirming that they are producing the expected results. As you grow in your career as a scientific analyst, you will find that a small number of errors is responsible for the vast majority of bugs.

DOI: 10.1201/9781003132233-8

Out of bound indexes

C and C++ do not perform array bounds checking and the programmer is responsible for making sure that indexes are valid. As an example, a 3D array allocated with

```
double vec[3];
```

contains components `vec[0]`, `vec[1]`, and `vec[2]`. Attempting to write to `vec[3]` leads to memory corruption as this addresses the non-existent 4-th item. This issue often arises when converting code from a one-indexed language such as MATLAB® or Fortran into C. Consider the following Fortran allocation:

```
real, dimension(3) :: vec
```

References to `vec(1)` need to be replaced as `vec[0]` and so on. This also implies that we need to pay attention to ranges of `for` loops, as well as the calculation of node index from (i, j, k) indexes. A MATLAB or Fortran version

```
u = (k-1)*ni*nj + (j-1)*ni + i
```

becomes

```
u = k*ni*nj + j*ni + i
```

when converted to Python, C/C++, or Java.

Use of uninitialized data

Consider the following code to build an identity matrix

```
1  double **I = new double*[n];
2  for (int i=0;i<n;i++) I[i] = new double[n];   // I[n][n] array
3  for (int i=0;i<n;i++) I[i][i] = 1.0;
```

While we do indeed end up with ones on the diagonal, what happens to the off-diagonal elements is not defined. Conceptually, the **new** operator returns an address of a memory block large enough to hold n double precision values. This block already holds whatever data happened to be written to that location by some other prior process. Without explicitly clearing (setting to zero) these values, our matrix starts with arbitrary "garbage" data. Where things get more complicated is that the **new** operator is really a hook to an operating system function. In our experience, some systems clear the newly allocated data while others do not. As such, it is possible that your code may work fine on one system, but produce incorrect results when compiled on a different machine. It is important to assume that data is left uninitialized and manually clear all values to zero! A good indicator of uninitialized variables being used is the presence of extremely tiny or huge floating point values such as 10^{-310} or 10^{68}.

Syntax Peculiarities

C and C++ are strongly typed languages. Among the most prominent demonstrations is the difference between an integer and a floating point division. Writing

$z = 5/2$

in MATLAB produces the expected 2.5. The almost identical code in C and C++

double $z = 5/2$;

gives us 2. This is unexpected. Since both 5 and 2 are integers, the compiler generates instructions for the faster integer division in which the fractional part is discarded (and not rounded to the nearest). In order to obtain the real-valued output, we need at least one operand to be a non-integer type. This conversion can be accomplished by including a trailing dot (or a .0) after whole numbers, including a **f** suffix to convert to a single precision float, or by explicitly *casting* to a real-valued type. Here are two examples

```
1  double z = 5/2.;  // trailing dot turns denominator to 2.0
2  double w  = 5/(double)2;   // explicit cast to a double
```

The use of **=** instead of the equality comparison operator **==** is another source of bugs. Consider the following code:

```
1  int x = 0;
2  if (x = 42) cout<<"x is 42";
```

You may be surprised to see **x is 42** printed despite the variable clearly initialized to a different value. Due to the use of the wrong operator, the second line actually translates into the following two instructions:

```
1  x = 42;       // x is assigned a new value
2  if (x) {... }  // boolean comparison, if (x!=0)
```

We are left with a boolean test. This test evaluates to true as long as **x** is non-zero. Since this is indeed the case here, the true branch of the **if** statement is executed, leading to the message being printed.

Math Errors

In order to improve code performance, it is customary to avoid checking inputs of math operations. This focus on speed can lead to errors if unexpected data in encountered. For instance, consider the following implementation of data averaging,

```
1  for (size_t c=0; n<num_cells; n++)
2    ave[c] = wp_vel_sum[c] / wp_sum[c];
```

This is the code you would use to compute mesh-averaged flow velocity in particle codes discussed in Chapter 5. This calculation results in a division by a zero in any cell without particles. The obvious correction is to include a conditional check,

```
1  for (size_t c=0; n<num_cells; n++)
2    if (wp_sum[c]>0) ave[c] = wp_vel_sum / wp_sum[c];
3    else ave[c] = 0;
```

which can be shortened as

```
1  for (size_t c=0; n<num_cells; n++)
2    ave[c] = (wp_sum[c]>0) ? (wp_vel_sum / wp_sum[c]) : 0;
```

Calculations can also fail if they lead to a numerical overflow. This is often the case with expressions in which large values cancel out mathematically "with pen and paper". For example, the following two lines should produce an identical output:

```
1  cout<<exp(10)<<endl;              // 22026.5
2  cout<<exp(1000)*exp(-990)<<endl;  // -nan
```

since $e^a e^b = e^{a+b}$. This is not the case. The computer attempts to evaluate the two operands independently which results in an overflow. It may sometimes be beneficial to utilize an approximation based on the Taylor series expansion when working with large values.

Incorrect use of pointers

Pointers are a common source of frustration for new programmers as they can easily lead to data corruption or segmentation fault errors. Pointer-based errors can be divided into three broad categories. First, we may be utilizing a pointer before it is assigned a valid memory address. For example,

```
1  double *ptr;
2  *ptr = 1.23;    // ptr not assigned a valid address
```

results in the code attempting to store the 1.23 value at some arbitrary memory location. This operation will lead to a segmentation fault or at least data corruption. While the former may seem like the worse outcome, it is actually preferred. It is easier to identify that pointers are being used incorrectly when their use leads to an instant crash. The other possibility is that the value of some variable is modified and the corrupted calculation does not show up until some later time.

Pointers are also typically used to access dynamically allocated arrays. Here we need to be cognizant of indexes, as has already been discussed. For a memory block allocated with

```
double *data = new double[N];
```

the valid indexes range from [0] to [N-1]. Using a for loop, we initialize all values using

```
for (size_t n=0; n<N; n++) data[n] = 0.0;
```

Using n<=N for the exit condition would result in the code setting *(data+N), which is located outside of the allocated memory space.

Dangling pointers are another source of error. A pointer is really just a variable storing a numeric address. Freeing a dynamically allocated array releases the memory block back to the operating system but does not "inactivate" the pointer that was used to access it. The following code is syntactically correct, but is faulty nevertheless:

```
1  delete[] data;   // deallocate the memory space from above
2  for (size_t n=0; n<N; n++) data[n] = 1.0;   // set some values
```

We should always invalidate a pointer after the deallocation by assigning to it the 0x0 nullptr address,

```
data = nullptr;
```

Of course, you likely will not write a code as shown above on purpose. However, there are situations that inadvertently lead to dangling pointers. One example includes returning the address of a local variable from a function. The memory block associated with the variable is deleted once the function exits. Another situation involves copying custom containers containing dynamic data. Consider the following structure:

```
1  struct MyCont {
2      MyCont() {};      // default constructor
3      MyCont(size_t N) {data = new double[N];}
4      ~MyCont() {delete[] data; data=nullptr;}
5  protected:
6      double *data = nullptr;
7  };such
```

Next consider the following code used for initialization

```
1  MyCont cont;
2  if (a<0.7) cont = MyCont(100); else cont = MyCont(50);
```

Here we perform a *shallow copy*, where only the address from the temporary `MyCont::data` array is copied over to `cont::data`. As the temporary object goes out of scope (at the end of the `if-else` clause), its destructor is called, with deallocates the memory. Hence, `cont::data` now holds a pointer to a freed memory block. Writing to it will lead to data corruption. To avoid this error, the `MyCont` object needs to implement a custom copy assignment operator,

```
1  struct MyCont {
2      MyCont& operator=(const MyCont &o) {
3          data = new double[o.N];
4          memcpy(data,o.data,sizeof(double)*N);}
5  }
```

One way to check for memory corruption is to take advantage of the compiler support to deploy 'canaries'. Adding `-fstack-protector` option to gcc compilation leads to data blocks getting padded with an extra buffer initialized with known data. Code is also included to check the consistency of these values. Any change indicates a buffer overflow, in the same way that a canary falling over in a coal mine indicates to a miner that the air is no longer safe to breathe.

8.1.2 PRINT STATEMENTS

With this brief overview of common bugs, let's now explore strategies for locating them. A typical simulation code reports only the main results. Screen, and especially file, output is slow compared to the speed at which the CPU operates. Including extraneous `cout` or `printf` statements leads to a reduced code performance. But by skipping the intermediate output, you may not realize that a calculation is not proceeding as planned until the final result is generated. The first step in debugging then becomes identifying the code section responsible for the incorrect result. Let's say that you find your output file contains NaN (not a number) values. You could modify an algorithm consisting of some three hypothetical steps to read:

```
1  doStepA ( field );
2  checkValues ( field , "A" );
3
4  doStepB ( field );
5  checkValues ( field , "B" );
6
7  doStepC ( field );
8  checkValues ( field , "C" );
```

where the **checkValues** function may be defined as

```
1  bool checkValues ( Field &field , const string &step ) {
2      for ( double val : field ) {
3          if ( ! isfinite ( val )) { cerr << "Found NaN at step " << step <<
               endl ;
4              return false ; }
5          return true ;
6  }
```

The code may generate a message such as **Found NaN at step** C. Since a similar message was not printed after step B, clearly something inside the **doStepC** function introduced the error. Subsequently, you would dive into that function and introduce similar instrumentation around its internal algorithms to further localize the error. Similar "print" statements can be used to determine the location of a segmentation fault crash.

Since it is likely that these messages are not desired in the production code release, you will subsequently need to delete these **checkValues** calls once debugging is complete. Alternatively, we can keep them, but have them execute only when the code is run for debugging purposes. C++ implements this functionality through the **assert** macro. Here is an example:

```
1  #include <cassert>
2  doStepA ( field );
3  assert ( checkValues ( field , "A" ));
```

The argument to the macro is any boolean expression. If it evaluates to **false**, the macro generates a screen output and calls **abort()** to terminate the program. The execution of assertions is however disabled when the code is compiled with the **NDEBUG** defined either through

#define NDEBUG

or by adding **-DNDEBUG** to the list of compiler arguments.

8.1.3 DEBUGGERS

Debugging using screen output can get tedious, especially when examining fine details of the calculation. Adding additional output requires re-compiling and re-running the code, which can be time-consuming. Furthermore, the actual change to the source code can lead to re-arrangement of memory storage and code that used to crash no longer doing so. This is where *debuggers* come in. A debugger is a program that launches another code as a child process

Figure 8.1 Use of the integrated debugger within Eclipse.

and allows the user to execute it line by line. The user can set *breakpoints* to pause the program at a specific line. Conditional breakpoints add further control by breaking only when a certain condition, such as a negative value, is met. A debugger typically also pauses whenever an exception is thrown. The exceptions likely originate deep inside the operating system function calls. The call that led to them can be identified by moving up the *stack frame*.

On Linux, you can enable stricter floating point exception checking using the `feenableexcept()` that is included in the `fenv.h` header. This command enables throwing exceptions when math errors such as division by zero (`FE_DIVBYZERO`), inexact rounding (`FE_INEXACT`), invalid function domain (`FE_INVALID`), overflow (`FE_OVERFLOW`), and underflow (`FE_UNDERFLOW`) are encountered. The following code will halt with the `SIGFPE` (Arithmetic exception) exception.

```
1  #include <fenv.h>
2
3  int main() {
4      // Enable division by zero and overflow exceptions
5      feenableexcept(FE_DIVBYZERO|FE_OVERFLOW);
6
7      double val = 1.0/0.0;
8      return 0;
9  }
```

Debuggers are directly integrated into common IDEs. Figure 8.1 shows a debugger used within Eclipse on the example code from Chapter 5. In order to utilize the debugger, we launch the program in the Debug mode. In the case of a C++ code, this involves using the executable built in the Debug profile. Whenever you create a new C++ project in Eclipse, NetBeans, or Microsoft Visual Studio, the IDE automatically add profiles for building a Debug and a Release version. The Release mode includes various optimizations to make the code run faster. These optimization lead to the binary no longer matching the source code. Some function calls may even get inlined to eliminate the overhead associated function calls. Some loops may get unrolled to eliminate the code for incrementing the index and checking the loop termination condition. Furthermore, a Release version lacks the full names of code objects such as variables and functions. They are not necessary for the program to run, since data access and function calls are performed via their respective memory addresses. Retaining symbol names makes it easier for a third party to decompile the application (to obtain the approximate source code from the binary) which may not be desired if the solver contains intellectual property you prefer to keep hidden. The Debug mode turns off optimization and retains the symbol names to simplify code review. The downside is that due to the lack of optimization, the Debug version runs several times slower than the Release one.

The IDE debugger interface provides access to several important tools. Firstly, it allows setting line breakpoints. Breakpoints are toggled by clicking along the left boundary of the editor window at the corresponding line. They are indicated by a small circle, as visible on line 24 in the above mentioned image. The code, launched in the Debug mode, suspends when it reaches a line containing a breakpoint. Breakpoints can optionally be made conditional in which case the code suspends only when the provided expression is true. Running code can also be suspended using the Suspend toolbar button. Suspending the code allows us to examine the values of variables in the local scope. This can be done by hovering the mouse over the variable of interest, or by adding "watch expressions", as illustrated by the window below the editor section. Here we can type in the name of a variable to investigate. The window also supports basic expressions, as shown in the figure. The first line shows a function call, while the second expression is used to investigate the data within the particle data structure. Suspended code can be resumed by clicking the Resume toolbar button. It can also be stepped line-by-line using Step Into and Step Over commands. These differ in their handling of function calls. The former jumps the debugger into a function call, where the execution continues to run line by line. The latter executes the function in a single step. There is also a Step Return option which executes the rest of the current function. Execution suspends on the following line of the calling function. The call stack can be examined in the Debug view, visible on the bottom left. This call stack lets us explore how the currently executed function got called. We can also explore the calling functions, and the values of variables within them, by moving up the call stack. Other useful views include Disassembly which allows you to see the compiled machine code, Registers which lists the values of CPU registers, Memory which lets you directly explore data according to the provided address, and OS Resources for listing system resources such as file handles.

8.1.4   COMMAND LINE DEBUGGING

Now let's say that you are connected remotely to a Linux cluster and don't have access to a graphical IDE. Linux offers a powerful command line debugger called **gdb**. In fact, Eclipse internally uses **gdb**. The graphical front-end interactions discussed in the prior section get translated behind the scene into **gdb** commands. In order to debug your code from the command line, first make sure to compile without optimization (-O0) and with symbols included (-g), i.e.

```
$ g++ -g *.cpp -o my_code
```

Next, launch the code using

```
$ gdb ./my_code
```

If your code requires command line arguments, use

```
$ gdb -args ./my_code arg1 arg2
```

You will subsequently see the following console output

```
$ gdb my_code
GNU gdb (Ubuntu 13.1-2ubuntu2) 13.1
Copyright (C) 2023 Free Software Foundation, Inc.
...
Reading symbols from my_code...
(gdb)
```

The (gdb) line indicates a prompt for entering commands. Type in **run** followed by Enter to start the program. This command can also be passed in through the command line using **-ex run** to start execution automatically. While the program is running, you can press Ctrl+C to break it. The **bt** command, for backtrace, shows the call stack letting you see which part of the code is being debugged. For instance, using the free molecular particle tracing code from Chapter 5, we may obtain output such as

```
ts: 243  0:183000
ts: 244  0:183750
^C
Program received signal SIGINT, Interrupt.
0x0000555555560b4b in operator*<double> at Field.h:120
120      return vec3<T>(a[0]*s, a[1]*s, a[2]*s);}
(gdb) bt
#0  0x0000555.... in operator*<double> at Field.h:120
#1  0x0000555... in Field_<vec3<double> >::scatter at Field.h:272
#2  0x0000555... in Species::sampleMoments at Species.cpp:166
#3  0x0000555... in main (argc=1, args=0x7fff..) at Main.cpp:47
(gdb)
```

You can also got up and down the call stack using up and down enabling you to inspect variables in the local context of the function call stack. Values of variables are printed using p,

```
(gdb) up
#2  0x000055... in Species::sampleMoments at Species.cpp:166
166         nv_sum.scatter(lc,part.mpw*part.vel);
(gdb) p lc
$3 = {d = {1.4000000000000006, 18.097333279855075}}
```

The source around the currently executed line can be listed using list (or l for short). l main,30 lists the main function from its start to file line 30. l Main.cpp:40 instead shows the code around line 40 of the specified file. b is then used to set a breakpoint,

```
b Main.cpp:41
```

Existing breakpoints can be printed using info break. Breakpoints can be deleted using d # where # is the breakpoint number as shown by the prior call. With the program suspended, either due to hitting a breakpoint or by pressing Ctrl-C, use next (or n), step, and finish to go to the next line, jumping over function calls, step into the function call, or run until the end of the current function, respectively. Another function stepi allows stepping over multiple instructions even if they are on the same line. cont (or c is used to restart (continue) the execution until another breakpoint is hit. Use quit to exit out of debugging.

8.1.5 MEMORY LEAKS

It is possible for a code to produce the correct results but to include an internal *memory leak* that leads to a degraded performance. Unlike languages such as Java, C++ does not include a built-in *garbage collection*. It is the responsibility of the programmer to deallocate (free) any dynamically allocated memory that is no longer in use. While the operating system automatically clears all data on program exit, a program with a particularly large leak can exhaust all system memory while still running. This leads to the operating system either shutting your program down, or more likely, the entire computer grinding to a halt as the O/S tries to accommodate the memory demand by swapping data between RAM and the slow hard disk.

Here is an example of a memory leak:

```
1  #include <iostream>
2  #include <thread>
3  using namespace std;
4
5  struct MyObject{              // some custom object
6      MyObject(int i) {
7          size_t N=1024*1024*20;
8          data = new double[N];  // some large buffer
9          for (int n=0;n<N;n++) data[n]=i;
```

```
10      }
11      double val() {return data[0];}
12  protected:
13      double *data;
14  };
15
16  int main() {
17      for (int i=0;i<10;i++) {
18          MyObject object(i);
19          cout<<object.val()<<endl;   // placeholder for some action
20          this_thread::sleep_for(3s);// wait for 3 second before
                continuing
21      }
22      return 0;
23  }
```

This code performs some placeholder action 10×. On each iteration, it utilizes a temporary local variable of type MyObject. This object uses an internal 160 Mb buffer, which is allocated by the constructor. The local variable object is deleted at the end of each for block. While this deletion frees the 8 bytes occupied by the data* pointer, it does not release the memory block at the stored address. As far as the operating system is concerned, the allocated buffer remains in use. The deletion of the pointer means that we have lost the "handle" to this data, and any subsequent hope of releasing it is gone. After 10 iteration, we lose 1.6 Gb of RAM capacity. The sleep_for command was added to allow you to watch the depletion of available system resources in real time.

In order to avoid memory leaks, it is important to get in the habit of always including a delete for every new added to the code. Fixing the above program simply involves adding a custom destructor, such as

```
~MyObject() {delete[] data; data = nullptr;}
```

Yet, even with this precaution, some memory leaks may still sneak in. We may have forgotten to label the base destructor in a polymorphic class with virtual, leading to an incomplete deletion. Or, our custom move or copy operators may not be working as intended. This is where dedicated memory checkers such as Valgrind come in. Valgrind is a command line tool for exploring program memory usage. It is specifically helpful for finding the use of unallocated variables, memory leaks, and buffer overruns. After compiling the code with the -g included to retain symbol names, we run our application using

```
$ valgrind -s --leak-check=full ./leak
```

The optional flags add an output of the error list and enable additional memory checks.

8.2   LARGE PROJECTS

8.2.1   MAKE FILES

A large simulation project may easily consist of thousands of lines of code distributed among tens of different source files. It does not make sense to recompile all the source files if only one of them is modified. A tool called **make** compares the file modification time stamp between the source and the object files to determine which files need to be rebuilt. A file called `Makefile` describes the build rules. IDEs, including Visual Studio and Eclipse, also implement their own project management systems, although they also support the use of external Makefiles. A sample Makefile is listed below:

```
 1  CC = g++
 2  LINK = g++
 3  CFLAGS = -O2
 4  LDFLAGS = -lpthread
 5  INCS = -I ./ -I include
 6  SOURCES := $(shell find $(./) -name '*.cpp' -or -name '*.cu')
 7  OBJECTS := $(SOURCES:.cpp=.o)
 8  OBJECTS := $(OBJECTS:.cu=.o)
 9  TARGET = my_code
10
11  all: $(TARGET) .depend
12  $(TARGET): $(OBJECTS)
13      $(LINK) $(OBJECTS) $(LDFLAGS) -o $@
14  .depend: $(SOURCES)
15      rm -f ./.depend
16      $(CC) $(CFLAGS) $(INCS) -MM $^ -MF ./.depend;
17  %.o: %.cpp
18      $(CC) $(CFLAGS) $(INCS) -c $< -o $@
19  %.o: %.cu
20      $(CUDA) $(CFLAGS) $(INCS) -c $< -o $@
21  clean:
22      rm -f -R $(OBJECTS)
```

The first 9 lines define various variables. The `:=` assignment operator allows executing expressions. A shell script, executed using `$(shell)`, finds all files with the `.cpp` or `.cu` extension. These files are stored in the `SOURCES` variable. Next, a substitution rule changes these extensions to `.o`. These names are stored in the `OBJECTS` variable. Variables are accessed using `$(name)` syntax.

The next section specifies the build rules. It consists of an arbitrary number of sections with the following syntax:

```
target: dependencies
    build instructions
```

The `all` target is the default one that is built if a target is not explicitly specified on the command line. Using the above file, we can "clean" the build, which involves deleting all object files to force a rebuild, by running

```
make clean
```

Another commonly encountered target `install` (not found here) is used to copy the compiled executable from the local "build" folder to a system directory, such as `\usr\local\bin` or `\opt` on Linux systems.

As we can see, the `all` rule depends on `my_code` (the value of the variable `TARGET`) as well as the contents of the `.depend` file. This file captures changes in the header files that would otherwise be skipped over since the build is based solely off the `.cpp` and `.cu` source files. The target is built using the command in the `LINK` variable using the provided list of object files and linker flags. The `$@` expression evaluates to the target name. In other words,

```
$(LINK) $(OBJECTS) $(LDFLAGS) -o $@
```

evaluates to the customary

```
g++ file1.o file2.o file3.o -lpthread -o my_code
```

The `.depend` target uses the g++ preprocessor to generate a list of dependencies to the .depend file. The `$^` variable holds the list of all target prerequisites. The next two targets specify rules for generating an .o file from a .cpp and .cu file, respectively. The only difference in the two rules is the use of the `CC` variable for the former and the `CUDA` variable for the latter to switch between the compiler in use. We discuss CUDA in Chapter 9. In both cases, the `$<` variable is used to obtain the rule's input and `$@` for the output file name. The actual list of files to build comes from the `OBJECTS` variable which is specified as the requirement for the `all` target.

8.2.2 CONFIGURATION SCRIPTS

Linux programs have historically been distributed as a collection of source files that had to be built on the local system. Typical build steps were

```
./configure
make
make install
```

The `configure` script probes the operating system and compiler for available features and include path directories, and sets the appropriate options for the Makefile. This script can be generated using `GNU Autotools`. A popular alternative is `CMake` from Kitware, the developers of Paraview and VTK. CMake offers a graphical as well as console-based editor that can be used to set program options interactively. CMake rules are stored in files called `CMakeLists.txt`. The console-based CMake interface is shown in Figure 8.2.

This custom build process is still recommended whenever code performance is critical since the code can be optimized using the most appropriate hardware settings. However, building code from scratch can be quite time consuming, with some libraries easily requiring hours to complete. As such, it is more typical to download prebuilt binary packages using tools such as `apt` on Ubuntu.

```
sudo apt install paraview
```

Mac supports a similar package management system called homebrew.

Figure 8.2 CMake console-based interface for setting build parameters.

8.2.3 LIBRARIES

The examples considered so far involved compiling all source files into an executable. It is alternatively possible to combine the source files into a *library*. A library can utilize static or dynamics (.dll or .so) linkage. Static libraries are saved in files with .lib or .a extension on Windows and Linux, respectively. Dynamical libraries use .dll and .so. The code from static libraries is included in the executable during the link step. Multiple executables utilizing the same static library functions each contain a copy of the library's compiled code. This leads to an increased total disk usage, which is perhaps of lesser concern nowadays than it was historically. Dynamic libraries are loaded and linked against during the program start-up. The single dynamic library file is shared among all programs utilizing it. The downside of this approach is that the code is no longer self-contained, and will not run if the appropriate shared library cannot be found. On Linux, the search path for dynamic libraries is specified by the LD_LIBRARY_PATH environmental variable.

We will now illustrate how to create and use a static library. Let's assume that you have developed some solver that you would like to share with the community, but are hesitant to release the source code to protect your intellectual property. This can be accomplished (but only to some extent due to the existence of decompilers) by distributing the compiled version of your solver. It involves compiling the relevant source code into a library and distributing it along with the header files containing function prototypes and class definitions. Codes utilizing the library are built by including the header files and linking against the library. Libraries are generated by first compiling-only (without linking) the source files to generate the .o object files. These files are subsequently archived into the library. The following steps are used on Linux:

```
$ g++ -c *.cpp
$ ar -crs libtoolkit.a solver.o output.o
$ g++ driver.cpp -ltoolkit -o my_sim
```

Here we assume that the library consists of two source files, `solver.cpp` and `output.cpp`. The first step compiles (due to the `-c` flag) these to `solver.o` and `output.o` object files. The second step uses the archiver `ar` program to combine these two files into the `libsolver.a` static library. Finally, in the third line, we demonstrate building an executable `my_sim` that links to the library. The library is specified using the `-l` parameter followed by the library name. Note that neither the `lib` prefix nor the `.a` extension is included. This call assumes that the library file is located in the current directory, or in the folders listed in the `LD_LIBRARY_PATH` search path. Other locations can be specified with `-L`. On Windows, one would use Microsoft Visual Studio to create a library-type project. The produced library is then added to the linker options for the application project that depends on it. Similar process is also available in other IDEs, including Eclipse.

8.3 USEFUL LIBRARIES

8.3.1 BLAS AND LAPACK

The prior chapters introduced numerical algorithms for matrix solving, data fitting, or signal processing. While you could implement these algorithms from scratch, it is often easier to take advantage of existing highly-optimized libraries such as LAPACK (Linear Algebra Package) and Basic Linear Algebra Subprogram (or BLAS) that provide this functionality. Both packages were initially written in Fortran, with the recent version ported to Fortran 90, however, they are compatible with C and C++ programs. The implemented algorithms are highly optimized, especially if compiled on the target system using local architecture specific settings. BLAS provides support for vector-vector, matrix-vector, and matrix-matrix operations such as addition, multiplication, and so on. LAPACK provides multiple routines for factoring and solving linear systems, with the routine specifically optimized for different matrix types, such as banded, tridiagonal, etc. optimized for a particular type (banded, tridiagonal, etc.). More information about LAPACK is available at `www.netlib.org/lapack/`.

These libraries can be built from scratch by downloading the source code. They can also be installed using a prebuilt binary. The prebuilt binary will likely lack the optimization that would be possible by compiling the code for the CPU architecture in your system. On Ubuntu, the library files are installed via

```
$ sudo apt install libblas-dev liblapack-devsudo liblapacke-dev
```

A sample LAPACK code is given below:

```
1  #include <lapacke.h>
2  double A[5][3] = {1,1,1,2,3,4,3,5,2,4,2,5,5,4,3};
3  double b[5][2] = {-10,-3,12,14,14,12,16,16,18,16};
4  lapack_int info,m,n,lda,ldb,nrhs;
5
```

```
6   m = 5;
7   n = 3;
8   nrhs = 2;
9   lda = 3;
10  ldb = 2;
11
12  // Solve a least squares problem
13  info = LAPACKE_dgels(LAPACK_ROW_MAJOR, 'N',m,n,nrhs,*A,lda,*b,ldb)
        ;
```

The code is then compiled using the following:

```
$ mpic++ lapack.cpp -o main -llapacke -llapack -lblas -lm -Wall
```

8.3.2 PETSC

PETSc is a library for solving matrix problems using parallel architectures, such as MPI or GPUs. It was developed at the Argonne National Lab in 1991, with more information available at https://www.mcs.anl.gov/petsc/index.html. PETSc supports parallel vectors and parallel sparse matrices, implements numerous preconditioners, direct and iterative Krylov-space methods, Newton-based non-linear solvers, and parallel time-stepping ODE solvers. It also provides an optimization library called Tao. An example of its use (based on a sample from https://www.mcs.anl.gov/petsc/documentation/tutorials/HandsOnExercise.html) is shown below:

```
1   KSP              ksp;         // custom object types
2   DM               da;
3   UserContext      user;
4
5   PetscInitialize(&argc,&argv,(char*)0,help); if (ierr) return ierr;
6   KSPCreate(PETSC_COMM_WORLD,&ksp);
7   DMDACreate2d(PETSC_COMM_WORLD, DM_BOUNDARY_NONE, DM_BOUNDARY_NONE
        ,
8                   DMDA_STENCIL_STAR,11,11,PETSC_DECIDE,PETSC_DECIDE
                      ,1,1,NULL,NULL,&da);
9
10  KSPSetComputeRHS(ksp,ComputeRHS,&user);
11  KSPSetComputeOperators(ksp,ComputeJacobian,&user);
12  KSPSolve(ksp,NULL,NULL);
13
14  DMDestroy(&da);
```

8.3.3 OPENFOAM

OpenFOAM (or "Open-source Field Operation And Manipulation") is a popular open-source C++ toolbox of numerical solvers. It was developed in the 90s at Imperial College London. OpenFOAM was historically mainly used for computational fluid dynamics (CFD), but volunteer developers have contributed packages for rarefied gas dynamics, plasma physics, and more. In addition, some commercial CFD software packages are essentially graphical user interfaces built on top of OpenFOAM. OpenFOAM can be downloaded from https://openfoam.com/, and tutorials are found at https://wiki.openfoam.com/Tutorials.

OpenFOAM makes heavy use of C++ operator overloading, leading at times to unusual syntax. For instance, solver parameters can be specified as shown below:

```
1  solvers {
2      p {
3          solver          PCG;
4          preconditioner  DIC;
5          tolerance       1e-06;
6          relTol          0.05;
7      }
8  }
```

The overloaded operators also make it possible to combine different field operators to construct governing equations. For example, the momentum conservation equation can be written as

```
1  fvVectorMatrix UEqn   (
2        fvm::ddt(rho, U)
3      + fvm::div(phi, U)
4      - fvm::laplacian(mu, U)
5  );
6  solve(UEqn == -fvc::grad(p));
```

8.3.4 VTK

The Visualization Toolkit (VTK) is a C++ library for processing and rendering 3D data. VTK was developed in 1993 by Will Schroeder, Ken Martin and Bill Lorensen in 1998, with this work eventually leading to the formation of the Kitware company. Paraview, which we have been utilizing in this book for data visualization, is essentially a GUI built on top of VTK. VTK is also commonly included in open-source computer-aided design (CAD) and finite element analysis (FEA) tools. We have already encountered VTK output formats in Chapter 5. Utilizing VTK directly as a library allows one to add visualization capability directly into the simulation code, with bindings available for popular languages including C++, Python, and Java. VTK is built around a "visualization pipeline". The pipeline starts with some data sources, which in many cases will be a file reader for one of the several supported formats. This source is then piped through arbitrary filters that, for example, extract isosurfaces or threshold the data. These are visualized by a mapper, which brings in a specified color translation lookup table. The mapper is assigned to an actor, which is added to the render window. A render window interactor handles the user keyboard and mouse interactions. The incomplete example below, taken from https://kitware.github.io/vtk-examples/site/Cxx/StructuredGrid/StructuredGrid/ illustrates the use of VTK. The code brings up a window in which the user can interact with a multicolored rectangular brick, as shown in Figure 8.3.

```
1  vtkNew<vtkLookupTable> lut;
```

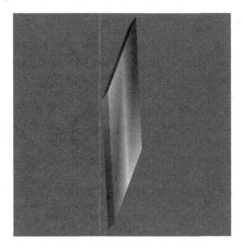

Figure 8.3 Visualization Toolkit render window.

```
2   lut->SetNumberOfTableValues(dataSize);
3   lut->Build();
4
5   // Create a mapper and actor
6   vtkNew<vtkDataSetMapper> mapper;
7   mapper->SetInputData(structuredGrid);
8   mapper->SetLookupTable(lut);
9   mapper->SetScalarRange(0, dataSize - 1);
10  mapper->ScalarVisibilityOn();
11
12  vtkNew<vtkActor> actor;
13  actor->SetMapper(mapper);
14
15  // Create a renderer, render window, and interactor
16  vtkNew<vtkRenderer> ren;
17  vtkNew<vtkRenderWindow> renWin;
18  renWin->AddRenderer(ren);
19  vtkNew<vtkRenderWindowInteractor> iRen;
20  iRen->SetRenderWindow(renWin);
```

8.3.5 BOOST

Boost is a set of C++ libraries providing support for a wide range of platform-independent tasks, such as multithreading, random number generation, image processing, regular expressions. Many of Boost's founders are on the C++ standards committee and many features originally available only in Boost have been migrated to the C++ standard library. For instance, prior to C++11, Boost was the only way to utilize platform-independent multithreading support. Multithreading, along with other aspects of Boost such as random number generation, is now part of the C++ standard library.

8.3.6 CYTHON

Finally, while not actually a library, another useful tool is Cython available from https://cython.org/. Cython compiles C++ code into Python modules that can then be imported from regular Python code. It allows computationally intensive algorithms of a Python program be ported to C, while still letting the code to retain linear algebra support of NumPy or plotting capabilities of pyplot. Cython is commonly used in mixed programming, in which a Python "front-end" is used to initialize and drive the simulation, with the actual number crunching happens in the compiled package.

8.3.7 GRAPHICAL USER INTERFACE FRAMEWORKS

Chapter 7 discussed the development of interactive applications running inside web browsers. We also noted that one challenge with this approach involves reading and writing local files. While Section 7.4 offered some workarounds, they are not particularly useful for consumer-level applications which may need to save and load a multitude of files, and even execute programs, as may be the case with a graphical user interface driving a command-line solver. For this reason, client-side HTML applications are not particularly well suited for developing general end user front-ends.

C and C++ programming language standard libraries do not include any support for graphical outputs. All codes that we have developed so far are examples of a *console application* that run through the command line. In order to develop a *windowed application*, the program needs to link against operating system libraries (SDK) to utilize platform-specific functions to open and populate windows. Since there is no commonality in these function calls across the operating systems, the developer needs to include platform-specific code for each supported operating system. As you can imagine, this can become quite tedious. For this reason, GUIs are typically developed by utilizing *cross-platform* frameworks. Some popular frameworks include Qt, wxWidgets, Swing, tkInter, and, to some extent, the already mentioned Visualization Toolkit. QT is a commercial C++ product with a large user base. QT examples can be found at https://doc.qt.io/qt-5/qtexamplesandtutorials.html. While the entire interface can be built solely using source code, QT also includes a graphical designer for constructing applications by interactively adding and repositioning appropriate elements. wxWidgets is an open-source C++ library for cross-platform GUI design. Swing and tkInter (and their derivatives) are the GUI frameworks built into Java and Python, respectively. Finally, the Visualization Toolkit (VTK), discussed above, offers the ability to launch the GUI window in which the rendering is made and where the user can interact with the data. Alternatively, this render space can be included as a widget inside another window developed using the aforementioned frameworks.

8.4   CODING STANDARDS

You may be currently working on a graduate degree and scrambling to complete your code in order to graduate on time. Software engineering is likely not high on the list of pressing concerns, unless that happens to be the subject of your thesis. However, there is a non-negligible chance that your research code ends up being used for years or decades to come. Many analysis tools used by the industry are nothing more than a graphical user interface wrapped around a legacy Fortran code dating to the early days of computing. Even if your code does not become widely adopted by external users, it may need to be expanded by other students who come after you. It is crucial to get in the habit of programming with *readability* and *maintainability* in mind.

Software companies tend to devise internal coding styles that developers are required to follow. Google's can found at `https://google.github.io/styleguide/cppguide.html`. Requiring all developers to adhere to a common set of principles makes the code not only more readable and maintainable, but also makes it easier for a new developer to pick up after the original author leaves the company, voluntarily or not. Below is a concise, and incomplete, summary of Google's coding standards:

- Optimize for the reader, not the writer
- Leave comments in tricky sections
- Be consistent with existing code and the broader C++ community
- Avoid surprising or dangerous constructs
- Be mindful of scale
- Concede to optimization when necessary

Here we instead present our recommendations, based on our experience working on multiple scientific simulation applications:

- *Be consistent*: Pick a naming and indentation convention and use it throughout the code. The approach utilized throughout this book is to name all class types with a leading capital letter. Variables and function names begin with a lower-case letter, and camel-cases is used to separate words:

```
1   class World {
2     /* ... */
3   };
4   World world;    // world is a variable of type World
5   performSimulation(world);  // call a function
```

- *Be organized*: Group your code into functions and files that make logical sense. Think of your code as a collection of black boxes that can be connected in a network. The function name and the input and output arguments should make it immediately obvious what inputs the function requires and what outputs are generated. One way to achieve this is by starting with a *skeleton code*. Write your main loop as a series of calls to yet undefined functions, each relying on data computed by functions preceding it. This approach will help clarify the required function arguments. Avoid using global variables, except for accessing common shared objects, such as a random number generator. Global variables otherwise tamper with the clear input/output interface offered by function arguments. Furthermore, sometimes it may also be tempting to just "stick" some code in an unrelated function instead of writing a dedicated function for it. This is a recipe for errors and hours wasted by debugging. You will likely forget the code was already added, possibly duplicating the math in another algorithm. Future developers may not even consider checking the function, since its name may suggest it to be benign.

- *Be clear*: Write code that is easy to understand. Use descriptive variable names and simple constructs. Of course, there will be some exceptions such as using `ni` instead of the more verbose `numberOfNodesInX`. Opt for clearer, perhaps more verbose syntax whenever possible. The compiler will optimize the code anyway. As an example, instead of writing

```
c = (b!=0?a/b:0);
```
it is clearer to write

```
if (b!=0) c = a/b; else c = 0;
```
The two lines should produce nearly identical code when compiled, but the second option is much easier to understand for someone not familiar with the C++ ?: conditional operator.[1]

- *Be team-oriented*: Use version control to keep track of changes, incorporate unit tests, and include sufficient comments. Code documentation is usually an afterthought, but even if you don't anticipate others using your program, there will be times when you revisit an old code to only find out that you have no idea what some function does (speaking from personal experience). At the minimum, reference sources for all equations that would not be obvious to a casual reader in your field. Also, develop a set of example inputs illustrating how to use the code, and include them with instructions and plots or tables of expected results.

8.5 CODE TESTING

In order to convince yourself and others that your numerical results can be trusted, your program should be put through a testing campaign. It consists of the following checks:

- *Verification:* Did we build the thing right?

---

[1] When in doubt, you can review the generated machine code by either adding `-S` compiler argument (this will generate a .S file containing the generated assembly code), or by using `objdump -d` with a binary file. Alternatively, during debugging in `gdb`, you can use `disas` command to display the assembly code.

- *Validation:* Did we build the right thing?
- *Uncertainty Analysis:* Given the same set of inputs, how much do the results vary between simulation runs?
- *Sensitivity Analysis:* How does a change in inputs affect the outputs?
- *Convergence Studies:* Do the results change if different numerical discretization is used?

8.5.1 VERIFICATION AND VALIDATIONS

Verification and Validation, typically shortened as V&V or IV&V for "independent", is the most frequently encountered form of code testing. *Verification* involves making sure that the developed product meets the prescribed specifications. Verification is of great importance to commercial software companies that need to ensure that all contractual obligations are being met by the delivered product. Yet, it is also comes up in the smaller settings. If a professors asks you to write a Navier-Stokes solver but you write code integrating the wave equation, then you did not "build the thing right" and most likely won't be receiving a pleasant grade on your submission.

Validation on the other hand involves checking whether the model correctly reproduces the physical process. Even if you write the Navier-Stokes solver as requested, you may find that computed results disagree with experimental observations. In this case, you may not have "built the right thing", since the physical system may be governed by a different set of laws.

Code validation is performed by comparing simulation results to known (true) answers. There are three approaches at our disposal. First, there is *analytical* validation. In this case, we use the code to solve a reduced problem with a known analytical solution. For example, if developing a fluid solver, we can simulate laminar flow in a cylindrical pipe and compare the velocity profile to the Hagen–Poiseuille equation. Alternatively, we can work backwards and use a prescribed solution to obtain the inputs needed to achieve that solution. This technique is referred to as the Method of Manufactured Solutions. Consider a one-dimensional Poisson solver, $d^2\psi/dx^2 = b$. The solver can be tested by assigning $\psi = f(x)$ where $f(x)$ is some at-least twice differentiable function (trigonometric or exponential functions work well). We then assign the values of the second derivative to the forcing vector b, along with appropriate boundary conditions. The solver solution should match $f(x)$.

We can also perform *numerical* validation. One possibility is to compare code results to that from a different, industry-accepted code. Another option is to conduct a "numerical experiment". This involves simulating a complex problem for which a direct analytical solution does not exist, while instrumenting the numerical world with virtual probes used for performing subsequent calculations. As an example, let's say that we have a box containing some initial free-molecular gas at a known number density and temperature. The box also contains an opening. We can add a virtual surface to extract the flux of molecules escaping the box. We then compare this numerically captured value to that predicted by the Langmuir flux model.

Finally, there is *experimental* validation. Experimental validation is often viewed as the panacea, but we need to be aware of experimental uncertainties (error bars) and details of the setup. One challenge with experimental validation is that, due to cost, many experiments are performed in direct support of hardware projects. In the space industry, these may include vacuum chamber characterizations of an entire assembled spacecraft. Utilizing data from such a test may prove to be logistically challenging due to legal limits on data release. Furthermore, even if the data can be obtained, the experimental complexity makes the telemetry unsuitable for code validation. The spacecraft itself may be too detailed to model, and the vacuum chamber may be filled with a multitude of ground support equipment that is not captured by the simulation. It then becomes challenging to distinguish discrepancies arising from incorrect code algorithms from those arising simply from the lack of included model fidelity. This is where university collaboration comes in handy, as universities are better suited for performing "basic-research" studies. Interested readers are referred to Oberkampf and Roy 2010 for an in-depth overview of these topics.

8.5.2 UNCERTAINTY ANALYSIS

Stochastic, Monte Carlo simulations will inherently contain noise due to random sampling. Such models are best suited for cases in which a *steady state* is achieved since we can then use the subsequent averaging to reduce the numerical noise. While averaging helps, it does not completely eliminate the noise. Furthermore, not all processes reach a true steady state, and may instead exhibit an oscillatory behavior. Uncertainty analysis helps to characterize the influence of noise. The general approach is to run the simulation multiple times with a different initial starting random number generator seed, to obtain a sequence of N results ψ_i. We next compute the *coefficient of variation*,

$$c.v.(\psi) = \left(\sqrt{\frac{1}{N} \sum_{i}^{N} \left(\psi_i - \bar{\psi} \right)^2} \right) / |\bar{\psi}| \qquad (8.1)$$

where

$$\hat{\psi} = \left(\sum_{i=1}^{N} \psi_i \right) / N \qquad (8.2)$$

This is essentially the standard deviation normalized by the magnitude. It allows us to characterize which parts of the solution can be "trusted".

8.5.3 SENSITIVITY ANALYSIS

The objective of a sensitivity analysis is to determine the relative importance of various inputs on the simulation results. In an analysis of regolith transport, we need to set the coefficient of restitution that controls the amount of energy lost on surface impact. The actual impact behavior is affected by the material types as well as the local roughness and the grain shape. These properties are likely not known for each individual grain, but we possibly estimate their applicable ranges. Performing the analysis over the expected possible input values allows us to bracket the solution. It also allows us to characterize which inputs are the most important.

Sensitivity analysis is a field with a strong mathematical interest, resulting in formulation of numerous schemes for performing it. At the heart of the challenge is the realization that a typical simulation requires a multitude of inputs which interact with each other through complex, often non-linear, relationships. The simplest method for performing sensitivity analysis is *One At a Time* or OAT. With this approach, we vary just a single variable at the time and compare the attained results to the baseline. The primary downside of this method is that it fails to capture the coupling between different inputs. Doing so requires running the code over the entire discrete spectrum of input combinations. This direct sampling quickly stops being practical, as the total number of runs is given by the product of unique values n_j to test for each parameter j, $n_{runs} = \prod_j n_j$. A more efficient alternative is to perform a Monte Carlo analysis by randomly picking values for all inputs from a uniform or a Gaussian distribution covering the expected data range. Other approaches involve utilizing *derivative* and *variance* based methods, and *regression analysis*. Tools such as Dakota (`dakota.sandia.gov/`) can automate these tests. These methods are described in some detail in Chapter 10.

8.5.4 CONVERGENCE STUDIES

Another important aspects of code testing are convergence studies. Numerical simulations rely on discrete intervals to capture the physical and time domains. It is important to confirm that the choice of time step and the spatial discretization is not affecting results outside the expected loss of resolution with a coarser computational grid. We generally consider *mesh*, *integration time step*, and *particle count* (for Monte Carlo methods) independence. The general approach is to run the code with a finer mesh, smaller time steps, and more particles to verify that results are identical within the resolution limits and the variation arising from random sampling. When it is not feasible to use a higher resolution due to computational limits, we can alternatively checking against a coarser mesh.

Note that some physical processes are inherently *chaotic* in nature. In these cases, small errors are amplified due to physical instability within the governing equations as is the case for problems such as turbulence and 3-body (or higher) interactions like gravitation. Though mathematically *deterministic*, arbitrarily small changes can grow to unbounded discrepancies. While discrete floating point arithmetic can reproduce exactly identical results, even relatively minor changes like reordering mathematically equivalent operations can result in differences in roundoff errors and divergent results. In these cases, convergence can only be studied in terms of statistical properties of the resulting outputs similar to the issues encountered in the uncertainty analysis section.

8.6 VERSION CONTROL

A *version control* system is a software tool for tracking code revisions. It is tremendously important in software teams consisting of multiple developers. But even when working solo, a version control system helps to maintain a log of revisions. Furthermore, if something in the code"breaks", a version control system allows reverting to a prior known working version.

Conceptually all version control systems contain a repository (or repo for short) which acts as a database of revisions for all tracked files. The repo is first initialized and then files to be tracked are added to it Generally only the source files, important configuration scripts, and example input files are included. Graphical programs may also need additional binary files such as images used for icons. The repo should contain all files needed to build the application. The developer commits (or checks in) code revisions either on a regular basis (such as nightly) or whenever a new code feature is implemented. Each commit includes a change log. The changes may be committed to a distinct branch. The developers may use a "develop" branch that contains the latest code. Once the new features have gone through software testing, the development branch is merged back into the "release" branch. The commit may also include tags to indicate special events, such as a release of a particular code version, or the finalization of a code feature.

The changes are then pushed to the repository. Over the years, different version control systems have come and gone. The earlier popular systems such as CVS (central version system) and SVN (subversion) have been surpassed by Mercurial and Git. Among the differences is the location of the repository. CVS and SVN utilized a centralized storage. Mercurial and Git support a distributed system, in which each developer has a local repository that tracks local edits. This system is attractive for teams spread out across multiple organizations where the developers do not share access to the same network. With the distributed system each group tracks its local revisions, and once a major milestone is reached (or as part of a contractual delivery), the local repository is transmitted to the client to be merged with their version. Alternatively, one can utilize a cloud-based service such as Github and Gitlab that offer a centralized storage location.

The local code is updated by performing a pull from the repository. A pull results in the latest revision fetched from the remote server. The version control software then attempts to automatically merge the remote and the local version. This automatic merge is prone to failures, especially when the same lines are modified both locally and remotely. Code with merge conflicts needs to be merged manually by reading through the file and looking tags such as

```
<<<<<<< HEAD
(your code)
=======
(external code)
>>>>>>> branch
```

While each file can be edited manually, it is easier to resolve merge conflicts using tools such as `meld`. Still, the best way to avoid merge conflicts is to avoid having multiple developers working on the same file at the same time.

8.7   UNIT TESTING

Unit tests are small algorithms included directly in the source code that check that functions produce the expected results. An example of a simple unit test may be one testing that $\sqrt{4.0}$ is indeed 2. The major strength of unit tests is their automation: after instrumenting the code, the test suite can be run on a regular basis to verify that the code is still functioning as expected. This helps us to catch a code change, typically implemented as part of some other feature improvement, that breaks functionality in another section of the program. These tests can be linked to repository code commits. This is known as *regression testing*.

There are many frameworks for performing unit testing, including JUnit for Java, and Google Test (GTest) for C++. In this section, we illustrate the use of GTest. The first step is to install the header and source files, which can be found as of writing at `https://github.com/google/googletest/blob/master/googletest/docs/primer.md`.

To use Google Test, we first add `#include <gtest/gtest.h>`. In order for the compiler to find this file, the `gtest` folder needs to be placed in a common path, such as `/usr/include/` on Linux. Alternatively, we add the path to compiler settings either graphically under "Additional Include Directories" in the development environment, or by adding the `-I` flag to `g++`. This header defines two macros: `TEST` and `TEST_F`. The former defines a standalone test, while the latter specifies a "test fixture". It allows for running more complicated routines that involve initializing objects. For instance, a test of a matrix solver likely requires the construction of a simplified computational mesh and the assignment of boundary conditions.

The listing below illustrates the use of unit tests. We first define a basic test called "Square" that belongs to a suite named "MathTests". We specify two checks, one testing that $\sqrt{784} = 28$, and the other one comparing $(-2)^2$ to 4. We also add another test called "AbsValue" belonging to the same suite. This one checks that the output of the absolute value function is ≥ 0. As can be deducted from this example, the actual comparisons are performed with macros of type `EXPECT_OP` and `ASSERT_OP` where the OP can be one of EQ, NE, LT, LE, GT, GE for numbers, TRUE and FALSE for booleans, and STREQ, STRNE, STRCASEEQ, and STRCASENE for strings. Macros of type EXPECT continue the test even when the test fails, while ASSERT terminates on failure. Therefore, assertions should be used to check inputs needed by subsequent tests. For example, there is no point in continuing a test suite if the input file used for comparison could not found.

```
1   #include <gtest/gtest.h>
2   #include <math.h>
3
4   TEST (MathTests, Square) {
5       EXPECT_EQ (sqrt(784.0), 28.0);
6       EXPECT_EQ (pow(-2,2),4.0))
7   }
8
9   TEST (MathTests, AbsValue) {
10      ASSERT_GE (abs(-1), 0);
11  }
12
13  // class used as a text fixture
14  class SolverTest : public :: testing :: Test {
15  protected:
16      SolverTest () { n=0;}  // initialization
17      ~SolverTest () {} // clean up
18
19      // called prior to each test
20      void SetUp() override {  n++; }
21
22      // called after each test
23      void TearDown() override { }
24      int n;
25  };
26
27  TEST_F(SolverTest , Run) {
28      EXPECT_EQ(n, 1);
29  }
```

Line 14 defines a class to be used with the test fixture on line 27. The first argument to the TEST_F macro is the name of the test. The second argument is the class used for testing. The macro constructs the object using the default constructor. Then, the member function SetUp() is called. As such, initialization can be done either in the constructor or inside this member function. We illustrate both approaches by first setting n to zero and subsequently incrementing it. In a production code, this set up may involve instantiating and initializing a computational mesh, and similarly instantiating the corresponding solver. The algorithm in the SetUp function needs to produce some output that can be tested using the EXPECT or ASSERT macros. The TearDown() member function is called on the test exit. This function would nominally clean up dynamically allocated memory, if any. These functions do not need to be defined if not needed.

The code is built by linking in gtest and gtest_main libraries, such as

```
$ g++ tests.cpp -lgtest -lgtest_main
```

Note that there is no main function in the above source file. It is instead provided by the `gtest_main` library. It may be preferred to define it directly in your code to allow using the same set of input files for production runs and for regression testing. We just need to add a call to `::testing::InitGoogleTest(&argc, argv);` followed by `return RUN_ALL_TESTS();` (the return statement is required). As an example, we can write

```
1   #include <gtest/gtest.h>
2
3   int main(int argc, char **argv) {
4   #ifdef TESTING
5       ::testing::InitGoogleTest(&argc, argv);
6       return RUN_ALL_TESTS();
7   #endif
8       // normal startup here
9       return 0;
10  }
```

The code is compiled using

```
$ g++ *.cpp -lgtest
```

for a normal run and

```
$ g++ *.cpp -DTESTING -lgtest
```

for the unit test run. Running the example code produces

```
[==========] Running 3 tests from 2 test suites.
[----------] Global test environment set-up.
[----------] 2 tests from MathTests
[ RUN      ] MathTests.SquareRoot
[       OK ] MathTests.SquareRoot (0 ms)
[ RUN      ] MathTests.AbsValue
[       OK ] MathTests.AbsValue (0 ms)
[----------] 2 tests from MathTests (0 ms total)

[----------] 1 test from SolverTest
[ RUN      ] SolverTest.Run
[       OK ] SolverTest.Run (0 ms)
[----------] 1 test from SolverTest (0 ms total)

[----------] Global test environment tear-down
[==========] 3 tests from 2 test suites ran. (0 ms total)
[  PASSED  ] 3 tests.
```

8.8   CODE DOCUMENTATION

As was already mentioned in the discussion of coding standards, code documentation is important to enable other developers to expand on your work and to remind yourself of the sources of implemented algorithms. Documentation can be divided into three main categories: *inline comments, developers guide*, and a *user guide*. Inline comments are small descriptive messages placed directly in the source code. We have already seen how these work. The syntax is language specific. In C++, multiline comments are included with /* and */, while a // comments out the rest of the line. This convention is used by a number of other languages including Java, PHP, and Javascript. In Python, we use % to include a single line comment and %%% to start a multiline comment block. Fortran uses the c character placed in the first column to comment out the entire line or ! to ignore the following text. MATLAB uses % and %(... %) for single and multi-line comments, respectively. Your variable and function names should be descriptive enough so that not every line needs documentation. Essentially, your code should be "self-documenting". However, there is always a need to use comments to document references of utilized equations, and to describe logic that may not be otherwise obvious. as The next important documentation type is the developer's guide. The purpose of this guide is to summarize the overall code organization and to specify the application programming interface (API) for the implemented functions. This document is of crucial importance when developing a library to be utilized by third parties who may not have access to the source code. Tools such as Doxygen can automate much of this effort. The output in nominally stored as a collection of .html webpages, as visualized in Figure 8.4, but other formats are also supported. Doxygen parses your code looking for function declarations. It then uses the `graphviz` package to generate call graph images.

Doxygen also parses code comments to generate the function description. It expects function definitions to start with a multiline comment. The first line is used to generate a brief description, while the subsequent lines form a detailed writeup. The comments can also include keywords such as `@param` and `@return` that describe function arguments and return values. Equations written in LATEX syntax are also supported as illustrated below.

```
1   /**
2    *  Computes a vector dot product
3    *
4    *  This function computes the dot product
5       \f[
6        d = \sum_{i=0}^{n-1} a_i b_i
7       \f]
8    *  The lenght of "a" and "b" needs to be the same
9    *  @param a first vector
10   *  @param b second vector
11   *  @return vector dot product */
12   double  dot(const vector<double> &a, const vector<double> &b);
```

Class member functions and variables are described with //!< as in

```
1   //! matrix base class, specifies the
2   // interface to be implemented
3   class Mat {
4   public:
5       double& virtual operator()(int i, int j); //!< access operator
6       const size_t nu; //!< number of rows (unknowns)
7   };
```

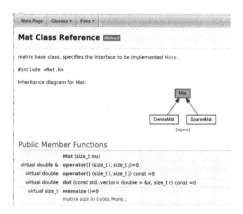

Figure 8.4 Doxygen-generated developer's guide.

Finally, every code needs a user guide. This is a document typically written using tools such as Microsoft Word or LaTeXthat describes the process of downloading, installing or building, and running the code. It should also summarize the implemented physics, and guide the end user through tutorials introducing program capabilities. Here it is a good idea to include some example outputs. The guide also needs to describe the syntax of the input file (if any) and guide the user through the use of a GUI environment, if utilized.

8.9 LATEX

LaTeX is a powerful tool for writing technical documents. Unlike graphical text editors such as Microsoft Word or Libre Office Writer, it is not a "what you see is what you get" environment. Instead, LaTeX is essentially a programming language for typesetting. One or more text files are "compiled" together into a single output .pdf document. Popular distributions include MikTex and TexLive. There is also an online collaborative platform called Overleaf. Since a LaTeX document is just a collection of text files, you can utilize version control systems such as git to track modifications made by your co-authors.

What makes writing in LaTeX unique is that you generally do not have to worry about the visual layout. Conference, journal, and book publishers provide *document class* template files that generate the desired formatting. The class file serves a purpose similar to that provided by the cascading style sheet (.css) does for an HTML file (see Section 7.1.2). By including a different class file you can completely change the look of your output document while having to make only minimal changes to the text itself. These modifications will be limited to changing out some class-specific commands. Examples of document classes include `article`, `book`, `IEEEtran`, and and `elsearticle`. The `beamer` class produces a presentation slide deck. Here is an example of a bare-bones .tex file:

```
1  \documentclass[draftcls,journal]{IEEEtran}
2  \usepackage{graphicx}
3  \begin{document}
4  % content
5  \end{document}
```

As you can see, the document starts with a header section which specifies the class file, along with any optional parameters inside the [] brackets. LaTeX commands follow format

```
\command{name}{required arguments}[optional arguments]
```

where the arguments are given by comma-delimited `key = value` pairs. It also loads *packages*. These packages extend the core functionality with custom code and serve the same purpose as `import` in Python and `#include` in C++. The list of available packages can be found at `https://www.ctan.org/pkg/` and a good overview of the most common features can be found at `https://en.wikibooks.org/wiki/LaTeX`. LaTeXdistributions generally come with a package manager, such as `tlmgr` for TexLive, that automates their download and installation.

8.9.1 SECTIONS AND REFERENCES

The main document sits inside the document environment. Class files implement custom commands for specifying the document title, author list, and so on. Document structure is introduced using commands such as `\chapter`, `\section`, and `\subsection`. These can be followed by a `\label` command to place a label that can be referenced from other parts of the document using `\ref`.

8.9.2 ENVIRONMENTS

Besides single commands, LaTeXalso utilizes *environments*. Each environment starts with a `\begin{name}` and ends with a `\end{name}`. You have already seen one environment, `document`, in use. Other common environments include `figure` and `table` for adding *floats* that get positioned automatically to reduce white space, and `minipage` for creating sections with individual margins. Here is an example:

```
1  \begin{figure}
2  \centering
3  \includegraphics[width=0.4\linewidth]{sim_results}
4  \caption{Simulaton results.}
5  \label{f:results}
6  \end{figure}
```

8.9.3 EQUATIONS

The ease with which equations can be written is one of LaTeX's greatest strengths. Inline equations are included with `$ ... $`. The `equation` environment produces a nicely formatted display-style equation. For example, the following text

```
1   The $\nabla \cdot \vec{f}$ term can be rewritten utilizing
2   \begin{equation}
3   \int_V \nabla \cdot \vec{f} dV = \oint_S \vec{f}\cdot\hat{n}dA
4   \label{e:div_th}
5   \end{equation}
6   where Eq. \ref{e:div_th} is the \emph{Divergence Theorem}.
```
yields

> The $\nabla \cdot \vec{f}$ term can be rewritten utilizing
>
> $$\int_V \nabla \cdot \vec{f}dV = \oint_S \vec{f} \cdot \hat{n}dA \qquad (1)$$
>
> where Eq. 1 is the *Divergence Theorem*.

Another useful math environment is `align`. It allows us to write a system of equations, such as

> $$y = (x + a)(x + a)$$
> $$= x^2 + 2ax + a^2 \qquad (1)$$

This output is generated using

```
1   \begin{align}
2   y &= (x+a)(x+a) \nonumber\\
3     &= x^2 +2ax + a^2
4   \end{align}
```

Notice the use of the & character to specify the alignment location. The command prevents printing of the equation number on the specified line. We can suppress all numbering by utilizing `equation*` or `align*` for the environment name.

You can easily find references for LaTeX commands online (for example at `wikibooks.org/wiki/LaTeX/Command_Glossary`). Below is a very brief summary of some common commands to get started.

`=, <, \le, >, \ge:`	$=, <, \le, >, \ge$
`\approx, \sim:`	$\approx, \sim$
`\left., \right.:`	start/end brackets, . one of $([\{)]\}.\|$
`\big,\Big:`	large and larger size
`\int_{x^2}^{a}, \oint:`	$\int_{x^2}^{a}, \oint$
`\text{if} \quad x\in[0,1):`	if $x \in [0, 1)$
`\sum_{i=1}^{N} \phi_i:`	$\sum_{i=1}^{N} \phi_i$

8.9.4 BIBLIOGRAPHY

LATEX also comes with a legacy bibliography manager called BibTex and a newer system called Biber. They use a text .bib file as the source of references. An example .bib file is shown below

```
1  @article{somebody,
2      author = {A. Somebody},
3      title = {Some Title},
4      journal = {Journal of Scientific Computing},
5      year = {2020},
6      volume = {10},
7      number = {456},
8  }
```

The file consists of multiple sections, each indicating the reference type. Some common types include `article` for peer-reviewed journals, `inproceedings` for conference papers, `book` for books, and `masterthesis` and `phdthesis` for theses, and `misc` for other sources such as web pages. References are cited in the document using \cite.

The concepts from the prior subsections are demonstrated in the following example.

```
1  \documentclass[%draftcls, journal
2  ]{IEEEtran}
3
4  \usepackage{graphicx}
5  \usepackage{hyperref}
6  \usepackage[caption=false, font=footnotesize]{subfig}
7  \usepackage{listings}
8
9  %set listings defaults
10 \lstset{
11   basicstyle=\small,
12   showstringspaces=false,
13   language=c++
14 }
15
16 \begin{document}
17
18 \title{Homework 8 Report}
19 \author{Jane~Doe%
20 \thanks{Undergraduate Student, Department of Astronautical
         Engineering}%
21 }
22 \markboth{Homework 8}{}%
23 \maketitle
24
25 \begin{abstract}
26 This report summarizes work performed for HW 8 in Class XYZ.
27 \end{abstract}
28
29 %————————————————
30 \section{Introduction}
31 Here is a quick summary of what was done on this homework
         assignment. We started by writing code to perform some matrix
```

```
         operation on vectors $\vec{a}$ and $\vec{b}$. We defined the
         \emph{dot} operation
32
33  \begin{equation}
34  \vec{a}\cdot\vec{b} = \sum_{i=0}^2 a_i b_i = a_0b_0 + a_1b_1 + a
         _2b_2
35  \end{equation}
36
37  We also defined the \emph{mag} operation ,
38  \begin{equation}
39  \left |\vec{a}\right| = \sqrt{\vec{a}\cdot\vec{a}}
40  \label{e:mag}
41  \end{equation}
42  The equation shown in \ref{e:mag} is coded up as
43
44  \begin{lstlisting}
45  friend double mag(const _vec3<T>&a) {
46      return sqrt(dot(a,a));
47  }
48  \end{lstlisting}
49
50  \section{Source Control}
51  We setup a Github repo, as shown in Figure \ref{f:repo}.
52
53  \begin{figure}[h]
54  \centering
55  {\includegraphics[width=0.8\linewidth]{github}}    %this requires
         github.png
56  \caption{Screenshot of the brand new Github repo}
57  \label{f:repo}
58  \end{figure}
59
60  We also learned how to use \href{http://doxygen.nl/manual/
         docblocks.html}{Doxygen}, \texttt{Google Test}, and \LaTeX.
         These tools are summarized in Table \ref{t:techs}.
61  Article by A. Somebody provides a good overview \cite{somebody}
         of these tools.
62
63  \begin{table}[t]
64  \centering
65  \caption{Summary of technologies}
66  \label{t:techs}
67  \begin{tabular}{l|l}
68  \textbf{Technology} & \textbf{Use}\\ %add new line
69  \hline
70  Doxygen & Developer guide documentation system \\
71  GTest & Unit Testing for C++ \\
72  Github & Source Control Repository \\
73  LaTeX & Technical paper writing language
74  \end{tabular}
75  \end{table}
76
77  \section{Conclusion}
78  There was no programming in this homework so it was very easy!
79
80  \section*{Acknowledgment}
```

81   I took advantage of the professor's office hour for help on this
          assignment.
82
83   \bibliographystyle{ieeetr}
84   \bibliography{ex8}
85   \end{document}

Compiling this file using

```
$ pdflatex ex8
$ bibtex ex8
$ pdflatex ex8
```

produces the `ex8.pdf` PDF document visualized in Figure 8.5. The second call to `pdflatex` is needed to incorporate the references produced by BibTeX.

Homework 8 Report

Jane Doe

Abstract—This report summarizes work performed for HW 8 in Class XYZ.

I. INTRODUCTION

Here is a quick summary of what was done on this homework assignment. We started by writing code to perform some matrix operation on vectors $\vec{a}$ and $\vec{b}$. We defined the *dot* operation

$$\vec{a} \cdot \vec{b} = \sum_{i=0}^{2} a_i b_i = a_0 b_0 + a_1 b_1 + a_2 b_2 \quad (1)$$

We also defined the *mag* operation,

$$|\vec{a}| = \sqrt{\vec{a} \cdot \vec{a}} \quad (2)$$

The equation shown in 2 is coded up as **friend double** mag (**const** _vec3 <T>&a) { **return** sqrt (dot (a, a)); }

II. SOURCE CONTROL

We setup a Github repo, as shown in Figure 1.

Fig. 1. Screenshot of the brand new Github repo

We also learned how to use Doxygen, Google Test, and LATEX. These tools are summarized in Table I. Article by A. Somebody provides a good overview [1] of these tools.

III. CONCLUSION

There was no programming in this homework so it was very easy!

ACKNOWLEDGMENT

I took advantage of the professor's office hour for help on this assignment.

Undergraduate Student, Department of Astronautical Engineering

TABLE 1
SUMMARY OF TECHNOLOGIES

Technology	Use
Doxygen	Developer guide documentation system
GTest	Unit Testing for C++
Github	Source Control Repository
LaTeX	Technical paper writing language

REFERENCES

[1] A. Somebody, "Some title," *Journal of Scientific Computing*, vol. 10, no. 456, 2020.

Figure 8.5 Output generated by compiling the example L^AT_EX script.

9 High-Performance Computing

This chapter introduces techniques for accelerating computer simulations. We begin by discussing profiling of serial programs to identify time-consuming algorithms. We then learn how to reduce run time by performing multiple computations in parallel using techniques such as multithreading, distributed computing with MPI, and the use of graphical processing units (GPUs) with CUDA. We also briefly discuss the use of graphics cards for screen rendering with OpenGL.

9.1 INTRODUCTION

As the complexity of numerical codes grows so does the simulation run time. While purely explicit solvers that rely only on a finite number of operations performed on local data (and similar inherently local algorithms such as the Thomas Algorithm) may scale with the number of elements n, algorithms that involve non-local data often scale with $n \log(n)$ *at best*. Example algorithms that potentially achieve this sort of scaling include staples such as the fast Fourier transform, quicksort, and Multigrid. The additional memory storage required by larger problems further reduces performance because of memory bandwidth limits and cache misses. It is not uncommon for production-level runs to require days, or even weeks, to complete. This is where parallel processing comes in. However, while it may be tempting to "throw the program onto a cluster", the first step ought to involve optimizing the serial (single processor) performance. Only then should one start focusing on parallelization. In particular, optimizing memory access patterns to avoid random jumps through memory tends to provide significant performance improvements in both serial and parallel implementations. If there is a significant chance that problem sizes may eventually exceed practical serial performance, some care should be taken early in selecting algorithms that are potentially amenable to parallelization without significant code redesign. Often such algorithms are not particularly more challenging to code in their serial implementations. This effort itself is not monolithic with multiple technologies at our disposal. This chapter introduces parallelization using multithreading, domain decomposition, and the use of graphics cards.

9.2 SERIAL OPTIMIZATION

9.2.1 PROFILING

The goal of serial optimization is to identify code sections that are responsible for the majority of computational time. These sections can then be rewritten using a different algorithm, ported to a faster programming language (as

DOI: 10.1201/9781003132233-9

would be the case with moving critical section from Python to C++), or sped up by improving memory access patterns. Identifying these sections involves measuring the amount of time spent in each function. This can be done by manually sampling the system clock before and after each function call. In, MATLAB®, the time taken to execute a function is obtained using a `tic`-`toc` pair,

```
1  tic
2  someFunction ()
3  toc
```

The `toc` function displays the time elapsed since the previous `tic` in seconds. In Python, we write

```
1  import time
2  start = time.perf_counter()
3  someFunction();
4  end = time.perf_counter()
5  print("Function took %.4g seconds"%(end-start))
```

C++11 provides access to a high resolution clock using the `<chrono>` header. The above algorithm can then be written as

```
1  #include <chrono>
2  #include <iostream>
3  using namespace std;
4
5  int main() {
6     auto start = chrono::high_resolution_clock::now();
7     someFunction();
8     auto end = chrono::high_resolution_clock::now();
9
10    chrono::duration<double> delta = end-start;
11    cout<<"Function took "<<delta.count()<<" seconds"<<endl;
12    return 0;
13 }
```

Timing an entire program in this manner can get really impractical. Luckily, there exist tools, called profilers, that perform this timing for us. Profilers fall into two categories: instrumenting and sampling. The first category "instruments" the code by wrapping each function call with a timing measurement, as demonstrated above. This can happen at the source level or can be done using the compiled executable. The process of recording the elapsed time leads to a noticeable slowdown of the running program. This is problematic for programs that already require a long time to complete. In that case, we may prefer to utilize a sampling profiler. These tools use operating system interrupts to pause the program at a regular frequency. They then inspect the program's call stack to determine which function is currently being executed. This stochastic approach is bound to suffer from noise. On the other hand, sampling profilers do not modify the code and generally lead to faster run times.

Development environments such as Microsoft Visual Studio or Eclipse contain built-in support for profiling. This includes recompiling the code with instrumentation enabled, launching it in the Profile mode, and presenting the timing data in a graphical manner. Figure 9.1 illustrates timing results from

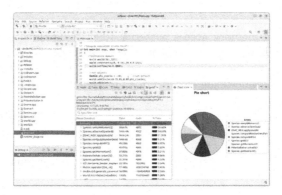

Figure 9.1   Code profiling using Eclipse.

the PIC simulation discussed in Chapter 5.

From the output we see that the majority of time is spent sampling velocity moments, followed closely by the function responsible for integrating particle positions and velocities. In the third place, at about 12% of the total time comes the DSMC collision algorithm. Field solver, which utilizes the Gauss-Seidel algorithm, does not show up until the 6th place at 4.7%. All other functions appearing at the top of the list involve particle operations, such as the computation of kinetic energy, momentum, and sampling of densities and macroparticle counts. The run time of this simulation can be seen to be driven by particle operations. This can be expected since the utilized plasma density is low enough such that there is very little change in plasma potential between time steps. Thus, much of the time measured for the field solver actually corresponds to the initial solution during the first step.

A popular Linux command line profiler is gprof. In fact, the output in Figure 9.1 was produced using this program. Eclipse simply acts as a front end that enables the appropriate compilation flags, launches the code, and on completion visualizes the collected telemetry. We can replicate this behavior with the following steps. First, the code needs to be compiled with -pg flag. This flag adds the gprof instrumentation. We then run the program normally. On completion, a file called gmon.out is generated. It contains the collected telemetry stored in a binary format. In order to convert it to a human-readable form, we run gprof with the executable file as an argument. The file can then be listed using Linux commands such as more, less, cat, or opened in graphical text editors. The listing below shows the entire process.

```
$ g++ -pg *.cpp -o my_code
$ ./my_code
$ gprof ./my_code > prof.txt
$ less prof.txt
```

The level of exported detail is controlled with optional arguments, but the default output is generally sufficient. The flat profile, shown in an abbreviated form below, is a good place to start. As we can see it shows the same information reported by Eclipse.

```
Flat profile:
Each sample counts as 0.01 seconds.
  %   cumulative   self               1
 time   seconds   seconds  ms/call  name
26.66   189.87    189.87    47.44   Species::sampleMoments()
14.63   294.06    104.19    26.03   Species::advance(Species&)
11.91   378.87     84.81    42.38   DSMC_MEX::apply(double)
```

Another useful section is the Call Graph. It shows how much time each function took to run its "own" code, and how much time was spent in "children" functions executed from within.

9.2.2 ALGORITHM MODIFICATION

Once the primary time sinks are identified, the next step involves determining if the logic could be implemented using an alternate, faster algorithm. We have already explored several algorithms for solving linear systems, including Jacobi and Gauss-Seidel in Chapter 2 and Preconditioned Conjugate Gradient and Multigrid methods in Chapter 3. For some general system, Gauss-Seidel may require thousands of iterations to converge, while PCG and Multigrid achieve convergence in just a few tens or hundreds of iterations. The algorithmic complexity of these solvers greatly exceeds that of Gauss-Seidel. However combined with the drastically improved convergence rate, the total time to solution becomes greatly reduced.

The overall benefit of such a modification will be very much problem specific. For example, we saw in the prior section that for the Chapter 5 PIC example, only about 4.7% of total run time is due to the field solve. Even fully eliminating this line item will have only a negligible impact on the total run time. As noted above, this atypical finding arises from the simulation running with low plasma density. At higher densities we may find the potential solver taking up the majority of run time. Profiling can thus help identify functions that will benefit the most from algorithm modification for the specific problem at hand. In this example, it is clear that any such changes should be focused on the algorithm responsible for particle pushing. For this simple example, perhaps not a whole lot of rework is feasible outside of streamlining memory access. However, in codes involving detailed geometries represented by tessellated surface meshes, much improvement can be achieved in optimizing the surface impact checking algorithm. This rework may involve utilizing line-triangle intersection functions with a reduced number of operations, as well as using efficient data structures such as octrees to limit the check to only triangles located in the particle vicinity. This algorithm shares similarity with ray tracing approach used in computer graphics. One may thus want to review tools developed in non-scientific disciplines such as computer graphics for their applicability to scientific numerical simulations.

9.2.3   CODE OPTIMIZATION

Once you tackle the "low-hanging fruit" by rewriting time consuming algorithms using an alternate approach, the next step of code optimization involves actually modifying the source code to produce more computationally efficient instructions. It is imperative to establish a timing baseline as some changes may not lead to the expected benefits. The compiler utilizes an optimizer (enabled in g++ using -O1 through -O3 settings) that already attempts to perform the steps described here. Aggressive manual optimization could lead to the optimizer failing to understand the code logic and skipping its own improvements. There are three main areas to pay attention to. These include: avoiding unnecessary calculations, avoiding jumps, and reducing cache misses.

To illustrate the first point, let's consider an algorithm for testing if a point is within a circle of radius r. One option is to write

$$\text{if } \sqrt{(\vec{x} - \vec{x}_0) \cdot (\vec{x}_0 - \vec{x}_0)} \leq r : \tag{9.1}$$

This requires computing the square root, which is computationally "expensive". It is also completely unnecessary. We can just compare the squared distance

$$\text{if } (\vec{x}_1 - \vec{x}_2) \cdot (\vec{x}_1 - \vec{x}_2) \leq r^2 : \tag{9.2}$$

where the value of r^2 is precomputed and saved. Similarly, multiplication is less expensive than division. A code performing repeated division by the same divisor may benefit by first evaluating the inverse, and subsequently multiplying the nominator by this value. Instead of

```
1   for (i=0; i<100; i++)
2       val[i] = f(i)/dx;
```

we may write

```
1   double idx = 1/dx;
2   for (i=0; i<100; i++)
3       val[i] = f(i)*idx;
```

Similar optimizations are also available when using trigonometric functions. A dot product of two unit vectors leads to the cosine of the angle between them. Let's say you need the sine to compute some other quantity. Instead of writing

```
1   double cos_theta = dot(vec_a, vec_b);
2   double theta = acos(cos_theta);
3   double sin_theta = sin(theta);
```

you can just use the $\cos^2 \theta + \sin^2 \theta = 1$ identity

```
1   double cos_theta = dot(vec_a, vec_b);
2   double sin_theta = sqrt(1-cos_theta*cos_theta);
```

which completely eliminates the need to evaluate trigonometric functions.

The next common trick, which may also be automatically applied by the optimizer, is to write conditional statements in a way that reduces the number of branches. Conditional expressions are implemented by the x86 family CPUs through a variety of "jump" instructions, such as JE (jump if equal) or JL (jump if less) that start executing operations at an alternate memory location if the condition is met. Jumps are expensive operations as they require loading the new instructions from RAM into the instruction pipeline. The memory address where the execution should return upon encountering a RET also needs to be stored. Let's say that you want to assign some coefficient to one value 10% of the time, and another value 90% of the time. If R is a uniformly distributed random number in $[0, 1)$, it may be better to write

```
1    int  coeff = <expr A>;
2    if  (R>=0.9)  val = <expr B>;
```
than
```
1    double  val;
2    if  (R<0.9)  val = <expr A>;
3    else <expr B>;
```
where **<expr A>** is some generic calculation or a function call. In the first example, 10% of cases will lead to the unnecessary evaluation of **<expr A>**, while avoiding the jump in 90% of cases. Depending on the time needed to perform this extra work, you may find that the first variant runs faster. If avoiding the **else** clause is not practical, then the if branches should be ordered by their likelihood of occurrence.

Conditional jumps are also "hidden" in a loop statement. Consider the following **for** loop:

```
1    for (int  i=0;i<N; i++) {
2       doCalculation (data [ i ]) ;       // some function
3    }
```

This loop implements the following logic

```
1    int  i=0;
2    start :                        // a label
3       doCalculation (data [ i ]) ;        // some function
4       i++;
5       if (i<N) goto start ;       // jump to the label ' start '
```
It is unusual to encounter **goto** in modern C++ codes, but logic like this was prevalent in legacy Basic and Fortran codes. As can be seen, each loop iteration involves a conditional check followed by a jump. If N is evenly divisible by 4, the number of these jumps can be reduced by rewriting the loop as

```
1    for (int  i=0;i<N; i+=4) {
2       doCalculation (data [ i ]) ;        // some function
3       doCalculation (data [ i +1]) ;     // some function
4       doCalculation (data [ i +2]) ;     // some function
5       doCalculation (data [ i +3]) ;     // some function
6    }
```

This is known as *loop unrolling*. While it was customary to make this optimization "by hand", modern compilers tend to incorporate it during the optimization phase.

9.2.4   CACHE MISSES

While these strategies lead to some benefit, they will pale in comparison with writing a cache-friendly code. The CPU does not directly operate on the data in RAM. Instead, it works with data stored within memory banks located on the CPU itself. These storage containers are called *registers*. Before a CPU operation can be performed, the data needs to be moved from RAM to the registers. The time required for this memory transfer is significantly greater than the actual time spent performing the mathematical calculation. For this reason, CPUs contain onboard memory storage called cache which can be accessed faster than the main memory. There are actually multiple levels of cache, with a smaller L1 cache located right on the microprocessor core, while a larger L3 cache residing on the motherboard. Whenever data is requested from RAM, the CPU first checks if the value already resides in the cache. If not, it is pulled from RAM. But instead of grabbing just the data itself, the memory controller also copies data from surrounding locations. This is done in the hope that the next memory request will ask for one of these values that have already been copied over. It is thus imperative to pay attention to data locality. Subsequent calculations should request data in the order it is stored in the memory. Let's consider code to duplicate values in a two-dimensional $N \times N$ matrix. The matrix is allocated as

```
1   double **A = new double*[N];
2   for (int r=0;r<N;r++) A[r] = new double[N];
```
We can then write

```
1   for (int r=0; r<N; r++)    // loop over rows
2     for (int c=0; c<N; c++)    // loop over columns
3       A[r][c] *= 2.0;
```
with the process itself repeated 500 times to reduce statistical noise. For $N = 4000$, we obtain the following:

```
$ g++ -O2 timing.cpp -o timing
$ ./timing
Took 7.967 seconds.
```

Now reverse the loop order to read

```
1   for (int c=0; c<N; c++)    // loop over columns
2     for (int r=0; r<N; r++)    // loop over rows
3       A[r][c] *= 2.0;
```
Again compile and run. We now obtain

```
$ ./timing
Took 71.807 seconds.
```

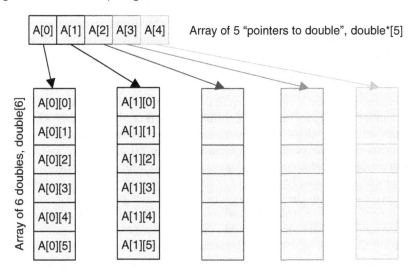

Figure 9.2 Memory storage pattern associated with C++ multidimensional array allocation.

This single change in loop ordering resulted in code running 9 × slower! The reason is that because of the allocation pattern, the column (c index) data is located in a consecutive memory block. The data in `A[r][c+1]` will likely already be in the cache after accessing `A[r][c]`. On the other hand, the data corresponding to the rows is scattered through RAM. The value `A[r+1][0]` is located at an arbitrary memory location that is not related to `A[r][0]` due to A being an array of pointers. Each A[r] entry points to an arbitrary location corresponding to the memory location where the **new** operator was able to acquire an array of N doubles. Due to this lack of data locality, the second version leads to repeated cache misses and much slower performance. The memory allocation is visualized in Figure 9.2. The approach used to allocate memory in C++ is called *row-major ordering*. This is in contrast with the *column-major ordering* used by languages such as Fortran or MATLAB. This is an important point to keep in mind when converting legacy code. The **do-loop** ordering from Fortran needs to be inverted for an optimal performance in C++. Paying attention to memory access patterns is important since nowadays application performance is often *memory-bandwidth limited* due to the fast CPU processing speeds. The CPU is mostly sitting idle waiting for data to become available. Using a CPU with a faster clock speed will not be beneficial for reducing simulation run time.

9.3 PARALLEL PROCESSING

Once we are satisfied with improvements to the serial code, the next step for reducing run time involves taking advantage of parallel processing. There are several technologies at our disposal: multithreading, distributed computing, the use of graphics cards, and dedicated hardware-based acceleration. Let's consider one of our office workstations. It contains an Intel i7-8700 CPU featuring 6 cores. Another of our workstations has the Intel Xeon W-1290 CPU with 10 cores. The CPU cores are independent processing units capable of performing calculations in parallel. This parallelization can be exploited by dividing the algorithm into multiple concurrent functional blocks, called *threads*, which are then distributed among the processors. These threads are belong to the same process and thus share the same memory space. While multithreading is simple to implement, the maximum speed-up will be limited by the fairly small number of cores. As noted in the prior section, these cores typically share the same memory bus and thus performance may be further hindered by bandwidth limitations. Large scientific simulations may also require more memory than available on a single system.

Additional scale up can then be accomplished by connecting multiple computers via network cables and having each computer work only on a subset of the global simulation. This collection of interconnected computers is called a cluster, supercomputer, data center, or "the cloud". In this distributed approach, each machine works only with a subset of the entire computational domain. Since these simulations are running on individual computers, they no longer have access to the same memory. Shared data needs to be communicated over the local network with the help of the Message Passing Interface (MPI) library. Computer clusters are typically built using rack-mounted nodes, with each node possibly containing multiple multi-threaded CPUs. As an example of scale, one system that we regularly utilize consists of 4,500 32-core compute nodes, thus providing a total of over 150,000 cores and 300 TB of RAM. Generally you will have access only to a subset of the total computational power as the cluster is shared among multiple researchers.

We can also utilize graphical processing units (GPUs). The complex geometrical shapes we are accustomed to interacting with in computer-aided design applications, digital art production, and video games, are internally represented by triangulated surface meshes. As was already introduced in Section 7.5, visualizing the model involves first translating vertex coordinates to the screen position by applying appropriate rotation, translation, and perspective transformations. This boils down to a matrix-vector multiplication,

$$\vec{x}_{screen} = \mathbf{PTR}\vec{x}_{model} \qquad (9.3)$$

The pixels making up each triangle need to be assigned a color. Added realism is accomplished by sampling colors from a texture map, which is an image of the target surface material. These steps require applying the same operation (matrix-vector multiplication and interpolation using triangle coordinates) to a large set of data (all mesh vertices and all visible triangle pixels). GPUs were specifically designed for this Single Instruction Multiple Data (SIMD) processing. As such they are the modern descendants of vector computers (such as Cray X-MP) from the early days of computing. A GPU contains hundreds

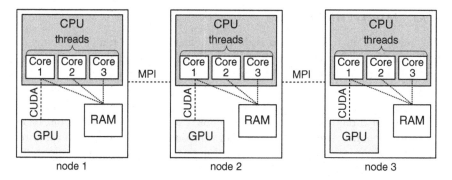

Figure 9.3 Coarse and fine grain parallel decomposition.

or even thousands of "shading units", although these processor cores tend to be slower than what is found on the CPUs. As an example, the entry-level NVIDIA Quadro P1000 card, which may be found in a desktop workstation, features 640 cores operating at 1.3 GHz base frequency and contains 4 Gb of onboard memory. On the other end of the spectrum you have a card such as NVIDIA GH100 which contains 18,432 cores operating at up to 1.6 Ghz frequency and storage capacity for 80 Gb of data. Theoretically, even the entry-level P1000 card could provide a 43× speed-up over a 6-core CPU operating at 3.2 GHz, although a 5 − 10 × improvement is more realistic in practice. Not all scientific computing algorithms are compatible with SIMD processing. For example, the stochastic nature of particle simulations implies divergence in the logic applied to some particles. GPUs have their own RAM, and any data to be processed on the GPU needs to be copied there first. This memory transfer can be a major bottleneck, often completely negating any performance gained by the rapid GPU processing, although it can be partly mitigated by utilizing concurrent streams.

Finally, an emerging field in high-performance computing involves the use of dedicated hardware such as Field Programmable Gate Arrays (FPGAs) or Application-Specific Integration Circuits (ASICs). These devices allow us to develop custom circuits that directly operate on the electrical signals associated with data bits. Introduction to FPGA programming can be found in Chapter 11. These technologies can be utilized individually or in combination. For example, it is not uncommon to begin with *coarse-grain* parallelization, in which the computational domain is distributed among individual machines of a supercomputer cluster. Then on each node, we deploy additional *fine-grain* parallelization by taking advantage of multithreading or GPU computing. This is visualized in Figure 9.3.

9.3.1 MULTITHREADING

We begin our introduction to code parallelization with multithreading due to its smallest software and hardware requirements. Support for multithreading is built into C++ compilers since the C++11 language and all modern CPUs contain multiple computational cores. Note that C++ compilers also implement a legacy parallelization paradigm called OpenMP which utilizes preprocessor directives to flag code sections, such as loops, that should be automatically parallelized by the compiler. We limit our discussion to the direct thread creation as it offers more control over the running code using and, in our opinion, a simpler interface.

Let's say that your code needs to add two large vectors, $\vec{c} = \vec{a} + \vec{b}$. A serial code may be written as

```
1  for (size_t i=0; i<N; i++)
2      c[i] = a[i] + b[i];
```

where N is the number of array elements. But we could just as well loop through the array backwards,

```
1  for (size_t i=N; i>=0; i--)
2      c[i] = a[i] + b[i];
```

We could even implement an algorithm that accesses the vector elements in a completely arbitrary order. The order of operations is not important; we just require that each element is accessed once, and only once. This makes this problem *trivially parallelizable*. The computer needs to perform a large number of operations, but the order in which they are performed is arbitrary. These types of algorithms are great candidates for parallelization.

A computer program is essentially a long sequence of instructions, quite like beads on a thread. The CPU simply executes the instructions one "bead" at a time. Here we have a sequence consisting of N additions (along with other steps involved in the for loop). There is nothing preventing us from dividing this single sequence into several blocks, such as:

```
1  block[0]:  for (size_t i=0*N/4; i<1*N/4; i++) c[i] = a[i] + b[i];
2  block[1]:  for (size_t i=1*N/4; i<2*N/4; i++) c[i] = a[i] + b[i];
3  block[2]:  for (size_t i=2*N/4; i<3*N/4; i++) c[i] = a[i] + b[i];
4  block[3]:  for (size_t i=3*N/4; i<4*N/4; i++) c[i] = a[i] + b[i];
```

Here we have four blocks, each processing one quarter of the vector. Given serial processing, there is essentially no difference between the above algorithm and the single for loop. However, given modern multicore architecture, we can assign each block to a different core. In that case, the calculation will continue in parallel, theoretically offering a 4× speedup over the serial implementation. Each of these blocks represents a unique thread. The operating system automatically shuffles threads among available computational cores. It is up to us to specify the instructions to be executed within each thread.

9.3.2 THREAD CREATION

Creating new threads requires interfacing with the operating system kernel using platform-specific instructions. Prior to C++11, one had to use Boost++ libraries to obtain access to cross-platform multithreading support. Nowadays, creating threads in C++ is trivial. We just need to instantiate a new object of type `thread`. The class definition is provided by the `<thread>` header. The source code for these functions is found in the `pthread` library, which historically had to be linked in using `-lpthread`; however, this no longer seems to be needed with recent versions of g++. Let's consider the following example:

```
1   #include <thread>
2   #include <iostream>
3   #include <sstream>
4   using namespace std;
5
6   //function to be executed from thread
7   void f(int i) {
8       //this_thread::sleep_for(chrono::milliseconds(5000));
9       stringstream ss;
10      ss<<"Hello from f("<<i<<")"<<endl;
11      cout<<ss.str();
12  }
13
14  int main() {
15      thread t1(f,1);
16      thread t2(f,2);
17
18      cout<<"Hello from main()"<<endl;
19
20      // wait for threads to finish
21      t1.join();
22      t2.join();
23
24      return 0;
25  }
```

Compile and run the code using (with `-lpthread` being optional),

```
$ g++ -O2 simple.cpp -lpthread && ./a.out
Hello from main()
Hello from f(2)
Hello from f(1)
```

The ordering of output messages is arbitrary, as you can prove to yourself by running the program a few more times.

The code in **main** begins with a creation of two objects of type `std::thread` called `t1` and `t2`. The first argument of the constructor is the function to execute in parallel, `f`. Subsequent arguments must match the argument list of the function being executed. In this case, this is just a single integer per `f(int i)`. The first object thus runs `f(1)`, while the second thread runs `f(2)`. The `f` function prints out a message to the screen, with

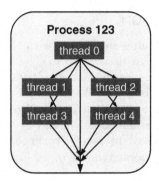

Figure 9.4 Program execution with multithreading.

stringstream used to prevent jumbled output. Directly writing to cout could result in the output from multiple threads intermixing after each « operator. The parallel execution begins immediately upon thread creation. There is no "start()" function to be called. The execution return back to main as soon as the thread start-up instructions are dispatched to the operating system. In other words, the execution returns before the specified function had a chance to complete. Waiting for such a completion would completely eliminate any parallel speedup.

This however also implies that the code in the main thread continues to run. The next instruction involves a screen output. Absent of the next two lines of code, main would subsequently process return 0; which would terminate the program. It is however quite likely that the child threads are still executing. This is not desired. For one, it would leave behind orphan threads, and second, there is no guarantee that whatever work the children threads were tasked with actually got completed. You can demonstrate this is indeed the case by uncommenting the sleep_for function to extend the f function run time and commenting the join calls. The program will end with a crash. We thus need main to wait for the threads to finish. This is done using the join function called on each thread. This function *blocks* (does not return) until the thread object it was called on terminates. Calling t1.join() from main makes the main thread wait for t1 to end. Program flow is visualized in Figure 9.4.

9.3.3 PARALLEL VECTOR ADDITION

Let's now return to our original vector addition example. We begin by defining the "worker" function, which adds two vectors. The important characteristic is that the array indexes to operate over are provided by input arguments:

```
1  void add(double *a, double *b, double *c, long start, long end) {
2      for (long i=start;i<end;i++)
3          c[i] = a[i] + b[i];
4  }
```

Next, in `main` we launch multiple copies of this function using threads. While it may be acceptable to hardcode the launch such as

```
1   thread t1(add,a,b,c,0*N/4,1*N/4);
2   thread t2(add,a,b,c,1*N/4,2*N/4);
3   thread t3(add,a,b,c,2*N/4,3*N/4);
4   thread t4(add,a,b,c,3*N/4,4*N/4);
```

there is no guarantee that the system is optimally suited for running with four cores. Instead, we may want to give the user the ability to specify the number of processors to utilize. The maximum number of available cores supported by the operating system can be queried using

```
size_t max_threads = thread::hardware_concurrency();
```

Note that modern CPUs split cores among "performance" and "efficiency", with "performance" cores supporting "hyperthreading" that makes a single core appear to support two threads. As such, the number of threads reported by the above function does not necessarily correspond to the number of concurrent computations that can be executed in parallel. From experience, we generally expect a speed up only up to `max_threads/2`. We let the user override the number of threads to use through a command line argument,

```
1    int main(int n_args, char *args[]) {
2        size_t max_threads = thread::hardware_concurrency();
3        int num_threads = max_threads;   //default value
4
5        // check for command line arguments
6        if (n_args>1) {
7            num_threads = atoi(args[1]);
8        }
9
10       cout<<"Running with "<<num_threads<<" of "<<max_threads<<"
             maximum concurrent threads"<<endl;
```

The example code, found in `add_threads.cpp` next allocates and initializes the data vectors. Next, we create the threads. The threads are stored using an array of pointers. We first initialize the vector,

```
1    thread **threads = new thread*[num_threads];
```

Subsequently, we launch the threads by dynamically allocating a `thread` object using `new`,

```
1    long chunk = N/num_threads+1;
2    for (int i=0;i<num_threads;i++) {
3        long start = i*chunk;
4        long end = (i+1)*chunk;
5        if (end>N) end=N;
6        threads[i] = new thread(add,a,b,c,start,end);
7    }
```

Note that this initialization could also be accomplished using the STL `vector` storage container. The end index is compared to vector bounds to make sure that the `add` function does not exceed data array bounds. This check is needed since N will generally not be evenly divisible by the number of threads. Finally, we wait for the threads to finish by calling `join` on all,

```
1   for (int i=0;i<num_threads;i++)
2       threads[i]->join();
```

we also need to clean up the dynamically allocated memory,

```
1   for (int i=0;i<num_threads;i++)
2       delete threads[i];
```

The above algorithm is wrapped within the chrono high-resolution clock calls to obtain timing. Running the code a few times results in:

```
$ ./add_threads 1
Running with 1 of 12 maximum concurrent threads
Calculation took 474ms
$ ./add_threads 2
Running with 2 of 12 maximum concurrent threads
Calculation took 267ms
$ ./add_threads 3
Running with 3 of 12 maximum concurrent threads
Calculation took 232ms
$ ./add_threads 4
Running with 4 of 12 maximum concurrent threads
Calculation took 202ms
```

9.3.4   PARALLEL EFFICIENCY

Clearly, some speed up is observed. Parallel efficiency can be characterized with

$$\eta_{parallel} = \frac{t_{serial}}{N_{cores} t_{parallel}} \tag{9.4}$$

The previously listed timings indicate efficiencies of 88.7%, 68.4%, and 58.6% for $N_{cores} = 2$, 3, and 4, respectively. Such diminishing returns are typical of multithreaded applications. There are two reasons why. First, only a subset of the serial code has been parallelized. Data initialization runs serially thus introducing a fixed overhead that cannot be eliminated even if the multithreaded algorithm itself runs at 100% efficiency. Second, as already discussed, much of CPU processing time is subject to memory bandwidth limitations. The amount of computation in this example is minimal (just a single addition), and thus much of the processing actually involves moving data between RAM and the CPU. The concurrent computations are competing with each other for the limited amount of memory transfer capacity.

There is another issue to consider. High parallel efficiency also requires uniform load balancing. It does not help to assign more cores at a problem if the added CPUs will be sitting mostly idle. To demonstrate load balancing, let's consider another example motivated by Sanders and Kandrot 2011. Julia set, visualized in Figure 9.5, is the collection of points $a = x + yi$ in the complex plane that satisfy the condition that a series given by

$$a^{k+1} = (a^k)^2 + c \tag{9.5}$$

converges to a finite number. We let $c = -0.8 + 0.156i$. The set is visualized by computing the number of terms the sum requires to diverge (which in practical terms means attaining value exceeding some threshold). This computation is performed using a `juliaValue` function,

```
1  int juliaValue(int i, int j, int ni, int nj) {
2      double fi = -1.0 + 2.0*i/ni;    // fi = [-1:1]
3      double fj = -1.0 + 2.0*j/nj;    // fj = [-1:1]
4
5      Complex c(-0.8, 0.156);     // coefficient for the image
6      Complex a(fi, fj);          // pixel pos as a complex number
7
8      int k;                      // iteration counter
9      for( k = 0; k < 200; k++) {
10         a = a * a + c;
11         if (a.magnitude2() > 1000) break;   // check for divergence
12     }
13     return k;
14 }
```

where `Complex` is a custom data type providing support for complex number arithmetic operations.

Let's say that we want to produce this image using a 4000 × 4000 grid. We let the horizontal i component correspond to the real value of a, while the vertical j component maps to the imaginary value. A serial version has us looping over all the nodes (pixels) in order:

```
1  for (int i=0;i<ni;i++)
2      for (int j=0;j<nj;j++)
3          julia[i][j] = juliaValue(i,j,ni, nj);
```

where `ni` and `nj` are 4000. This is clearly another example of a trivially parallelizable problem. We need to evaluate Julia value 16 million times, but the individual calculations are completely independent of each other.

One option to accelerate the code is to split up the outer loop into several horizontal blocks. We move the loop into a standalone function,

```
1  void calculatePixels(int i_start, int i_end, int ni, int nj, int
       **julia, int thread_id) {
2      for (int i=i_start;i<i_end;i++)
3          for (int j=0;j<nj;j++)
4              julia[i][j] = juliaValue(i,j,ni, nj);
5  }
```

and subsequently launch it in separate threads as

```
1  vector<thread> threads;
2  threads.reserve(num_threads);
3
4  for (int t=0;t<num_threads;t++ ) {
5      int i_start = t*(ni/num_threads);
6      int i_end = (t+1)*(ni/num_threads);
7      if (t==num_threads-1) i_end=ni; // correct the upper limit
8
9      threads.emplace_back(calculatePixels, i_start, i_end, ni, nj,
           julia,t);
10 }
```

Figure 9.5 Parallelization of Julia set computation using 3 blocks.

```
11
12   for (thread &t:threads) t.join();  // blocks until t finishes
```

Here we demonstrate the use of C++11 vector to store threads instead of the dynamically allocated array of pointers from before. Otherwise, this algorithm's logic is identical to the vector addition example. Running the code, found in `MT/julia_blocks.cpp`, with an increasing number of threads, we may obtain timing data such as

N_{cores}	1	2	3	4	5	6	7	8
time (s)	1.74	0.95	0.94	0.65	0.72	0.56	0.57	0.50
η_{par} (%)	100%	92%	62%	67%	48%	52%	44%	44%

While some speed up is seen, parallel efficiency is not optimal. In fact, there is essentially no reduction in run time from 2 threads to 3. Going from 4 threads to 5 actually increases the run time. These findings illustrate that simply "throwing more resources at a problem" does not necessarily lead to faster execution. To see what's going on, we can have the `calculatePixels` function set a thread id,

```
julia[i][j] = thread_id;
```

instead of assigning the Julia value, as was done before. Storing these results in a separate file, such as `thread_ids.vti`, let's us visualize which thread is responsible for which pixel. This visualization is accomplished by overlaying the two plots, as shown for 3 threads in Figure 9.5. The Julia set shading indicates the number of terms of Equation 9.5 that had to be evaluated for that pixel. Darker regions (given the utilized white-to-black color map) required more computations. It is obvious that the middle thread is doing much more work than the other two threads as it contains more dark area. This is an example of improper *load balancing*.

Figure 9.6 Computation of the Julia set using strides.

We can improve performance by assigning pixels to threads in a round-robin fashion. Instead of dividing the horizontal index into blocks, we assign threads to pixels in a circular pattern: first pixel goes to the first thread, second pixel goes to the second thread, third pixel to the third one, fourth pixel back to the first one, and so on, assuming three threads. This assignment is performed by utilizing the number of threads as the stride with the thread id number used as the offset and the pixel index linearized per $n = j \cdot ni + i$. Using our 2D data structure, we loop over all nodes but perform the computation only if $(n - \mathrm{id}_{thread})\%(n_{threads}) = 0$. This is written as

```
1  void calculatePixels(int ni, int nj, int *julia, int thread_id,
2  int num_threads) {
3      int n=0;
4      for (int i=0;i<ni;i++)
5          for (int j=0;j<nj;j++) {
6              if ((n-thread_id)%num_threads == 0)  // if belongs to me
7                  julia[i][j] = juliaValue(i,j,ni, nj);
8              n++;  // 1D node index
9          }
10 }
```

We can again alternatively have the `calculatePixels` function assign the thread id. Overlaying it we obtain the visualization in Figure 9.6. Using this round-robin assignment, we obtain $> 70\%$ parallel efficiency for up to 4 cores, as the following table demonstrates:

N_{cores}	1	2	3	4	5	6	7	8
time (s)	1.74	0.98	0.75	0.61	0.52	0.44	0.42	0.38
η_{par} (%)	100%	89%	77%	71%	67%	66%	59%	57%

This data was collected on a laptop using a 10-core Intel i5-1230U CPU. Similar timing using one of our workstations indicates nearly uniform 90% efficiency.

9.3.5 THREADS WITH CLASSES

The above sections demonstrated launching threads using standalone functions. Required inputs were passed in as function arguments. Results were stored at a memory address that was also passed in using a data pointer.

Such an approach is not always practical. We may prefer to use custom class objects to provide these inputs and storage locations. Unfortunately, it is not possible to launch a class method directly using the thread interface. We can however accomplish a similar outcome with the help of static functions. Static functions are not tied to a particular instance of the class. Consider the following class definition:

```
1  class Worker {
2  public:
3      // constructor, initializes parameters and launches new thread
4      Worker(double *a, double *b, double *c, long start, long end):
5          a(a), b(b), c(c), start(start), end(end) {
6          finished = false;
7          thr = new thread(run, this); // create new thread
8      }
```

The class constructor stores the provided data pointers as class member variables and also initializes a variable used to monitor status completion. It then creates a new thread that launches a static class member function defined as

```
static void run(Worker *p)    {p->add();}
```

This function receives as its input a pointer to an instance of the Worker class on which to start the calculation. The constructor passes this pointer using its **this** pointer. The add function performs the usual addition over a defined range,

```
1  void add() {
2      for (long i=start; i<end; i++)
3          c[i] = a[i] + b[i];
4      finished = true;
5  }
```

The threads are disposed of in the destructor,

```
1  ~Worker() {
2      thr->join(); //rejoin thread with the main code
3      delete thr;  //free memory
4  }
```

In the main thread, we use an array of Worker pointers to store the instances,

```
1  Worker **workers = new Worker*[num_threads];
2  ...
3  workers[i] = new Worker(a,b,c,start,end);
```

We can then wait for completion by monitoring the **finished** boolean in the worker classes,

```
1  bool finished;
2  do {
3      finished = true;
4      for (int i=0; i<num_threads; i++) {
5          finished &= workers[i]->finished; //boolean and
6      }
7
8      this_thread::sleep_for(chrono::milliseconds(5));
9  } while (!finished);
```

This variable is initialized as false during the construction of the object and is then set to true once the calculation completes. The sleep call is needed to yield the execution to other running threads, otherwise, the loop checking for thread completion will fully occupy the computational cycles of a single CPU core. Subsequently, we dispose of the workers using

```
1  // delete workers, calls destructor
2    for (int i=0;i<num_threads;i++) {delete workers[i];}
```

This approach offers the added flexibility of being able to monitor the data being operated on while the parallel code is running, unlike directly calling the blocking `join`. With some additional code modification, it is also possible to rewrite the code such that the thread continues running once its main calculation is completed. Here we could take advantage of `condition_variable` which allows the thread to suspend until the state of the monitored variable changes. We use this variable to signal to the thread that new data is available. This approach may be beneficial in codes in which we otherwise would need to create and dispose threads over and over. Thread creation is not "free". In past measurements using one of our particle tracing codes, we observed that the time needed to create a thread is comparable to the time needed to push several thousand particles. This overhead makes use of multithreading impractical to cases without a huge number of particles, if the threads are created on each time step. However, by creating the threads just once, during program start up, may make it feasible to utilize multithreading even in such simulation.

9.3.6 DOT PRODUCT

Let's now consider another example: the computation of a dot production,

$$\vec{a} \cdot \vec{b} = \sum_{i=0}^{N-1} a_i b_i \tag{9.6}$$

Here is a possible serial version:

```
1  double dot = 0;
2  for (size_t i=0;i<n;i++) dot += a[i]*b[i];
3  cout<<"Serial dot product = "<<dot<<endl;
```

A naive multithreaded implementation may look as follows

```
1  void dot(double *a, double *b, int i1, int i2, double *res) {
2      for (int i=i1;i<i2;i++)
3          *res += a[i]*b[i];
4  }
```

which is launched using

```
1  double res = 0;
2  for (int i=0;i<num_threads;i++) {
3      ...
4      threads[i] = thread(dot,a,b,i1,i2,&res);
5  }
```

This example can be found in `MT/dot_threads_bad.cpp`. For vectors initialized as

$$a_i = (i + 1)/N \tag{9.7}$$
$$b_i = 1/a_i \tag{9.8}$$

the expected value of the dot product is 1. However, running the code we observe a different value. Furthermore, the produced value is not constant and varies from run to run. For example:

```
$ ./dot_threads_bad
Result Parallel: 0.142047
$ ./dot_threads_bad
Result Parallel: 0.113404
$ ./dot_threads_bad
Result Parallel: 0.156082
```

One benefit of multithreading is that all threads have access to the same memory space. As such, it is not necessary to transmit data to each thread. While this reduces the computational overhead, we need to be extremely careful to avoid multiple threads *writing* to the same memory location as this can lead to a *race condition*. Race condition is a scenario in which code results are nondeterministic and vary according to which thread happens to reach a certain logic first. Let's consider a simplified example in which we use two threads to increment a counter,

```
1   void doWork(int *counter) {
2       *counter += 1;
3   }
4
5   int main() {
6       int counter = 0;
7       thread t1(doWork, &counter);
8       thread t2(doWork, &counter);
9       t1.join();
10      t2.join();
11      cout<<counter<<endl;
```

We expect this counter to be 2 but this is not always the case. The actual accumulation on line 2 actually involves three CPU steps:

1. Copy the current value of **counter** to a CPU register.
2. Increment the register value with the value corresponding to A.
3. copy the computed value from the register to the memory location corresponding to **counter**.

These steps are needed since the CPU does not operate directly on data in system RAM. Avoiding data corruption requires these three steps to be performed without interruption, such as

1. Value of counter = 0 is copied to CPU 1 register

2. CPU 1 increments its copy, value in register $= 1$
3. CPU 1 copies its result to RAM, counter $= 1$
4. Value of counter $= 1$ is copied to CPU 2 register
5. CPU 2 increments its copy, value in register $= 2$
6. CPU 2 copies its result to RAM, counter $= 2$

However, due to concurrent computation, we can have the following sequence of events:

1. Value of counter $= 0$ is copied to CPU 1 register
2. Value of counter $= 0$ is copied to CPU 2 register
3. CPU 1 increments its copy, value in register $= 1$
4. CPU 2 increments its copy, value in register $= 1$
5. CPU 1 copies its result to RAM, counter $= 1$
6. CPU 2 copies its result to RAM, counter $= 1$

This is in the nutshell what is happening with the prior dot product example. The preferred approach for avoiding the race condition is to avoid having multiple threads write to the same memory space. This can be accomplished by having each thread store its data in a thread-local buffer. These local results are then combined (reduced) together *in serial* using the main thread. Here is an example of how this can be implemented:

```
1   vector<double> local_dot(num_threads);
2
3   for (int i=0;i<num_threads;i++) {
4       ...
5       threads[i] = thread(dot,a,b,i1,i2, &local_dot[i]);
6   }
7
8   for (int i=0;i<num_threads;i++) threads[i].join();
9
10  double res = 0;
11  for (int i=0;i<num_threads;i++) res+=local_dot[i];
```

This code now produces the expected value of 1.

9.3.7 LOCKS AND MUTEXES

The previously discussed approach is not always practical, such as when the target data structure is too large to be duplicated for each thread. In this case, we can serialize critical sections using mutex-es. Mutex is an object, often implemented in hardware, that can be locked by a thread. When a thread attempts to lock an already locked mutex, it suspends until the lock is released. In other words, only one thread can lock a mutex at a time. While a mutex can be locked and unlocked manually, it is customary to wrap it in a lock object such as unique_lock that performs this unlocking automatically as it goes out of scope. It is important not to overuse mutexes as the critical code runs serially, thus eliminating the benefits of parallel processing. The use of mutexes and locks is demonstrated using the following example:

```
1  #include <mutex>
2  std :: mutex mtx;
3  {
4      std :: unique_lock lock(mtx);   // lock mtx, hangs if already
           locked
5      /* some critical code here */
6  } // mutex is unlocked here automatically
```

When developing multithreaded applications, it is important to be cognizant of hidden serialization in standard library functions. Otherwise, the program may experience degraded parallel performance. An example of such serialization is found in the C++11 random number generators introduced in Chapter 4. These generators use state vectors that need to be updated after each call, which requires serialization in order ot avoid the race condition. We can avoid this performance hit by providing each thread with its own generator. This is easily accomplished if the generators are encapsulated into a class object as illustrated in Section 4.4.12.

9.3.8 ATOMICS

Another option for avoiding the race condition is to utilize atomic operations (or atomics). These are uninterruptable mathematical operations often implemented by hardware as unique CPU instructions. Here is an example of a thread-safe integer increment:

```
1  #include <atomic>
2  std :: atomic_int counter = 0;
3  counter++;      // thread−safe increment
```

Atomic operations are generally only available for integer types and hence are not applicable to algorithms that accumulate real values.

9.4 MESSAGE PASSING INTERFACE (MPI)

We are now ready to introduce the second parallelization option: the use of the Message Passing Interface or MPI. MPI defines a set of commands allowing independent processes to communicate with each other. This communication typically takes place over a network in a distributed computing environment (a cluster), but MPI code can also be run on a single machine in which case the different processes run as independent threads. Below is a basic "Hello World" MPI example:

```
1  #include <iostream>
2  #include <mpi.h>
3  using namespace std;
4
5  int main(int n_args, char **args) {
6      MPI_Init(&n_args, &args);
7
8      int mpi_size;
9      int mpi_rank;
10
```

```
11    MPI_Comm_size(MPI_COMM_WORLD, &mpi_size);
12    MPI_Comm_rank(MPI_COMM_WORLD, &mpi_rank);
13
14    char processor_name[MPI_MAX_PROCESSOR_NAME];
15    int name_len;
16    MPI_Get_processor_name(processor_name, &name_len);
17
18    cout <<"I am "<<mpi_rank<<" of "<<mpi_size <<" running on "<<
          processor_name<<endl;
19
20    MPI_Finalize();
21    return 0;
22  }
```

Support for MPI is not built into the C++ standard library. We instead need to install the MPI development environment. There are several distributions available including MPICH and OpenMPI. On Windows, Microsoft provides its own MPI implementation. The difference between the distributions is limited to the syntax of command line arguments and configuration files. There is no impact on the source code. On Ubuntu, we install MPI using

```
$ sudo apt install libmpich-dev
```

or

```
$ sudo apt install libopenmpi-dev
```

This installs the mpi.h header, the MPI library, as well as the mpic++ wrapper for g++ that automatically sets the correct include paths and libraries. Another program that gets installed is mpirun. This is the program used for launching MPI jobs.

Here is how to compile and launch the parallel job using 4 local processes:

```
1  $ mpic++ hello_mpi.cpp -o hello_mpi
2  $ mpirun -np 4 ./hell_mpi
3  I am 3 of 4 running on node1
4  I am 0 of 4 running on node1
5  I am 1 of 4 running on node1
6  I am 2 of 4 running on node1
```

If you have access to a private cluster with nodes that already support password-less public-key login via SSH, you could additionally create a file listing their node names or IP addresses, such as

```
node1    slots=2
node2    slots=4
```

where the slots parameter specifies the number of available cores on each node. You would then start the program using

```
$ mpirun -np 4 --hostfile my_hosts ./hello_mpi
```

This results in four copies of hello_mpi distributed among the two nodes.

All MPI programs start with a call to MPI_Init and end with a call to MPI_Finalize. The first command initializes the inter-process communication. The second command indicates to mpirun that the program terminated normally. A process terminating without calling this function is interpreted to have crashed. This then leads to `mpirun` terminating all the other running processes.

MPI running processes are grouped into one or more communicators. `MPI_COMM_WORLD` is the default one that encompasses all processes. The program can create custom communicators that could then be used to split up the processes according to assigned tasks. The number of processes belonging to a communicator give its size, which is queried by calling `MPI_Comm_size`. Note that the output is returned by writing to a memory address provided as the second function argument. This is how all MPI functions return data as the actual function return is reserved for status codes. Each process is also assigned a unique integer "rank" within the communicator. This is a consecutive integer running from 0 to `size-1`. The rank of the current process is obtained using `MPI_Comm_rank`.

Before continuing, it is important to review the difference between multi-threading and MPI. With multithreading, there is only a single program (process) executed by the O/S. This process "splits itself" into multiple threads. Processes operate within a walled-off memory space to prevent programs from accessing each others data. But since the threads belong to the same process, they share their memory space. This is why in the multithreaded examples, we could allocate data vectors in `main` and then pass their addresses to the calculation functions running in separate threads. An MPI launch actually launches the specified number of copies of the given program. This is analogous to "double clicking" the executable file multiple times in the file browser. These running processes share the same disc file, but as far as the operating system is concerned, they are independent processes. They don't have access to each other's memory space. This leads to the *distributed memory* model, visualized in Figure 9.7. The programs communicate by passing data to each other. This communication is accomplished by MPI.

9.4.1 MPI DOT PRODUCT

Let's again consider the computation of a dot product. Similarly to what was done previously with threads, we let each rank work on a section of the array. The primary difference is that instead of a "main thread" calculating the extents that each thread works on, each MPI process figures this out individually using its rank and the known communicator size. Let's consider this initial implementation:

```
1  #include <iostream>
2  #include <sstream>
3  #include <mpi.h>
4  using namespace std;
5
6  int main(int n_args, char **args) {
7      MPI_Init(&n_args, &args);
8
```

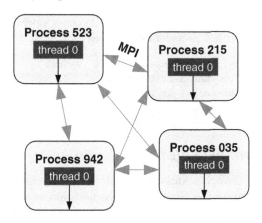

Figure 9.7 Distributed memory program architecture.

```
9       int mpi_size;
10      int mpi_rank;
11      MPI_Comm_size(MPI_COMM_WORLD, &mpi_size);
12      MPI_Comm_rank(MPI_COMM_WORLD, &mpi_rank);
13
14      // every rank allocates and initializes full array
15      size_t N = 1000000;
16      double *a = new double[N];
17      double *b = new double[N];
18
19      //set some values
20      for (size_t i=0;i<N;i++) {a[i] = i+1/(double)(N); b[i]=1/(a[i
            ]*N);}
21
22      int my_size = N/mpi_size+1; // number of nodes per processor
23      int i_start = mpi_rank*my_size; // our first node
24
25      if (mpi_rank==mpi_size-1)  // limit n on the last rank
26         my_size = N - i_start;
27
28      double dot_local = 0;
29      for (size_t i=i_start; i<i_start+my_size; i++) dot_local += a[
            i]*b[i];
30
31      stringstream ss;
32      ss<<"Local dot product on rank "<<mpi_rank<<" is "<<dot_local
            <<endl;
33      cout<<ss.str();
34
35      // wrap up and clean up
36      MPI_Finalize();
37      return 0;
38   }
```

Just as with threads, we use a **stringstream** to format a single string to be outputted using a single call to **cout**. This prevents garbled output that would arise otherwise due to pieces between « operators intermingling. Running the code with 4 processors produces

```
$ mpic++ -O2 dot_mpi1.cpp && mpirun -np 4 ./a.out
Local dot product on rank 3 is 0.249997
Local dot product on rank 1 is 0.250001
Local dot product on rank 0 is 0.250001
Local dot product on rank 2 is 0.250001
```

9.4.2 SEND AND RECEIVE

We can confirm, through manual addition, that the printed values do indeed add up to the expected 1. Such a manual addition is not ideal. We should let the computer do the work. This can be accomplished by having each rank communicate its local result instead of just printing it to the screen. This is where the "message passing" part of MPI comes in. We designate one processor, by convention the one with rank $= 0$, to act as the *root* that handles screen and file input and output. After each non-root process computes its local dot product, it sends its local result to the root. The root receives this data and uses it to increment its own data. This communication is performed using `MPI_Send` and `MPI_Recv` commands. Instead of outputting `dot_local` to the screen, we now have

```
1   if (mpi_rank!=0) {  // if not root
2       MPI_Send(&dot_local,1,MPI_DOUBLE,0,42,MPI_COMM_WORLD);
3   }
4   else { // if root, receive data from all other ranks
5       double dot_global = dot_local; // my contribution
6       for (int r=1;r<mpi_size;r++) { // start with rank 1
7           MPI_Status status;
8           double dot_remote;
9           MPI_Recv(&dot_remote,1,MPI_DOUBLE,r,42,MPI_COMM_WORLD,&
                status);
10          dot_global+=dot_remote; // add in the remote data
11      }
12      // only root is doing screen output
13      cout<<"Using "<<mpi_size<<" processors, dot product is "<<dot_
            global<<endl;
14  }
```

First, any non-root (rank > 0) process uses `MPI_Send` to send the local value to the root (rank 0). The first argument is the address of the memory buffer to send. It is followed by an integer indicating the number of items to send, and an MPI predefined constant indicating the data type. Here we are sending just a single double-precision value corresponding to `dot_local`. Normally one packs data of equal type into a buffer to be sent using a single call to reduce communication overhead. It is also possible to define custom data types that map to structures. The following argument indicates the rank to send to, such as the root rank 0. We next specify an arbitrary tag used to distinguish data packages of otherwise same recipient and size. Finally, we specify the communicator to use.

On the other hand, the rank 0 process initializes a `dot_global` variable with its already computed local value. It then loops over the remaining ranks (ranks 1 through size - 1) and uses `MPI_Recv` to receive a single double-precision real value from each processor. This value is placed into the `dot_remote` variable and is then used to increment `dot_global`. Finally, once the loop over all ranks completes, the accumulated value is printed to the screen. We obtain

```
1  $ mpic++ -O2 dot_mpi2.cpp && mpirun -np 4 ./a.out
2  Using 4 processors, dot product is 1
```

9.4.3 DEADLOCK

Whenever sending data, it is important to watch out for *deadlock*. `MPI_Send` and `MPI_Recv` are blocking operations (there are also non-blocking variants in `MPI_ISend` and `MPI_IRecv` but they require some additional function calls and are usually not necessary). These functions block (do not return) until the data being transferred is dispatched through the communication buffer. For all practical purposes, this means that a send command does not return until the data is "picked up" on the other end. As such, every send command needs a matching receive. Consider the following pseudocode exchanging two values between two processors:

```
1  if (rank==0) {
2     MPI_Send(&a,1);   // send a to 1
3     MPI_Recv(&b,1);   // recv b from 1
4  } else if (rank==1) {
5     MPI_Send(&a,0);   // send a to 0
6     MPI_Recv(&b,0);   // recv b from 0
7  }
```

This code deadlocks (assuming the data being sent is too large to fit in the transfer buffer). The deadlock occurs since both ranks execute a send operation at the same time and are both waiting on the other to perform the receive. The deadlock is eliminated by swapping the order of operations as,

```
1  if (rank==0) {
2     MPI_Send(&a,1);   // send a to 1
3     MPI_Recv(&b,1);   // recv b from 1
4  } else if (rank==1) {
5     MPI_Recv(&b,0);   // recv b from 0
6     MPI_Send(&a,0);   // send a to 0
7  }
```

to ensure there is a matching receive for each send. This situation of needing to exchange data between two processors is so common that there is a single `MPI_Sendrecv` instructions that performs this exchange without needing to worry about ordering of the send and receive calls. It is used as

```
1  MPI_Sendrecv(&send_data, send_count, send_type, dest_rank, send_tag,
2               &recv_data, recv_count, recv_type, src_rank, recv_tag,
3               communicator, status);
```

The blocking nature of these calls also implies that send and receive operations can be used for process synchronization. Perfect load balance cannot be achieved in practice and some processes will finish their workload before the others. Calling `MPI_Recv` to receive needed data will naturally lead to the calling process blocking until the data is made available. While not generally necessary, additional synchronization can be introduced using `MPI_Barrier`. Finally, other functions are available to simplify data transfer. Examples include `MPI_Scatter` and `MPI_Gather` for distributing and collecting buffer sections. Just as was the case with `MPI_Sendrecv`, these functions are form of syntactic sugar that offer simpler interface for functionality otherwise provided by `MPI_Send` and `MPI_Recv`.

9.4.4 REDUCTION

The operation described in Section 9.4.2 is an example of a *reduction*. Multiple values (the local dot products) are reduced (through addition) into a single value corresponding to the global dot product. Instead of doing the reduction manually, we can utilize `MPI_Reduce` or `MPI_Allreduce`. The difference between these two functions is that in the first variant, only the specified rank receives the reduced data. In the second, it is communicated to all. Quite often, the data is needed by just a single processor. There is no need to waste network resources transferring data to processors that won't use it. The entire code listing demonstrating the use of `MPI_Send` and `MPI_Recv` in the prior section can be replaced with

```
1   double  dot_global;
2   MPI_Reduce(&dot_local ,&dot_global ,1 ,MPI_DOUBLE,MPI_SUM, 0 ,MPI_COMM
        _WORLD) ;
3   if  (mpi_rank==0) cout <<"Using "<<mpi_size <<" processors , dot
        product  is  "
4   <<dot_global <<endl ;
```

The first argument provides the address of the data to reduce, while the second argument is the memory location where the reduction result should be saved. The data size is given as a single double-precision value. The reduction is performed through a summation, and the result is be provided only to rank 0. This means that `dot_global` will have a valid value only on the root. Finally, we specify the communicator to use.

9.4.5 DOMAIN DECOMPOSITION

Right now, each rank allocates and initializes the entire $\vec{a}$ and $\vec{b}$ vectors. This is inefficient and also wastes memory. Instead, each rank should load only the data needed for its local computation. This is known as domain decomposition. We demonstrate it by having each processor allocate a vector of just `my_size` entries instead of the global N,

```
1   double  *a = new double[my_size];
2   double  *b = new double[my_size];
```

We next assign the vector values. It is important to pay attention to the indices. The i in Equation 9.8 corresponds to the global index, but the value is stored locally in a smaller array with indexing starting at 0. We use our understanding that each processor handles a `my_size` large vector to calculate the starting offset. This offset is used to compute the global index, $i_{glob} =$ rank $\cdot$ my_size $+ i$,

```
1   // set some values
2   int i_start = mpi_rank*my_size; // our first node
3   for (size_t i=0;i<my_size;i++) {
4       int global_i = i+i_start; // index in the global vector
5       a[i] = global_i+1/(double)(N);  b[i]=1/(a[i]*N);}
```

The rest of the program remains unmodified.

This same concept is also used to parallelize general simulation codes. Consider a solver for some PDE operating on a two-dimensional Cartesian mesh. We can subdivide the global domain into as many blocks as there are processors. An example split is visualized in Figure 9.8(a). Note that this type of splitting restricts the number of MPI processes to the product of domain counts in each spatial dimension. The global computational domain now contains internal "processor" boundaries. Let's say that we are solving the Laplace equation. Nodes along the processor boundary are part of the global non-boundary domain and hence the standard five-point stencil for the second derivative applies there. But placing the stencil on the processor boundary leads to one "leg" extending out of the local domain into the neighbor processor domain. This issue is addressed by incorporating *ghost cells*. Ghost cells (or ghost nodes) are a buffer of cells or nodes that contain data from the neighbor as shown in Figure 9.8(b). The number of ghost layers will depend on the discretization scheme used. The ghost cells are populated with values from the neighbor processors prior to each solver iteration. This is implemented by copying data from the first internal layer to the transfer buffer, and using `MPI_Sendrecv` to exchange the data with the neighbor. The primary challenge with implementing ghost cells involves deciding just how to store them. One option is to expand the local computational domain so that the actual solution domain is found in $i \in [1 : n_i]$, $j \in [1 : n_j]$, and $i = 0$, $i = n_i + 1$ are the ghost nodes in the x direction.

A somewhat similar approach is used for parallelizing particle codes. Here we may again perform domain decomposition, in which case a processor handles all particles existing within a physical subdomain. Alternatively, we could distribute the particle population among the processors in which case each processor handles the full domain but has only a fraction of the particle population. In the first approach, particles crossing domain boundaries need to be collected to a transfer buffer and communicated to the neighbor. It is possible that particles leaving near domain corners need to undergo through multiple hops until the correct target domain is found. In the latter case, the locally computed number density needs to be combined with densities on the other domains using a reduction operation.

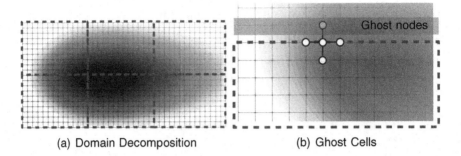

(a) Domain Decomposition   (b) Ghost Cells

Figure 9.8   Use of MPI in a parallel field solver.

9.4.6   REMOTE ACCESS

MPI allows launching instances on multiple physical computers connected by a network. While with multithreading we are limited by the total RAM and CPU-cores available on the single system, MPI allows for a conceptually unlimited computational resources, with the actual limitations only due to the available hardware budget. These resources are provided by computer clusters. We now briefly describe the typical cluster usage. A so-called "Beowulf" cluster consists of a head node that is connected to the outside world, and a massive number of compute nodes connected to the head. These clusters are described in more detail in Sterling 2002. The compute nodes are firewalled from the outside world and communicate with each other on their own private network. The connection to the cluster is accomplished using `ssh` (Secure Shell), such as

```
$ ssh user@host_name
```

Generally, SSH communication is using a text-only "command shell" but graphics (X Windows) forwarding can be enabled with a `-Y` argument if both server and client systems support it. This forwarding makes it possible to launch a GUI-based program on the server, but have the windows show up on the local machine. SSH connection is encrypted. The connection requires some form of authentication, which can be in the form of a password, a public key, or a Kerberos token.

The majority of clusters run Linux. There are several flavors of Linux, including Ubuntu, Debian, RedHat, and CentOs. While there are differences between these implementations when it comes to installing and configuring programs, they tend to use the same shell environment (such as `bash`). Some useful commands include

- `cd` changes the current directory, `mkdir` makes a new directory.
- `cp` copies file (`cp -r` recursive), `mv` moves files.
- `rm` removes files `rm -rf` recursive remove that also deletes directories. These files are immediately deleted (not placed in a recycle bin) and

cannot be easily recovered.

- `vi` is one option for editing files. Press "i" to enter interactive mode that allows you to type. Press "esc" to get out of it. ":" enters a command mode, ":wq" saves file and exits, ":h" shows help, ":q!" quits without writing. "w" outside interactive mode moves cursor one word at a type, and "?" can be used to search. Another popular editor is `emacs`.
- `chown` changes file user or group ownership, `chmod` changes file permissions.
- `man` shows the manual page for some command.
- `tar` and `gzip` are used to combine (tar) files and to compress them. These commands can be combined by passing `-z` argument to `tar`.
- `sshfs` mounts a remote directory over SSH. The mount point directory then works just like a regular local folder, except that that the data is stored at the remote location. This assumes that you are running a compatible O/S on your local system.
- `scp` is used to copy files over SSH. However, `rsync` provides an improved alternative by transferring only the files that have been modified.

Unless you are lucky to have access to a private cluster, you will not be able to run jobs directly by specifying node names in the host file. This is because the cluster is shared with hundreds or thousands of other researchers. Without program launch controls, the system resources would easily become saturated. Clusters utilize some *scheduler* such as `qsub` or `slurm`. These systems provide commands for submitting jobs to a queue and monitoring their status. The cluster user's guide will include a sample "batch" file used for job submission. This file will usually have you specify your account id, so the system knows which account to charge for the CPU time. It will specify the desired number of CPU cores and memory, as well as the expected maximum run time. The queue to use is also specified. There may be some high-priority queues with access limited to only select users. Other queues provide limited resources but shorter start up time to aid in code debugging. The scheduler starts the job once enough resources are available. It is important to request sufficient resource since jobs get killed if the runtime or memory allowance is exceeded. However, requesting excessive resource may lead to your job sitting in the queue for a long time. In the case of heavily utilized systems, it is not unusual to have a job stuck in a queue for days or even weeks. In such cases, you may be able to obtain results in less total time by utilizing smaller local resources.

9.4.7 PARALLEL DEBUGGING

We finally mention a few notes about parallel debugging. Debugging MPI programs is not trivial. The most common issue is deadlock, which can usually be identified by instrumenting the code with `print` statements as described in Section 8.1. However, there may be a time when the use of a debugger is warranted. Assuming you are running MPI code on a local system, or are connected remotely to a server via `ssh -Y` to enable XWindows forwarding, you can take advantage of an `xterm` trick to concurrently run multiple GDB sessions, each associated with a specific MPI rank, in individual terminal windows. Launch the code using

```
$ mpirun -np 4 xterm -e gdb -ex run ./my_sim
```

This command launches four graphical xterm terminals, with each running `my_sim` through the GDB debugger. The individual gdb sessions communicate with each other. Pressing Ctrl-C in any of the windows suspends the execution in the rest. One can then use typical gdb commands to examine variables or the call stack. Resuming the code in one window resumes the execution across the entire parallel ecosystem.

9.5 CUDA

Finally, we discuss code parallelization using graphics cards (GPUs). As was already mentioned in the introduction, graphics cards were originally developed to accelerate single instruction multiple data (SIMD) operations encountered in screen rendering. Early work with the use of GPUs for general computation involved "hacking" this screen output to produce calculation results. This is akin to the π WebGL computation example in Section 7.5. Later, two architectures emerged to directly facilitate GPU computation without needing to worry about image buffers. These architectures are OpenCL and CUDA. The former system works with a wide range of hardware architectures, while CUDA is supported only by NVIDIA graphics cards. However, from our experience, the use of CUDA is much more widespread, and hence we limit our discussion only to this system. CUDA is a language extension to C++ that introduces new syntax for identifying code running on the GPU. The GPU is referred to as the *device*. CUDA programs are typically given a .cu extension. They are compiled with |nvcc| which is obtained by installing the CUDA Toolkit available from **developer.nvidia.com/cuda-toolkit**. This compiler parses and builds only the GPU-specific sections. The rest of the CPU *host* code is processed by gcc.

To demonstrate the use of CUDA, let's consider the following example:

```
1  __global__ void add(float a, float b, float *c) {
2      *c = a + b;
3  }
4
5  int main() {
6      float *dev_c;   // pointer to data on the GPU
7      cudaMalloc((void**)&dev_c, sizeof(float));
8
9      // malloc C version of "new"
10     add<<<1,1>>>(1.1,2.2,dev_c);
11
12     float c;
```

```
13      cudaMemcpy(&c , dev_c , sizeof ( float ) , cudaMemcpyDeviceToHost ) ;
14
15      cout <<"GPU computed result is "<<c<<endl ;
16
17      cudaFree ( dev_c ) ;
18      return 0;
19  }
```

This code uses CUDA to add two numbers. The actual addition is performed using the function **add** which is declared with an unfamiliar **__global__** prefix. CUDA classifies all functions into one of three categories: host functions that run on the CPU, device functions that run on the GPU and are only callable by other GPU functions, and global functions that run on the GPU but are executed from the CPU. Functions of this last category are also referred to as kernels. Functions without a classifier are assumed to be of type **__host__**. Subsequently, we see that **main** begins by using the **cudaMalloc** function to allocate 4 bytes (size of a single precision float) of memory. This function shares a calling signature with the legacy C **malloc**, however unlike malloc, it allocates memory on the GPU. The memory address assigned to the **dev_c** pointer is not valid on the CPU side! Attempting to access it from **main** will lead to a program crash.

We instead pass this pointer to the **add** kernel. This is where the function stores the result of its calculation. Kernels are launched from host code with a <<<nb, tpb>>> syntax. The first value provides the number of requested *blocks*, each consisting of the number of threads per block given by the second parameter. These values can be scalars (as is the case here) or can be of type **dim3** and describing a three-dimensional grid. The total number of threads is given by cross-multiplying these parameters. Each kernel function has access to built-in variables **threadIdx**, **blockIdx**, and **blockDim** that it can use to determine which part of the problem to work on. In this case, we launch just a single instance. The code in kernel stores the computed value in GPU memory. The data needs to be copied to main system RAM in order to actually use it from the CPU side. This is accomplished with the **cudaMemcpy** function. This function follows the syntax of the legacy C **memcpy** function in which the first parameter is the memory address of the destination and the second argument is the address of the source. We also need to specify the number of bytes to transfer. The final parameter specifies transfer type. Here we copy from the GPU to the CPU, implying that the source pointer, **dev_c** needs to store a GPU memory address, while the destination address, given by &c needs to map to the main system memory. Finally, we display the result and free the dynamically allocated memory. The above code is compiled and run as

```
$ nvcc -O2 add_cuda.cu -o add_cuda
$ ./add_cuda
$ c = 3.3
```

The calculated result is indeed correct.

9.5.1   ONLINE COMPUTATIONAL RESOURCES

Running this example requires you to have access to a computer with an NVIDIA GPU. In the absence of such a system, you can, at least as of this writing, obtain free access to GPU resources using Google Colaboratory, colab.research.google.com/. This online environment is usually used to run Python code within the embedded Jupyter notebook. You can however also access the underlying Linux environment. This is done with special %% commands. Before starting, it is important to navigate to the Change Runtime Type screen found under the Runtime menu to select the GPU hardware accelerator option. By default, Google colab starts up on a system containing just a CPU. You can confirm that the system does indeed have access to a GPU by adding the following code block:

```
%%script bash
nvidia-smi
```

The first line indicates that this block contains commands to be executed using the system Bash shell. Alternatively, you can also write

```
!nvidia-smi
```

You should see a table listing hardware information of the available GPU along with a list of running processes (which should be empty). Next, write out your example code to file using

```
%%writefile add.cu
#include <iostream>
using namespace std;
...
```

Then compile and run the code as illustrated below:

```
%%script bash
nvcc add.cu -o add
./add
```

While this example program does not produce any file out, had you been running a code that does produce output files, you would subsequently download them to your computer using a Python code block containing

```
1  from google.colab import files
2  files.download("<filename>")
```

9.5.2   VECTOR ADDITION

After this brief introduction, let's see how to actually use CUDA to perform vector addition. The use of CUDA for code acceleration is analogous to multithreading, with the difference that the number of utilized threads is substantially larger. Specifically, we usually assign individual threads to each element index. Instead of the 4-block decomposition shown on page 350, with GPU we essentially execute

```
1   thread  0:  c[0]  =  a[0]  +  b[0];
2   thread  1:  c[1]  =  a[1]  +  b[1];
3   thread  2:  c[2]  =  a[2]  +  b[2];
4   ...
5   thread  N-1:  c[N-1]  =  a[N-1]  +  b[N-1];
```

Let's start with data allocation and initialization. This entire code is found in CUDA/add2.cu.

```
1   #include <iostream>
2   using namespace std;
3
4   int main() {
5       // allocate vectors
6       const int N = 1000;
7       float *a = new float[N];
8       float *b = new float[N];
9       float *c = new float[N];
10
11      // initialize values
12      for (int i=0;i<N;i++) {
13          a[i] = i;
14          b[i] = 2*i;
15      }
```

We are utilizing single precision floats instead of doubles since historically GPUs lacked support for double precision. This is no longer the case, however, computation with single precision is still faster. GPUs also tend to have smaller available memory and hence using single precision allows the program to store information about twice as many items. We next allocate memory on the GPU and copy the data from the CPU to the GPU:

```
1       float *dev_a;   // pointer to data on the GPU
2       float *dev_b;   // pointer to data on the GPU
3       float *dev_c;   // pointer to data on the GPU
4
5       cudaMalloc((void**)&dev_a, N*sizeof(float));
6       cudaMalloc((void**)&dev_b, N*sizeof(float));
7       cudaMalloc((void**)&dev_c, N*sizeof(float));
8
9       cudaMemcpy(dev_a,a,N*sizeof(float),cudaMemcpyHostToDevice);
10      cudaMemcpy(dev_b,b,N*sizeof(float),cudaMemcpyHostToDevice);
```

We then launch N copies of the **add** kernel. Subsequently, we copy the results from the GPU to the CPU using **cudaMemcpy**. Finally, we visualize the first 10 components to confirm that the calculation executed correctly:

```
1       add<<<1, N>>>(dev_a, dev_b, dev_c);
2       cudaMemcpy(c,dev_c,N*sizeof(float),cudaMemcpyDeviceToHost);
3
4       // make sure the results make sense
5       for (int i=0;i<10;i++)
6           cout<<a[i]<<" + "<<b[i]<<" = "<<c[i]<<endl;
7
8       delete[] a;
9       delete[] b;
10      delete[] c;
11
```

```
12      return 0;
13   }
```

The kernel is given below:

```
1   __global__ void add(float *a, float *b, float *c) {
2       int i = threadIdx.x;
3       c[i] = a[i] + b[i];
4   }
```

This code looks similar to our initial addition example, except that the inputs are single precision float array pointers instead of scalars. Every thread receives the same pointers and must subsequently decide which array item to operate on. This is done using the built-in **threadIdx** variable. It is of a CUDA type **dim3** (based on **uint3**, which is a 3-component integer array with components accessed using .x, .y, and .z. Any unspecified components default to 1. Given a "one-dimensional" launch, as is the case here, only the .x component is meaningful. It gives us the thread id, which is subsequently used for array indexing.

9.5.3 ERROR CHECKING

Now, let's say that you wanted to run the code with a bigger vector. While the code compiles and runs without any warning message with N increased to 10,000, the results are incorrect:

```
$ ./add2
0 + 0 = 0
1 + 2 = 0
2 + 4 = 0
```

CUDA functions return a non-zero value if they failed. As such, it is customary to wrap each CUDA call (or at least the ones you suspect of causing a problem) with a macro that checks the return value and prints out a message as needed. Such a macro can be defined using

```
#define CUDA_ERROR(f) {if (f!=cudaSuccess) {cerr <<
    cudaGetErrorString(f)<<" on line "<<__LINE__<<endl; exit(-1)
    ;}}
```

where **cudaSuccess** evaluates to 0 and **cudaErrorString** is used to translate the error message to an informative message. We use it as in

```
CUDA_ERROR(cudaMalloc((void**)&dev_a, N*sizeof(float)));
```

The only exception is the kernel launch. We need to use **cudaPeekAtLastError**, as in

```
1   add<<<1, N>>>(dev_a, dev_b, dev_c);
2   CUDA_ERROR(cudaPeekAtLastError());
```

Running this modified code (found in **CUDA/add2_ec.cu**, we obtain

```
invalid configuration argument on line 1
```

with the line corresponding to the kernel launch.

9.5.4 BLOCKS

The GPU limits how many threads can be launched per block. This maximum number can be obtained using

```
1  cudaDeviceProp cuda_props;
2  cudaGetDeviceProperties(&cuda_props,0);
3  cout<<"Max threads per block = "<<cuda_props.maxThreadsPerBlock<<
       endl;
```

which usually evaluates to 1024. As such, it is not possible to launch the add function, as written so far, using 10,000 threads. In order to process the larger array, we need to subdivide the launch into multiple blocks, such as

```
1  int threads_per_block = 512;
2  int num_blocks = (N+threads_per_block-1)/threads_per_block;
3  add<<<num_blocks, threads_per_block>>>(dev_a, dev_b, dev_c, N);
```

The number of threads per block becomes a user input and can in fact have noticeable impact on code run time. A non-negligible aspect of CUDA optimization involves trying different thread per block sizes to find the configuration that maximizes the performance for a given problem. The total *grid size* is given by the product of `num_blocks` and `threads_per_block`. Inside the kernel, we use two additional built-in variables to compute the thread id,

```
1  __global__ void add(float *a, float *b, float *c, int N) {
2      int i = blockIdx.x*blockDim.x+threadIdx.x;
3      if (i<N)
4          c[i] = a[i] + b[i];
5  }
```

Since the data size is generally not evenly divisible by the number of threads per block, the code ends up launching more copies of the kernel than there are data points. Some kernels will be launched with indexes exceeding the data size. Thus, as shown in the listing above, we need a bounds check to limit the calculation only on threads mapping to valid data indexes.

9.5.5 TIMING

The next version of this code, found in `add4.cu`, adds timing. We use

```
using type = float;
```

to easily allow switching between single and double precision, with all subsequent instances of `float` replaced with `type`. We also use

```
auto start = chrono::high_resolution_clock::now();
```

to time the CPU and GPU performance with a larger, 50 million items long vector. We grab time points before and after memory transfer,

```
1  auto t1 = chrono::high_resolution_clock::now();
2  cudaMemcpy(dev_a,a,N*sizeof(type),cudaMemcpyHostToDevice);
3  cudaMemcpy(dev_b,b,N*sizeof(type),cudaMemcpyHostToDevice);
4
5  auto t2 = chrono::high_resolution_clock::now();
6  add<<<num_blocks, threads_per_block>>>(dev_a, dev_b, dev_c, N);
7  cudaDeviceSynchronize();
```

```
 8   auto t3 = chrono :: high_resolution_clock :: now () ;
 9
10   cudaMemcpy ( c 2 , dev_c ,N* sizeof ( type ) , cudaMemcpyDeviceToHost ) ;
11   auto t4 = chrono :: high_resolution_clock :: now () ;
12
13   std :: chrono :: duration <double , std :: milli > duration1 = t4−t 1;
14   std :: chrono :: duration <double , std :: milli > duration2 = t3−t 2;
```

The cudaDeviceSynchronize call is used to wait for the calculations to finish. A kernel launch is not a blocking operation (the actual completion block happens in cudaMemcpy) and thus without this call, the time point t3 would be meaningless. The time difference between t2 and t3 corresponds to the time spent solely performing the calculation. The elapsed time from t1 to t4 captures the entire GPU processing including memory transfer. Using single precision, we obtain the following output:

```
$ nvcc -O2 add4.cu -o add4 && ./add4
CPU time 86.9048 ms
GPU time 8.69542 ms
GPU time with memory transfer 128.235 ms
```

With double precision we obtain:

```
$ nvcc -O2 add4.cu -o add4 && ./add4
CPU time 153.111 ms
GPU time 17.7095 ms
GPU time with memory transfer 254.745 ms
```

We can make several observations. First, the use of double precision is indeed slower by a factor of about 2. This factor shows up both on the CPU and the GPU side. We can also notice that it took the GPU only 8.7 ms to crunch through the vector, while it took the CPU 86.9 ms. This seems great. But once the memory transfer is included, the GPU processing time increased to 128.2 ms for the single precision case. This is 48% slower than just running on the CPU.

This finding is unfortunately not unexpected. While GPUs offer rapid computation, the corresponding memory transfer can be a major bottleneck. Yet, even by utilizing the tools discussed next, GPU code optimization is a non-trivial effort. CUDA threads are organized into groups of 32 called warps. The GPU itself contains multiple "streaming multiprocessors" or SMs with a scheduler that switches warps onto the SM once they are eligible to run. Each SM has a theoretical maximum number of warps it can handle. It can be determined using

cout ≪cuda_props . maxThreadsPerMultiProcessor / cuda_props . warpSize ;

which usually leads to the value of 64. NVIDIA CUDA toolkit also comes with the NSight IDE which integrates with Microsoft Visual Studio or Eclipse, and also provides a profiler that visualizes the warp efficiency. Running it on this example code, you will see that perhaps only as few as 14% of all warps are eligible for execution. The rest are stalled for a variety of reasons such as waiting for an instruction, or, as is usually the case, due to memory dependency. This implies that the processor is waiting for data. One way to alleviate this problem is by utilizing *shared memory* instead of the *global memory* we have been using so far. Threads running in the same block have access to a faster local memory space that can be used for intermediate computations. Subsequently, only the reduced data is placed in the global memory space that was previously allocated with `cudaMalloc`. We will see how to utilize shared memory in the next section on the dot product. Another important issue to pay attention to is the instruction dependence. The warp itself is split into two halves, and each half-warp (16 threads) must be executing the same instruction at the same time - this is the whole concept behind the SIMD architecture. This means that if the code diverges, due to the presence of conditional branches, the threads in the true branch run while those in the false branch are suspended, and vice versa. Thus, with GPU parallelization, it may be beneficial to perform extra calculations that are later discarded instead of trying to "optimize" the code by performing those calculations selectively.

9.5.6 PINNED MEMORY

Returning back to the memory transfer penalty, one reason for the long memory copy times is that the main system memory is *pageable*. RAM data is organized into virtual pages, which the operating system can move around or even offload to the disk when insufficient physical memory is available (which comes with a noticeable slowdown in system performance). Before memory is copied to the GPU, it is necessary to pin it in place so that a page swap does not inadvertently happen while the buffer is mid-copy. The `cudaMemcpy` function handles this by first allocating a temporary *pinned* buffer. The data to be copied to the GPU is copied into this temporary buffer before actually being transferred to the device. This is quite inefficient especially if the array is to be reused multiple times. For this reason, CUDA provides the `cudaHostAlloc` function that allows the program to allocate pinned memory on the CPU. This function is used in lieu of `new` or `malloc`. Its use is demonstrated in `cuda5.cu`. Instead of

```
1  float *a = new float [N];
2  float *b = new float [N];
3  float *c = new float [N];
```

we write

```
1  cudaHostAlloc(&a, sizeof(type)*N, cudaHostAllocDefault);
2  cudaHostAlloc(&b, sizeof(type)*N, cudaHostAllocDefault);
3  cudaHostAlloc(&c, sizeof(type)*N, cudaHostAllocDefault);
```

Just with this one change, the GPU addition time for the 50 million-item long single precision arrays dropped from 128.2 ms to 54.7 ms, which is now faster than the 86.9 ms time recorded on the CPU. Similarly, the time with double precision dropped from 254.7 ms to 109.7 ms, again beating the CPU time of 153.11 ms.

9.5.7 STREAMS

Yet, an even further improvement can be made. All modern CUDA-capable GPUs support concurrent copy and execution. In essence, data can be copied to (or from) the device while the GPU is running calculations. This is accomplished using streams. All operations within a single stream are executed in the specified order. Having multiple streams allows the GPU to interleave the operations. The use of streams is also demonstrated in add5.cu. We begin by creating two streams,

```
1  cudaStream_t  stream1, stream2;
2  cudaStreamCreate(&stream1);
3  cudaStreamCreate(&stream2);
```

We next divide our problem set into some arbitrary, but evenly divisible, number of chunks, 10 in this example. We use the chunk size to determine the number of thread blocks to launch,

```
1  const int CHUNK = N/10;
2  num_blocks = (CHUNK+threads_per_block−1)/threads_per_block;
```

We then loop over the chunks but with each loop iteration processing two consecutive chunks, as shown below

```
1  for (int i=0;i<N; i+=CHUNK*2) {
2      // schedule copies of chunk of "a" on streams 1 and 2
3      cudaMemcpyAsync(dev_a,a+i ,CHUNK*sizeof(type),
             cudaMemcpyHostToDevice,stream1);
4      cudaMemcpyAsync(dev_a, a+i+CHUNK,CHUNK*sizeof(type),
             cudaMemcpyHostToDevice,stream2);
5
6      // repeat for vector "b"
7      cudaMemcpyAsync(dev_b,b+i ,CHUNK*sizeof(type),
             cudaMemcpyHostToDevice,stream1));
8      cudaMemcpyAsync(dev_b, b+i+CHUNK,CHUNK*sizeof(type),
             cudaMemcpyHostToDevice,stream2));
9
10     // schedule kernel execution
11     add<<<nb,tpb,0,stream1>>>(dev_a, dev_b, dev_c, CHUNK);
12     add<<<nb,tpb,0,stream2>>>(dev_a, dev_b, dev_c, CHUNK);
13     cudaPeekAtLastError());
14
15     // schedule memory copy back
16     cudaMemcpyAsync(c+i , dev_c ,CHUNK*sizeof(type),
             cudaMemcpyDeviceToHost,stream1);
17     cudaMemcpyAsync(c+i+CHUNK, dev_c ,CHUNK*sizeof(type),
             cudaMemcpyDeviceToHost,stream2);
18  }
```

Inside the loop, we use cudaMemcpyAsync to schedule the memory copy of the vector blocks to the GPU. This is a non-blocking operation that returns control back to the calling CPU code right away. We schedule the copy of the a vector for the first chunk using the first stream, and also a copy of the a vector corresponding to the indexes for the second chunk, using the second stream. We similarly interleave the copy of the b vector, the kernel launch, and the subsequent copy back to the CPU. Note that the kernel is now launched using

4 parameters. The third parameter indicates the number of bytes of additional shared memory to dynamically allocate on top of the already available statically allocated memory space. The fourth parameter indicates the stream to use. It is important to realize that all of these operations are asynchronous. There is no guarantee that any of the specified operations finish or even start executing before we go on to the next loop iteration. The two streams end up with a long sequence of operations such as 'copy A to GPU', 'copy B to GPU', 'execute kernel', 'copy C to CPU', 'copy A to GPU', 'copy B to GPU' and so on. Outside the loop, we call `cudaStreamSynchronize` to wait for the stream operations to complete,

```
1  cudaStreamSynchronize ( stream1 ) ;
2  cudaStreamSynchronize ( stream2 ) ;
```

So how did we do? With streams (and pinned memory), the completion time reduces to 38.2 ms for single precision, and 78.6 ms for double precision. This corresponds to a $3.36\times$ improvement over the original version, and a $2.25\times$ improvement over the CPU code. For double precision, this timing corresponds to a $1.95\times$ improvement over the CPU.

9.5.8 DOT PRODUCT

Let's now consider the computation of a dot product. Again, let's assume that we have allocated two vectors on the CPU and have copied their contents to the GPU. We may then be tempted to write the kernel as

```
1  __global__ void dot_bad ( float *a , float *b , float *dot , int N) {
2      int i = blockIdx.x*blockDim.x+threadIdx.x;
3      if ( i<N)
4          *c += a[i]*b[i];
5  }
```

But as we already know from our prior discussion on the race condition, this calculation will lead to an erroneous result since we have many threads attempting to update the same memory space at once. To get around this race condition, we need to utilize another memory storage option. So far we have only explored using the GPU global memory. CUDA defines two additional memory types: *texture* and *shared*. Texture memory offers fast access for data interpolation, but we don't discuss it here. Shared memory on the other hand is quite useful, and in fact, in many cases necessary. It corresponds to memory that is shared between threads belonging to the same block. It is declared using `__shared__` keyword, such as

```
__shared__ float prod[1024];
```

Each thread block has its own "prod" array. It is declared to be size 1024 since per hardware limitations, that is the maximum number of threads per block. The thread can then use the memory at the index indicated by `threadIdx.x` as its own scratch space. Instead of having the threads update the global dot product value, we just let each thread store the multiplication result for its items in the appropriate shared memory location,

```
1   __global__ void dot(float *a, float *b, float *c, int N) {
2       __shared__ float prod[1024];
3
4       int i = blockIdx.x*blockDim.x+threadIdx.x;
5
6       // default
7       prod[threadIdx.x] = 0;
8
9       if (i<N)
10          prod[threadIdx.x] = a[i]*b[i];
```

While this avoids the race condition, we now have the multiplication results in a memory space that cannot be accessed outside the block. Furthermore, we care about the total dot product, not a collection of individual products. We thus need to reduce this data. While there are better algorithms (see Sanders and Kandrot 2011), here we just let the thread with id 0 act as the "root" that performs this summation serially. Before this can happen, we need to wait on all threads within the block to finish. This is accomplished using **__syncthreads__**. This result, corresponding to the dot product contribution just from this block, is then placed into another array, this one located in the global memory, using the block id as the index,

```
1       // wait for all threads in the block to finish
2       __syncthreads();
3
4       // only "root" performs the sum
5       if (threadIdx.x==0) {
6           float sum=0;
7           for (int i=0;i<blockDim.x;i++)
8               sum+=prod[i];
9               c[blockIdx.x] = sum;
10      }
11  }
```

The final sum over the blocks is performed on the CPU:

```
1   float *c = new float[num_blocks];
2   cudaMemcpy(c,dev_c,num_blocks*sizeof(float),
        cudaMemcpyDeviceToHost);
3
4   /*perform final sum on the CPU*/
5   float dot_gpu = 0;
6   for (int i=0;i<num_blocks;i++) dot_gpu+=c[i];
```

9.5.9 CONCURRENT MPI-CUDA COMPUTATION

Many supercomputer clusters offer GPU-accelerated nodes. To fully take advantage of such a system, it is necessary to combine MPI and CUDA parallelization. This subsequently requires compiling your application with support for both MPI and CUDA processing. One option is to place all CUDA code in separate .cu files and compile them with nvcc, with the rest of code compiled with mpic++. But it's probably easier to skip mpic++ altogether and just run the entire compilation through nvcc. mpic++ is just a shell script that adds the location of MPI headers and also links in the MPI runtime library. These

inputs can be provided to nvcc which itself also routes compilation though
g++. Using just a single step, we can write

```
$ nvcc -x cu mpi_cuda.cpp -I<path_to_MPI> -lmpi
```

Alternatively, we could perform the build using separate compile and link
steps,

```
$ nvcc -x cu -arch=sm_20 mpi_cuda.cpp -I<path_to_MPI> -dc
$ nvcc mpi_cuda.o -lmpi -o mpi_cuda
```

The -x cu flag instructs nvcc to treat the inputs as CUDA files despite
the .cpp extension. -dc means to compile and build the device code, and
-arch=sm_20 specifies the CUDA version to build for.

9.5.10   VISUALIZATION

The prior sections introduced the use of graphics cards for general computing.
But returning to their roots, what about using them for actual rendering? One
option is to combine the code with a visualization library such as the Visualization Toolkit. That approach is recommended whenever data manipulation,
such as computation of cutting planes or isosurfaces, is needed. However, let's
assume that your code produces results that already map to a 2D buffer. In
this case, we can directly visualize this data as a pixel map. 3D data can be
visualized using volume ray casting. We use openGL to perform this visualization. openGL is the low-level programming interface used for GPU rendering.
    The example in render1.cu introduces the use of openGL. Much of the
code found in main shares similarities with commands you are already familiar
with from the discussion of webGL in Section 7.5,

```
1   //initialize openGL using double buffer and RGBA mode
2   glutInit (&argc, argv);
3   glutInitDisplayMode (GLUT_DOUBLE|GLUT_RGBA);
4   glutInitWindowSize (DIM, DIM);
5   glutCreateWindow ("Render 1");
6   glewInit ();
7
8   //generate buffer object and bind it to pixel data
9   GLuint buffer_obj;
10  glGenBuffers (1, &buffer_obj);
11  glBindBuffer (GL_PIXEL_UNPACK_BUFFER_ARB, buffer_obj);
12
13  //allocate memory to hold pixel data, each pixel is 4 bytes (RGBA
        )
14  GLubyte *rgba_data = new GLubyte[DIM*DIM*4];
15  //call our placeholder function to generate data
16  generateData (rgba_data);
17
18  //copy our data to the openGL buffer on the GPU
19  glBufferData (GL_PIXEL_UNPACK_BUFFER_ARB, DIM*DIM*4,rgba_data,GL_
        DYNAMIC_DRAW_ARB);
20
```

```
21  //add hooks for processing keyboard and draw requests
22  glutKeyboardFunc(key_func);
23  glutDisplayFunc(draw_func);
24
25  //start main loop, leaves the window open and fires keyboard
       events
26  glutMainLoop();
27
28  //clean up
29  delete[] rgba_data;
30  glDeleteBuffers(1,&buffer_obj);
```

We start by initializing openGL and specify that we want to use double buffering and RGBA mode. We then create a GUI window with the specified dimensions and also give it a title. We also initialize the extension-management library GLEW. We next generate openGL buffers. Buffer creation consists of three steps. First, we obtain a handle to the buffer. Next, we associate this handle with some openGL data structure such as the pixel data used here. Finally, we copy the data to the buffer. We use a custom **generateData** function to produce data for visualization. In your simulation code, you would use the code results instead. We also add listeners for keyboard press and redraw events. These event handlers let the application respond to user interaction. We then start the glut (gl utility) main loop. This function does not return until the window is closed.

The following keyboard listener closes the window if the ESC (ASCII code 27) key is pressed,

```
1  static void key_func(unsigned char key, int x, int y) {
2      switch(key) {
3          case 27: exit(0);  // esc
4      }
5  }
```

The actual visualization happens inside the repaint listener,

```
1  static void draw_func(void) {
2      glDrawPixels(DIM, DIM, GL_RGBA, GL_UNSIGNED_BYTE, 0);
3      glutSwapBuffers();
4  }
```

We use **glDrawPixels** to draw pixel data. We could instead draw triangle buffers, as was shown in Chapter 7. The last argument is the pointer to the data, or 0 if openGL should use the buffer already bound to the pixel unpack buffer.

In the prior example, the redraw happens only as needed. However, let's say you wanted to animate some results. To do so, we register an "idle listener".

```
glutIdleFunc(idle_func);
```

This function gets called by the **glutMainLoop** when openGL is ready for repainting. We populate its body with code that was initially placed in main. First **generateData** is called to recompute data, the data buffer is then updated, and finally **glutPotRedisplay** is called to force a redraw.

```
1  static void idle_func(void) {
2      generateData();
```

```
3      glBufferData(GL_PIXEL_UNPACK_BUFFER_ARB, DIM*DIM*4,rgba_data,
           GL_DYNAMIC_DRAW_ARB);
4
5      // redisplay image
6      glutPostRedisplay();
7    }
```

These two examples only demonstrated the use of openGL, without utilizing any GPU computing. The third version, found in `render3.cu` moves
the data computation to the GPU. We essentially modify `generateData` to
launch a CUDA kernel to compute the individual pixels in parallel instead of
looping over them serially,

```
1    void generateData() {
2        static float t = 0;
3        static float dt = 0.1;
4
5        dim3 dims = {DIM,DIM,1};
6        gen_data_gpu<<<dims,1>>>(dev_data,t);
7        cudaMemcpy(rgba_data,dev_data,DIM*DIM*4*sizeof(GLubyte),
           cudaMemcpyDeviceToHost);
8
9        t+=dt;
10       if (t>25 || t<-25) dt=-dt;
11   }
```

Note that we launch one block per pixel. This is definitely not a good way to
write fast GPU code. This approach is used for simplicity.

Now, while this version works, it is rather silly. The code implements the
following logic:

1. Use GPU to compute some data
2. Copy the data to the CPU using `cudaMemcpy`
3. Copy the same data back to the GPU using `glBufferData`
4. Use the data on the GPU to redraw the screen image

Clearly, steps 2 and 3 should be eliminated. This is possible with CUDA
/ OpenGL interoperability. There are only a few changes required in the
code to make this happen. First, we need to include `cuda_gl_interop.h`.
We also need to use `cudaGLSetGLDevice` to tell CUDA that we want to use
the specified card for rendering. This function needs a device id parameter.
We start by using `cudaChooseDevice` to get the id of an available GPU that
meets the specified requirements (in this case supporting very old 1.1 compute
capability)

```
1    cudaDeviceProp prop;
2    int dev;
3    memset(&prop,0,sizeof(cudaDeviceProp));
4    prop.major = 1;
5    prop.minor = 1;
6    cudaChooseDevice(&dev,&prop);
7    cudaGLSetGLDevice(dev);
```

The prior code versions used a dynamically allocated `GLubyte` array to
store the pixel data. This is no longer need. We also initialize `glBufferData`

with NULL as the data source, indicating that no data is to be copied. Instead, we use

```
cudaGraphicsGLRegisterBuffer(&resource , buffer_obj ,
    cudaGraphicsMapFlagsNone) ;
```

to get a CUDA "resource id" corresponding to the openGL data buffer. Then in **generateData**, we use

```
1  cudaGraphicsMapResources(1,& resource ,NULL) ;
2  cudaGraphicsResourceGetMappedPointer (( void **)&dev_data ,& size ,
       resource) ;
```

to get the device memory address of the openGL buffer. This address is subsequently passed to the kernel to use for its calculation output. And that's it. These changes, found **render4.cu** allow the CUDA calculations to be rendered directly without involving the CPU.

10 Optimization and Machine Learning

This chapter covers two separate, but related topics. We first begin by introducing several numerical approaches for finding optimal input parameters that reduce the difference between the predicted and the expected (true) value. We next provide a very elementary introduction to machine learning, which involves optimization of the coefficient space based on a large set of training data. We develop a simple neural network for classifying real values.

10.1 INTRODUCTION

Our discussion so far has revolved around approaches for solving some governing equations given a set of *known* initial and boundary conditions. The reality is not so clear cut. Some physical properties are simply not available and we need to make educated guesses on their values. Other times, the physical model may include coefficient terms which do not have a well-defined magnitude. As an example, consider the issue of electron confinement encountered in fusion and plasma propulsion applications. In a sufficiently strong magnetic field, electrons are expected to orbit around the magnetic field lines until a scattering event such as a wall or an atomic collision kicks them off. The experimentally observed rate of these scattering events is greater than can be explained by classical theory. This led to the inclusion of an "anomalous diffusion" coefficient in governing equations for electron transport. This model was not based on any fundamental, first principle derivations. It was simply incorporated to improve the agreement with experimental results. Such a coefficient is an example of a simulation "knob" (or a fudge factor) that is ubiquitous in simulation codes. A good part of a numerical modeler's workload may involve running batches of simulations to find the set of inputs that produce the optimal agreement with experimental data. Different communities invoke a variety of different terminology for activities falling under this general umbrella such as optimization, parameter inference, inverse problems, data assimilation, or more generally some variant of "learning". They generally involve a process of searching through possible models in pursuit of improved or accelerated predictive power supporting some higher-level outer goal. In this chapter we introduce several approaches for automating this search.

DOI: 10.1201/9781003132233-10

10.1.1 COST FUNCTION

A numerical simulation is essentially a transfer function converting some inputs $\vec{x}$ into outputs $\vec{y}$,

$$\vec{y} = F(\vec{x}) \tag{10.1}$$

The input space can be further divided into the "known" values $\vec{k}$ and the unknown model parameters $\vec{m}$ discussed above, $\vec{x} = \vec{k} \cup \vec{m}$. We can define a *cost function* to characterize how well the model resolves the expected output,

$$e = \left[\frac{\sum_i^n (y_i - y_{true,i})^2}{n} \right]^{0.5} \tag{10.2}$$

This is just the L2 norm between the computed and expected values. Of course, this description assumes that the two solution vectors can be compared directly. In reality, we may first need to run both sets through a secondary algorithm to perform data reduction. For instance, the results may not be known at the same physical locations, necessitating the use of an interpolation scheme. In the case of systems that only achieve an oscillatory steady state, we may need to compute the spectral density or a phase portrait.

10.2 OPTIMIZATION APPROACHES

Holding the $\vec{k}$ known inputs fixed, our goal is to select $\vec{m}$ which minimizes e (or, alternatively, maximizes a *fitness function*, $f = 1 - e$). There are multiple approaches to chose from. The most obvious option is to assign ranges and discretization to the parameters of interest and compute e for every combination. Whenever we encounter a lower error than the current minimum, we update our "best known" fit. The challenge of this *brute force* scheme is that the number of combinations quickly spirals out of control as the number of input parameters increases. The total number of cases is given by the product $\prod_i N_i$, where N_i is the number of discrete values to check for the parameter i. Evaluating the $\vec{y}$ solution vector requires running a numerical simulation, which depending on the problem at hand, may require hours of run time. For more than a handful of coefficients, sampling a multidimensional parameter space becomes computationally prohibitive.

The brute force approach is however not without its use. The cost function may exhibit multiple *local minima*. The collection of model parameters that minimizes the error within certain lower and upper bounds may not be the true global minimum that would be found over a larger input space. A typical optimization run thus consists of two stages. First, during the *exploration* stage, we identify regions that look promising by exhibiting a low value of the cost function. In other words, we identify regions encompassing local minima. In the second *exploitation* stage, we investigate these regions in detail to find the actual global minimum.

The brute force scheme can be useful during the initial exploration stage to obtain an insight into the "lay of the land". However, if the gridding is too

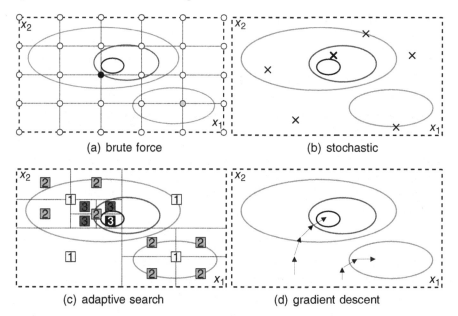

(a) brute force   (b) stochastic

(c) adaptive search   (d) gradient descent

Figure 10.1 Summary of common optimization approaches.

coarse, there exists the danger that a local minimum is completely skipped over if the error response is highly non-linear. An alternative approach is to utilize a *stochastic (Monte Carlo) search*. Here we select random values for each parameter from a distribution spanning the input space. We keep track of the input vector that produced the best fit. The sampling repeats for a prescribed number of samples or until convergence.

10.2.1 ADAPTIVE SEARCH

Once a promising parameter region is identified using the above approaches, we can use another technique to hone in on the optimal values. One option here is to use an adaptive binary search. This approach is illustrated in Figure 10.1(c). We start with a coarse grid surrounding the region of interest. We compute the error at each cell center. The cell with the lowest error is subdivided into equal-sized partitions in each parameter dimension. We again compute the error at each cell center and subdivide further, repeating the process until convergence. Convergence can be defined as

$$\left| \frac{e^L - e^{L+1}}{e^L} \right| < \epsilon_{tol} \tag{10.3}$$

where e^L is the error at subdivision level L. This approach is best suited for problems with a single minimum in the domain bounds.

10.2.2   GRADIENT DESCENT

The rate of change of the cost function $e(\vec{x})$ is given by its gradient,

$$\nabla e = \frac{\partial e}{\partial x_0}\hat{x}_0 + \frac{\partial e}{\partial x_1}\hat{x}_1 + \ldots \qquad (10.4)$$

Instead of selecting the next parameter by iterating over a discretized domain, we can try to minimize the error by moving in the $-\nabla e$ direction. This approach is known as *gradient descent*. Gradient descent is also used in linear solver algorithms such as the *conjugate gradient* introduced in Chapter 3.

In a simulation driven optimization we often will not have an analytical expression for the gradient and instead need to rely on a numerical approximation.[1] The gradient of a function $e(\vec{x})$ can be estimated as

$$\nabla e \approx \sum_i C_i \hat{x}_i \qquad (10.5)$$

where C_i is the central difference approximation of $\partial e/\partial x_i$. For example

$$C_1 = \frac{e(x_0, x_1 + \Delta x_1, x_2, \ldots) - e(x_0, x_1 - \Delta x_1, x_2, \ldots)}{2\Delta x_1} \qquad (10.6)$$

The new parameters are then given by

$$\vec{x}^{k+1} = \vec{x}^k - \nabla e \cdot \vec{x} \qquad (10.7)$$

An alternative approach involves selecting a single nearby random parameter space completely randomly, and use information just from these two data points to compute the gradient. This method is called the *random gradient-free method*, see Nesterov and Spokoiny 2017. We can also eliminate the computation of the gradient altogether, and just select several near-by random samples. We then continue the search with the parameters that minimized the error. Regardless of the implementation details, gradient descent is analogous to observing how a single marble rolls down an error hill. Additionally, instead of launching just a single marble, we could launch a swarm of such particles, and track them all in parallel. We can allow them to exchange information such that far-away particles are naturally guided toward the region that minimizes the solution. This approach is aptly known as a *particle swarm*.

10.2.3   EXAMPLES

The SciPy Python package includes an optimization library that implements a `minimize` method that can be used for general optimization. To illustrate,

---

[1]Note: In the case of very large-scale optimization problems with costly function evaluations, rewriting the simulation to compute this gradient information along with the function evaluation can often be much more computationally efficient than the repeated function evaluations necessary for its approximation. However, retrofitting legacy codes not designed with the feasibility of computing such gradients in mind is often impossible or impractical.

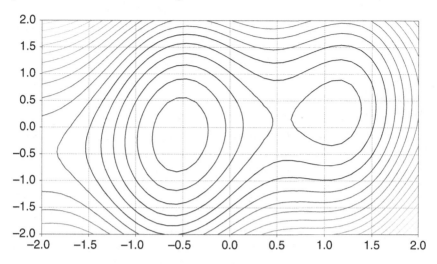

Figure 10.2 Contours of an example function to be minimized.

let's say that you want to minimize the following function (defined as a lambda expression):

```
fun = lambda x:(np.sin((x[0]+0.1)*np.pi)+(x[0]+0.2)**2-0.5*x[0]*x
    [1] + x[1]**2 + 1)
```

To find the minimum, we simply call

```
1  from scipy.optimize import minimize
2  res = minimize(fun, (-1.3, 0), method='CG')
3  print ("%.3g, %.3g -> %.3g"%(res.x[0],res.x[1],res.fun))
```

Running the code we find that the minimum value is 0.115 and is located at $(-0.538, -0.134)$. This result can be confirmed visually in the contour plot of Figure 10.2.

Let's now consider a C++ implementation. A brute force version that searches a 100×100 uniform grid could be written as

```
1  constexpr double PI = acos(-1.0);
2
3  double fun(double x[2]) {
4      return sin((x[0]+0.1)*PI) + (x[0]+0.2)*(x[0]+0.2) -0.5*x[0]*x
        [1] + x[1]*x[1] + 1;
5  }
6
7  void minimize(double (*f)(double[2]),double x1[2], double x2[2],
        double nn[2], double &min_val, double min_x[2]) {
8      min_val = 1e66;    // some large value
9      for (int i=0;i<nn[0];i++)
10         for (int j=0;j<nn[1];j++) {
11             double x[2];
12             x[0] = x1[0] + i*(x2[0]-x1[0])/(nn[0]-1);
13             x[1] = x1[1] + j*(x2[1]-x1[1])/(nn[1]-1);
14             double val = f(x);
```

```
15              if (val<min_val) {
16                  min_val = val;
17                  min_x[0]=x[0]; min_x[1]=x[1];
18              }
19          }
20 }
21
22 int main() {
23     double x1[2] = {-2,-2};
24     double x2[2] = {2,2};
25     double nn[2] = {100,100};
26     double min_val;
27     double min_x[2];
28
29     minimize(fun,x1,x2,nn,min_val,min_x);
30     cout<<"Minimum "<<min_val<<" at x=("<<min_x[0]<<","<<min_x
         [1]<<")"<<endl;
31     return 0;
32 }
```

The minimum is found to be 0.115415 at $(-0.545455, -0.141414)$, which is in good agreement with the SciPy result. The result could be refined by using a finer grid. However, a better use of computational resources involves utilizing adaptive search. This version reads as

```
1 double minimize_as(double (*f)(double[2]),double x1[2], double x2
         [2], double min_val, double min_x[2]) {
2     size_t sub_divs = 4;
3     double dh[2] ={(x2[0]-x1[0])/sub_divs,(x2[1]-x1[1])/sub_divs};
4     double df;
5
6     for (int i=0;i<sub_divs;i++)
7         for (int j=0;j<sub_divs;j++) {
8
9             // evaluate value at centroid
10            double x[2];
11            x[0] = x1[0] + (i+0.5)*dh[0];
12            x[1] = x1[1] + (j+0.5)*dh[1];
13            double val = f(x);
14
15            if (val<min_val) {
16                df = val-min_val;
17                min_val = val;
18                min_x[0]=x[0]; min_x[1]=x[1];
19            }
20        }
21    cout<<" f("<<min_x[0]<<","<<min_x[1]<<") = "<<min_val<<endl;
22    if (abs(df)>0.001) {
23        for (int i=0;i<2;i++) {
24            x1[i] = min_x[i]-0.5*dh[i];
25            x2[i] = min_x[i]+0.5*dh[i];
26        }
27        return minimize_as(f,x1,x2,min_val,min_x);
28    }
29    return min_val;
30 }
```

This function performs a similar brute force search but over a coarser grid containing just 4 subdivisions in each dimension. Instead of returning the position of the minimal value, it recursively calls itself with new domain bounds limited to the cell containing this minimal value. This process continues until the difference between the newly computed minimum and the one obtained at the prior, coarser resolution is greater than some tolerance. In this case, the algorithm converges after just 4 recursive calls, leading to an improved minimum of 0.115074 at $(-0.539062, -0.132812)$.

A Monte Carlo version could be written as given below:

```
void minimize_mc(double (*f)(double[2]),double x1[2], double x
    2[2], size_t num_samples, double &min_val, double min_x[2]) {
  min_val = 1e66;       // some large value
  for (size_t n=0;n<num_samples;n++) {
    double x[2];
    x[0] = x1[0]+rnd()*(x2[0]-x1[0]);
    x[1] = x1[1]+rnd()*(x2[1]-x1[1]);
    double val = f(x);
    if (val<min_val) {
      min_val = val;
      min_x[0]=x[0]; min_x[1]=x[1];
    }
  }
}
```

Using 1000 samples, we obtain 0.151938 at $(-0.560195, -0.324446)$. As noted earlier, this Monte Carlo-based approach may be preferred in problems containing multiple minimums.

Finally, a stochastic gradient descent version is given in the following example:

```
void minimize_sgd(double (*f)(double[2]), double x_guess[2],
    double &f0, double x[2]) {
  constexpr int max_iter = 1000;
  constexpr int num_samples = 5; // samples per point
  constexpr double R = 0.001;   // radius for point picking
  constexpr double w = 0.1;     // integration factor

  x[0] = x_guess[0];    // initialize
  x[1] = x_guess[1];
  f0 = f(x);
  double df;
  int iter = 0;

  do {
    double x1[2], x2[2];
    // point 1
    double r = (0.01+rnd())*R;    // sample random radius
    double theta = 2*PI*rnd();
    x1[0] = x[0] + r*cos(theta);
    x1[1] = x[1] + r*sin(theta);
    double df1 = f(x1)-f0;

    // point 2
    r = rnd()*R;   // sample another random radius
```

```
24          theta = 2*PI*rnd();
25          x2[0] = x[0] + r*cos(theta);
26          x2[1] = x[1] + r*sin(theta);
27          double df2 = f(x2)-f0;
28
29          double a = x1[0]-x[0];      // dx1
30          double b = x1[1]-x[1];      // dy1
31          double c = x2[0]-x[0];      // dx2
32          double d = x2[1]-x[1];      // dy2
33          double detA = a*d-b*c;      // det(A)
34
35          double invA[2][2] = {{d/detA, -b/detA},{-c/detA, a/detA}};
36          double df_dx = invA[0][0]*df1+invA[0][1]*df2;
37          double df_dy = invA[1][0]*df1+invA[1][1]*df2;
38          x[0] -= df_dx*w;
39          x[1] -= df_dy*w;
40
41          // calculate difference in f(x) from prior step
42          df = f0;          // save old value
43          f0 = f(x);        // update valule
44          df -= f0;         // difference
45          cout<<iter <<": f("<<x[0]<<","<<x[1]<<") = "<<f0<<endl;
46      } while (abs(df/f0)>1e-4 && ++iter<max_iter);
47 }
```

This function requires that an initial guess be provided. We use it to store the initial $f(\vec{x})$. Next, two points are randomly selected within a sphere of a user specified radius R and centered on the $\vec{x}$ position. These points give us the following system:

$$\begin{bmatrix} \Delta x_1 & \Delta y_1 \\ \Delta x_2 & \Delta y_2 \end{bmatrix} \begin{pmatrix} \Delta f/\Delta x \\ \Delta f/\Delta y \end{pmatrix} = \begin{pmatrix} \Delta f_1 \\ \Delta f_2 \end{pmatrix} \tag{10.8}$$

This system can be solved by computing the inverse analytically. We then move the current $\vec{x}$ position in the $(-w\Delta f/\Delta x, -w\Delta f/\Delta y))$ direction. Convergence is obtained after 29 steps, with minimum value of 0.115066 at $\vec{x} = (-0.538246, -0.136064)$. Note that in this particular implementation, the size of the search radius as well as the integration step size w are held constant. An optimized gradient descent would start with a larger step that shrinks as the minimum is approached. Several different algorithms exist, but they generally scale the step size according to the difference in gradient magnitude between the new and old point.

10.3 GENETIC ALGORITHMS

An interesting variation on the stochastic Monte Carlo approach gives us the *genetic algorithm*. A particular combination of N random $\vec{m}$ input values can be treated as a genetic sequence defining an "individual". These individuals can be classified according to their fitness, $1 - e$. We next select random pairs, $(\vec{m}_i, \vec{m}_j$ with the important caveat that fitter individuals have a higher chance of selection. These individuals mate, producing an offspring that inherits a genetic sequence (a strand of the parameter space) from parent 1 and another section from parent 2. We also introduce random mutations and allow for the possibility of no mating. The process repeats until a new generation of equal size is produced. We eventually converge on the optimal solution through the survival of the fittest. This approach requires a sufficiently diverse initial population so that the optimal solution vector can be constructed.

To illustrate, let's consider a program for finding a solution to the well-known *traveling salesman problem*. The objective of this problem is to find the shortest round trip path through a set of points. Each point needs to be visited once, and only once. The method implemented here is based on a Fullstack Academy Youtube video found at `youtube.com/watch?v=XP8R0yzAbdo`.

We consider a two-dimensional implementation in which the points are represented by (x, y) pairs. Next, we generate P random paths through the points. This involves filling a N-sized vector of integers with indices to the point array. The ordering indicates the sequence in which the points (towns) will be visited. These vectors provide the genetic sequence for each of the P individuals. We next compute the total path length for each individual,

$$L_i^2 = \sum_{j}^{N-1} \left(P_{x,j+1} - P_{x,i}\right)^2 + \left(P_{y,j+1} - P_{y,j}\right)^2 \tag{10.9}$$

The lengths are then normalized to obtain fitness value $\mathfrak{f} \in [0, 1]$. We then create N new paths by randomly selecting two unique paths, with the probability of selections scaling with $\mathfrak{f}$. Using Javascript, this is written as

```
1  do {
2    p1 = Math.floor(Math.random()*N)
3  } while (paths[p1].fitness<Math.random());
```

Let's assume that the two selected individuals p_1 and p_2 with the following genetic sequence:

p1:	0	1	**6**	**4**	**2**	**3**	5
p2:	1	4	2	6	5	0	3

We randomly select a strand of p_1's genetic sequence, as shown by the bold letters. We copy this sequence to the offspring. The rest is filled with the shaded data from p_2. Duplicates are however rejected, leading to blank cells. We thus obtain:

o:	1	-	6	4	2	3	-

Random unique values are then picked to fill the blanks:

o:	1	0	6	4	2	5	5

Finally, we add random mutations to a small fraction of offspring. The mutation in this case implies a random swap of two cities. Thus, assuming this child undergoes a mutation, the final genetic sequence may read

o	6	0	1	4	2	5	5

This example, found in `genetic.html`, is implemented using Javascript to leverage the interactive graphing capabilities of web browsers, as discussed in Chapter 7. The user interface lets the user select between initial points placed at uniform intervals on a circle or points sampled randomly within the rectangular canvas. The relevant code sections is listed below.

```
1    // produces new generation by mating random parents
2    function makeNewGeneration() {
3        var new_paths = [];
4        while (new_paths.length<pop_size) {
5            var p1,p2;
6            do {
7                p1 = Math.floor(Math.random()*pop_size);
8            } while (paths[p1].fitness<Math.random());
9
10           do {
11               p2 = Math.floor(Math.random()*pop_size);
12           } while (p1==p2 || paths[p2].fitness<Math.random());
13
14           // copy random section to copy from parent 1
15           var a = [];
16           for (var k=0;k<num_points;k++) a[k] = -1;
17           var j1 = Math.floor(Math.random()*num_points);
18           var j2 = j1+Math.floor(Math.random()*(num_points-j1));
19           for (var j=j1;j<j2;j++) a[j] = paths[p1].path[j];
20
21           //second param is j2+num_points-(j2-j1)
22           for (var j=j2;j<j1+num_points;j++) {
23               var jj = j%num_points; //wrap around
24
25               // do we have this point already?
26               var found=false;
27               for (var k=0;k<num_points;k++)
28                   if (a[k]==paths[p2].path[jj]) {found=true;break;}
29               if (!found)  a[jj] = paths[p2].path[jj];
30               else a[jj]=-1;
31           }
32
33           // replace any -1s
34           for (var k=0;k<num_points;k++) if (a[k]<0) {
35               var val=0;
36               do {
37                   val = Math.floor(Math.random()*num_points);
38                   for (var k2=0;k2<num_points;k2++) if (a[k2]==val) val
                          = -1;
39               } while(val<0);
40               a[k] = val;
41           }
42
43           // mutation, swap two points
44           if (Math.random()<0.05) {
45               var j1 = Math.floor(Math.random()*num_points);
46               var j2 = Math.floor(Math.random()*num_points);
47               var b = a[j1];
48               a[j1] = a[j2];
49               a[j2]=b;
```

```
50        }
51
52            new_paths.push({path:a});
53        }
54        paths = new_paths;
55  }
```

This code implements the logic described in the previous paragraphs. Figure 10.3 illustrates this algorithm applied to a system consisting of 7 points. The dark line shows the optimal solution (the shortest path), while the remaining lines trace the paths taken by the less fit individuals. The attained solution is indeed the correct one.

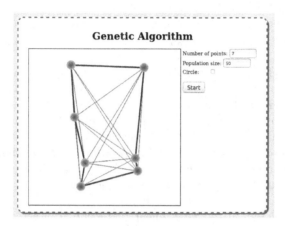

Figure 10.3 Illustration of the genetic algorithm to find the shortest path through a set of random points.

10.4 MACHINE LEARNING

Machine learning (ML) has seen explosive growth in both capabilities and usage over the recent years especially with the release of end-user generative tools such as ChatGPT and Midjourney. These tools may seem like magic, but the underlying principles tend to be surprisingly simple. ML is a subset of a broader category of Artificial Intelligence (or AI). Earlier approaches into AI involved coding the instructions the system should follow in response to various inputs. Machine learning is a catchall term for algorithms that attempt to extract relationships from data without such a prescribed response logic. Machine learning methods that attempt to approximate the network structure of the human brain are often referred to as artificial neural networks (ANNs). Structured to initially mimic the brain of a newborn baby, ANNs consist of a multitude of randomly weighted interconnected neurons. As the baby grows, some connections become stronger while others diminish. In ML, this corresponds to the 'training' phase where the system uses data to adapt its responses to improve the approximation of the expected output.

The networks can consist of few (shallow) or many (deep) layers and can have connections that are dense, sparse, or follow other problem-inspired structural patterns as in the case of convolutional networks. In the classical fully connected feed-forward scheme, described here, information propagates only in a single direction from layer to layer. The first layer provides the inputs while the last layer is the result of the calculation. In an image classification application, the input layer will contain the pixels of the flattened image. These pixels feed into several "hidden layers" that terminate in the output layer. The individual neurons of the output layer correspond to the probabilities of different categories. For example, in a program classifying images of animals, the output layer neurons may correspond to categories such as "cat", "dog", or "bird". Output layer with normalized values of $[0.85, 0.14, 0.01]$ indicates that the image is most likely a cat, although there is a slight chance it is actually a dog. However, it doesn't seem very likely that this is a picture of a bird.

A neural network can be envisioned as a general unknown 'universal function approximator' that maps input vectors to outputs. In its simplest form, this can be envisioned as an analogue to a linear system $\vec{y} = \mathbf{A}\vec{x}$. This matrix expression transforms input $\vec{x}$ to an output vector $\vec{y}$. Early in your linear algebra study you learned how to perform matrix-vector calculation so that $\vec{y}$ can be computed by knowing $\mathbf{A}$ and $\vec{x}$. Later, you learned different schemes for solving for $\vec{x}$ from $\mathbf{A}$ and $\vec{y}$, assuming A is full-rank and square. But what if we know $\vec{x}$ and $\vec{y}$ and want to determine the coefficients of matrix $\mathbf{A}$? This is essentially a simplified and generalized objective of the machine learning training phase. Indeed, fitting a linear regression to a set of sampled data points as we did in Chapter 3 could be considered among the most basic of machine learning algorithms. In ML, however, we are often referring to a much higher dimensional curve fitting than the scalar function examples. Here the input, $\vec{x}$, could be a flattened image and the expected output, $\vec{y}$, be the vector encoding one image label per dimension. The generalization arises from the fact that in ML the size of $\vec{x}$ and $\vec{y}$ does not need to be the same. For classification problems, the output layer is often much smaller than the input such that many inputs (i.e. think of all possible pictures of cats) map to the same label. The same thing occurs in linear algebra when the forward map is not full rank and invertible. Starting with a large collection of $\vec{x} \to \vec{y}$ pairs, known as training data, the objective of the training phase is to find the single transformation (i.e. the matrix $\mathbf{A}$ in this simplified description or more generally $\vec{y} = f(\vec{x})$) that provides good overall agreement over the entire population. Once the system is trained, it can be applied to a new set of test data inputs that were not included in the training set, and hopefully still produce a solution that is in good agreement with the expected outcome. As such, it is important to ensure that the training data broadly covers the entire input space that may be encountered during the subsequent use.

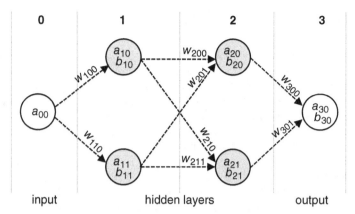

Figure 10.4 Visualization of a simple dense neural network with a single input and a single output neuron.

10.4.1 NEURAL NETWORKS

Figure 10.4 illustrates a simple neural network consisting of a single input neuron, two hidden layers of two neurons each, and a single output neuron. This is an example of a dense (fully-connected) network, since each neuron receives information from every single neuron in the preceding layer. The value of each neuron is given by summation of the values (called activations) of the input neurons scaled by some weights. In addition, we allow for a fixed bias. Thus we have

$$x_{i,j} = \sum_{k}^{n_{k-1}} w_{i,j,k} a_{i-1,k} + b_{i,j} \tag{10.10}$$

The i index corresponds to the layer, while the j indexes the neuron within that layer. n_k is the number of neurons in layer k. There are other application-specific neuron arrangements. For example, image processing typically begins with a convolution layer to blur small features and to account for imperfect alignment. Long short-term memory (LSTM) neural networks have backward feedback connections that make it possible to process data regardless of the specific ordering. These networks are frequently used for speech recognition.

10.4.2 ACTIVATION FUNCTIONS

In linear algebra, the 'families' of inputs for equivalent output points for degenerate matrix, A, correspond to lines, planes, and their higher dimensional generalizations. However, the $x_{i,j}$ values in Equation 10.10 corresponding to the same output can have quite arbitrary shapes that are not necessarily well approximated by these linear subspaces. By wrapping the outputs of the layers with some nonlinear "activation" (or squishing) functions, the networks acquire the ability to approximate arbitrary shaped functions. Historically, the logistic (Sigmoid) function was used

$$f_\sigma(x) = \frac{1}{1 + e^{-x}} \tag{10.11}$$

This function is nearly linear around the origin but asymptotes to -1 and 1 as one moves away from this linear region. These days it is more common to apply the Rectified Linear Units or *ReLU*,

$$f_{\text{ReLU}}(x) = \max(x, 0) \tag{10.12}$$

ReLUs are computationally inexpensive for computing both values and gradients[2], and result in sparsification of the network connections further accelerating computation. Indeed, as described in Balestriero and Baraniuk 2021, these ReLU activations effectively approximate arbitrary nonlinear functions as piecewise linear compositions providing good approximation power. Utilizing activation functions, we write the value stored on each neuron as

$$a_{i,j} = f(x_{i,j}) \tag{10.13}$$

where $x_{i,j}$ is the direct summation from Equation 10.10.

10.4.3 COST FUNCTION

While conceptually simple, a neural network consists of a huge number of weights and biases that need to be set. This is where the training data comes in. The neural net can be initialized with random values assigned to the weights and biases. We run each piece of training data through the network and compare the activation on each output layer neuron to the expected value (which is also included with the training data). This difference is used to compute the cost (or loss) function. Averaging the cost function over the entire training set (or possibly, a batch) lets us characterize how well the network is doing. The actual "learning" of a neural network involves adjusting the weights and biases to reduce the average error. There are different approaches for computing the cost function. Here we characterize the error using

$$E = \sum_n \frac{1}{2} (a_{o,n} - T_n)^2 \tag{10.14}$$

where o is the index of the output layer and T_n is the expected true value.

10.4.4 BACK PROPAGATION

The computed error is next used to adjust the network using a process called *back propagation*. It is essentially a form of gradient descent. For each unknown

---

[2]Note except at $x = 0$ where df/dx is undefined leading to some theoretical debates and analysis challenges that have resulted in the introduction of assorted smoothed versions of ReLU activation.

weight or a bias, we compute the derivative $\partial E / \partial w$. We subsequently update the value per

$$w = w - \eta \frac{\partial E}{\partial w} \tag{10.15}$$

where η is known as the learning rate.

The actual derivatives are computed with the help of the chain rule. Let's consider the example from Figure 10.4. This system depends on a total of 13 unknowns split between 8 weights and 5 biases. Let's consider the output layer. We have

$$a_{30} = f(x_{30}) \tag{10.16}$$

$$x_{30} = w_{300}a_{20} + w_{301}a_{21} + b_{30} \tag{10.17}$$

We are interested in computing $\partial E / \partial w_{300}$, $\partial E / \partial w_{301}$, and $\partial E / \partial b_{30}$. Using the chain rule, we can write

$$\frac{\partial E}{\partial w_{300}} = \frac{\partial E}{\partial a_{30}} \frac{\partial a_{30}}{\partial x_{30}} \frac{\partial x_{30}}{\partial w_{300}} \tag{10.18}$$

Per Equation 10.14, $\partial E / \partial a_{30} = (a_{30} - T_0)$. Next $\partial a_{30} / \partial x_{30} = f'(x_{30})$. For the logistic function

$$\frac{d}{dx} f_\sigma(x) = \frac{e^x}{(1 + e^x)^2} = f_\sigma(x)(1 - f_\sigma(x)) = a - a^2 \tag{10.19}$$

Finally, $\partial x_{30} / \partial w_{300} = a_{20}$. Collecting the terms, we have

$$\frac{\partial E}{\partial w_{300}} = \frac{\partial E}{\partial a_{30}} \frac{\partial a_{30}}{\partial x_{30}} \frac{\partial x_{30}}{\partial w_{300}} = (a_{30} - T_0)(a_{30} - a_{30}^2)(a_{20}) \tag{10.20}$$

and

$$\frac{\partial E}{\partial w_{301}} = \frac{\partial E}{\partial a_{30}} \frac{\partial a_{30}}{\partial x_{30}} \frac{\partial x_{30}}{\partial w_{301}} = (a_{30} - T_0) \left(a_{30} - a_{30}^2 \right) (a_{21}) \tag{10.21}$$

$$\frac{\partial E}{\partial b_{30}} = \frac{\partial E}{\partial a_{30}} \frac{\partial a_{30}}{\partial x_{30}} \frac{\partial x_{30}}{\partial w_{300}} = (a_{30} - T_0) \left(a_{30} - a_{30}^2 \right) (1) \tag{10.22}$$

The remaining terms are computed similarly. We need to pay attention to multiple paths that can be used to reach a neuron as is the case with a_{10} and a_{11}. For example,

$$\frac{\partial E}{\partial w_{100}} = \frac{\partial E}{\partial a_{30}} \frac{\partial a_{30}}{\partial x_{30}} \left[\frac{\partial x_{30}}{\partial w_{300}} \frac{\partial w_{300}}{\partial a_{20}} \frac{\partial a_{20}}{\partial x_{20}} \frac{\partial x_{20}}{\partial a_{10}} + \right.$$
$$\left. \frac{\partial x_{30}}{\partial w_{301}} \frac{\partial w_{301}}{\partial a_{21}} \frac{\partial a_{21}}{\partial x_{21}} \frac{\partial x_{21}}{\partial a_{10}} \right] \frac{\partial a_{10}}{\partial x_{10}} \frac{\partial x_{10}}{\partial w_{100}} \tag{10.23}$$

which evaluates to

$$\frac{\partial E}{\partial w_{100}} = (a_{30} - T_0)(a_{30} - a_{30}^2)\Bigg[(w_{300})(a_{20} - a_{20}^2)(w_{200})(a_{10} - a_{10}^2) +$$

$$(w_{301})(a_{21} - a_{21}^2)(w_{210})(a_{10} - a_{10}^2)\Bigg](a_{10} - a_{10}^2)(a_{01}) \qquad (10.24)$$

This math is not difficult, but it is very easy to get lost in the indexes. Deploying a similar algorithm to anything larger than our toy problem would require automation of this calculation.

10.4.5 SIMPLE NEURAL NET IN C++

Let's now demonstrate the power - and simplicity - of ML with a simple example. The entire code can be found in **nn.cpp**. We would like the computer to classify whether a number in $[0, 1)$ is larger or equal to 0.5. In other words, we would like to round an arbitrary real value in $[0, 1)$ to the nearest integer. We will utilize the neural network discussed above. The training data is a collection of pairs such as $(0.4423, 0)$, $(0.8033, 1)$, $(0.2101, 0)$, and so on. We begin by creating this training set:

```
1  int main() {
2      auto eval = [](double x) {return x>=0.5;};
3      // create training data
4      std::vector<std::pair<double,double>> training_data;
5
6      for (size_t i=0;i<10000;i++) {
7          double x = rnd();
8          training_data.emplace_back(x,eval(x));
9      }
10
11     // print training data
12     cout<<"—— TRAINING DATA ——"<<endl;
13     for (size_t n=0;n<10;n++) cout<<training_data[n].first<<" -> "
            <<training_data[n].second<<endl;
```

We next initialize the neural net. For simplicity, we are representing the network using statically-allocated multidimensional arrays corresponding the activations and weights. These values are stored inside of a structure of type NN,

```
1  template<int NL, int MN>
2  struct NN {
3      double w[NL][MN][MN];
4      double b[NL][MN];
5      double x[NL][MN];      // neuron value before squishing
6      double a[NL][MN];      // neuron value after squishing
7
8      //error sums
9      double dE_dw[NL][MN][MN];
10     double dE_db[NL][MN];
11
```

```
12        int num_neurons[NL];
13        int num_layers = NL;
14
15        // evaluate output layer using current weights
16        void evaluate(double (*squish)(double)) {
17            for (int l=1;l<NL;l++) {
18                for (int n=0;n<num_neurons[l];n++) {
19                    x[l][n] = 0;
20                    for (int m=0;m<num_neurons[l-1];m++) {
21                        x[l][n] += w[l][n][m]*a[l-1][m];
22                    }
23                    x[l][n] += b[l][n];
24                    a[l][n] = squish(x[l][n]);
25                }
26            }       // for layers
27        }   // evaluate
28   };
```

The **eval** function propagates information from the input layer using the currently set weights and biases. We instantiate the neural net as:

```
1   // build neural net: 1->2->2->1
2   const int NL = 4;
3   NN<4,2> nn;
4   nn.num_neurons[0] = 1;
5   nn.num_neurons[1] = 2;
6   nn.num_neurons[2] = 2;
7   nn.num_neurons[3] = 1;
```

We also assign random initial weights but set the biases to zero:

```
1   // generate random weights
2   for (int l=1;l<NL;l++)
3       for (int n=0;n<nn.num_neurons[l];n++) {
4           nn.b[l][n] = 0;
5           for (int m=0;m<nn.num_neurons[l-1];m++) {
6               nn.w[l][n][m] = -1 + 2*rnd();
7           }
8       }
```

Next, we assign the activation (squishing) function and its derivative,

```
1   // set squishing function (logistic)
2   auto squish = [](double x)->double{return 1/(1+exp(-x));};
3   auto dsf = [](double a)->double{return a*(1-a);};   // derivative
```

This brings us to the core of the calculation: the actual training. Due to the relatively small size, we don't utilize batching, and use the entire training data to compute the derivative. We repeat the process for a maximum of 500 epochs, but allow the loop to exit earlier if convergence is achieved. The **dE_dw** and **dE_db** arrays of the **NN** data structure are used to accumulate the error, and thus we begin by clearing them at the start of each epoch:

```
1   double eta = 10;        // learning rate
2   constexpr int max_epochs = 500;
3   double E_initial = 1e66;
4
5   // start training
6   for (int epoch = 0; epoch<max_epochs; epoch++) {
7       double E_sum = 0;
```

```
8
9      // clear error sums
10     for (int l=1;l<NL;l++)
11         for (int n=0;n<nn.num_neurons[l];n++) {
12             for (int m=0;m<nn.num_neurons[l-1];m++)
13                 nn.dE_dw[l][n][m] = 0;
14             nn.dE_db[l][n] = 0;
15         }
```

We then loop over the training data. We assign a value to the input neuron. We then propagate the activation through the network by calling the **evaluate** function. We then grab the expected result from the test data, and use it to compute the local error,

```
1   for (std::pair<double,double> &pair:training_data) {
2       nn.a[0][0] = pair.first;  // set input layer
3       nn.evaluate(squish);       // evaluate output layer
4
5       double T[1] = {pair.second};   // true solution, keeping as an
                array for generality
6
7       // local error
8       double E = 0.5*(nn.a[NL-1][0]-T[0])*(nn.a[NL-1][0]-T[0]);
9       E_sum += E;
```

This error calculation is used solely to track the system's performance. The actual processing takes place next in the hard-coded backpropagation algorithm,

```
1   // hardcoded error derivatives per chain rule
2   nn.dE_dw[3][0][0] += (nn.a[3][0]-T[0]) * dsf(nn.a[3][0]) * (nn.a
        [2][0]);
3   nn.dE_dw[3][0][1] += (nn.a[3][0]-T[0]) * dsf(nn.a[3][0]) * (nn.a
        [2][1]);
4   nn.dE_db[3][0] += (nn.a[3][0]-T[0]) * dsf(nn.a[3][0]) * (1);
5
6   nn.dE_dw[2][0][0] += (nn.a[3][0]-T[0]) * dsf(nn.a[3][0]) * (nn.w
        [3][0][0]) * dsf(nn.a[2][0]) * (nn.a[1][0]);
7   nn.dE_dw[2][0][1] += (nn.a[3][0]-T[0]) * dsf(nn.a[3][0]) * (nn.w
        [3][0][0]) * dsf(nn.a[2][0]) * (nn.a[1][1]);
8   nn.dE_db[2][0] += (nn.a[3][0]-T[0]) * dsf(nn.a[3][0]) * (nn.w
        [3][0][0]) * dsf(nn.a[2][0]) * (1);
9
10  nn.dE_dw[2][1][0] += (nn.a[3][0]-T[0]) * dsf(nn.a[3][0]) * (nn.w
        [3][0][1]) * dsf(nn.a[2][1]) * (nn.a[1][0]);
11  nn.dE_dw[2][1][1] += (nn.a[3][0]-T[0]) * dsf(nn.a[3][0]) * (nn.w
        [3][0][1]) * dsf(nn.a[2][1]) * (nn.a[1][1]);
12  nn.dE_db[2][1] += (nn.a[3][0]-T[0]) * dsf(nn.a[3][0]) * (nn.w
        [3][0][1]) * dsf(nn.a[2][1]) * (1);
13
14  double a = (nn.w[3][0][0]) * dsf(nn.a[2][0]) *(nn.w[2][0][0]) *
        dsf(nn.a[1][0]);
15  double b = (nn.w[3][0][1]) * dsf(nn.a[2][1]) *(nn.w[2][1][0]) *
        dsf(nn.a[1][0]);
16  nn.dE_dw[1][0][0] += (nn.a[3][0]-T[0]) * dsf(nn.a[3][0]) * (a+b)
        * dsf(nn.a[0][0]) *(nn.a[0][0]);
17  nn.dE_db[1][0] += (nn.a[3][0]-T[0]) * dsf(nn.a[3][0]) * (a+b) *
        dsf(nn.a[0][0]) *(1);
```

```
18
19   a = (nn.w[3][0][0]) * dsf(nn.a[2][0]) * (nn.w[2][0][1]) * dsf(nn.
         a[1][1]);
20   b = (nn.w[3][0][1]) * dsf(nn.a[2][1]) * (nn.w[2][1][1]) * dsf(nn.
         a[1][1]);
21   nn.dE_dw[1][0][1] += (nn.a[3][0]-T[0]) * dsf(nn.a[3][0]) * (a+b)
         * dsf(nn.a[0][1]) * (nn.a[0][1]);
22   nn.dE_db[1][1] += (nn.a[3][0]-T[0]) * dsf(nn.a[3][0]) * (a+b) *
         dsf(nn.a[0][1]) * (1);
23   }  // for training data
```

This is just an implementation of terms partially listed by Equations 10.20 through 10.24. These accumulated values are averaged once the loop over the population is completed,

```
1   // compute derivative averages over the test population
2   for (int l=1;l<NL;l++)
3       for (int n=0;n<nn.num_neurons[l];n++) {
4           for (int m=0;m<nn.num_neurons[l-1];m++)
5               nn.dE_dw[l][n][m] /= training_data.size();
6           nn.dE_db[l][n] /= training_data.size();
7       }
```

The averaged derivative terms are used to update the weights,

```
1   // apply correction to w and b terms
2   for (int l=1;l<NL;l++)
3       for (int n=0;n<nn.num_neurons[l];n++) {
4           for (int m=0;m<nn.num_neurons[l-1];m++)
5               nn.w[l][n][m] -= eta*nn.dE_dw[l][n][m];
6           nn.b[l][n] -= eta*nn.dE_db[l][n];
7       }
```

Finally, we display the error. The algorithm terminates once the error reduces to 10% of the initial value,

```
1   // display total error and check for convergence
2   double E = E_sum/training_data.size();
3   if (epoch>0) {
4       double E_frac = E/E_initial;   // in percent
5       cout<<epoch<<"\t"<<E_frac<<endl;
6       if (E_frac<0.1) break;
7   } else E_initial = E;
8   }  // epoch
```

And that's basically it. So how did we do? To determine this, we start by creating a new set of input-output pairs. This is a different set than what was used for training. We have

```
1   // generate test data
2   std::vector<std::pair<double,double>> test_data;
3   for (size_t i=0;i<1000;i++) {
4       double x = rnd();
5       test_data.emplace_back(x,eval(x));
6   }
```

We then simply loop over this test data set, and just as before, plug in the input into the neural net input layer, evaluate the network, and compare the output to the expected value from the test data. The comparison is performed by rounding the activation on the output layer to the nearest integer in order to match the format of the test data. A counter is incremented for each correct classification. We also display the first 10 calculations to the screen.

```
1   int num_correct = 0;
2   int item = 0;
3   for (std::pair<double,double> &pair:test_data) {
4       nn.a[0][0] = pair.first;    // set input layer
5       nn.evaluate(squish);
6
7       int o = (int)(nn.a[NL-1][0]+0.5);   // round to nearest (i.e.
            0.6 -> 1)
8       int T = (int)pair.second;
9       if (o==T) num_correct++;
10
11      // display first 10 entries
12      if (++item<10) cout<<nn.a[0][0]<<" -> "<<o<<" ("<<(o==T?'Y':'x
            ')<<")"<<endl;
13  }
14
15  cout<<"Sample size: "<<test_data.size()<<endl;
16  cout<<"Correct: "<<num_correct/(double)test_data.size()*100<<"%"
        <<endl;
```

Running the code, we obtain the following:

```
ch10$ g++ -O2 nn.cpp && ./a.out
...
0.918609 -> 1 (Y)
0.173283 -> 0 (Y)
0.684588 -> 1 (Y)
0.189842 -> 0 (Y)
0.358008 -> 0 (Y)
0.333293 -> 0 (Y)
0.380889 -> 0 (Y)
0.533627 -> 1 (Y)
0.317462 -> 0 (Y)
Sample size: 1000
Correct: 98.3%
```

Amazingly, our primitive 5 neuron system was able to correctly classify over 98% of the inputs!

10.4.6 LIBRARIES

While it is important to understand the underlying mathematics behind machine learning, the practical way to actually utilize ML in simulation programs is through existing libraries. One such library is TensorFlow (with PyTorch being a popular alternative). TensorFlow implements a GPU-accelerated algorithms for efficient matrix (or tensor) data manipulation, allowing results from one operation to "flow" into another. It was developed at Google in 2005 and comes with language bindings for C/C++ and Python. It utilizes Python library called Keras to build neural networks. What makes TensorFlow particularly appealing is its availability for microcontrollers such as Arduino

Nano. This makes it easy to add machine learning wearables and robotic projects. The basic usage is illustrated below. This example is a based on the library's tutorial website https://www.tensorflow.org/tutorials/keras/classification.

```python
#!/usr/bin/env python3
# -*- coding: utf-8 -*-

#TensorFlow example based on https://www.tensorflow.org/tutorials
                             /keras/classification
import tensorflow as tf
import matplotlib.pyplot as plt

mnist = tf.keras.datasets.mnist
(x_train, y_train), (x_test, y_test) = mnist.load_data()
x_train, x_test = x_train / 255.0, x_test / 255.0

model = tf.keras.models.Sequential([
tf.keras.layers.Flatten(input_shape=(28, 28)),
tf.keras.layers.Dense(128, activation='relu'),
tf.keras.layers.Dropout(0.2),
tf.keras.layers.Dense(10)
])

# specify the loss function (cross entropy)
loss_fn = tf.keras.losses.SparseCategoricalCrossentropy(
    from_logits=True)

# initialize model to start training
model.compile(optimizer='adam', loss=loss_fn, metrics=['accuracy'
    ])

# train the model and then run with test data
model.fit(x_train, y_train, epochs=3)
model.evaluate(x_test, y_test, verbose=2)

# visualize few example results
fig, axs = plt.subplots(1,4)
for i in range(4):
    axs[i].imshow(x_test[120+i],cmap='gray_r');
    axs[i].set_xlabel(y_test[120+i],size=22,weight='bold');
    axs[i].set_xticks([]); axs[i].set_yticks([])
plt.tight_layout();
plt.show(block=True);
```

The code begins by downloading a training set consisting of 28 × 28 pixel images of handwritten 0 through 9 digits. These images are stored in the x_train list. The corresponding integer classification is found in y_train. Next, the neural network is built. The image is first flattened into 784-item input layer. It feeds to a 128-neuron dense layer that uses the ReLU squishing function. Subsequently, a drop-out layer is added that randomly drops 20% of the input signals. These layers are frequently utilized in image classification to avoid overfitting. The final 10-neuron dense layer captures the image classification. Details of the loss function are then specified. This particular example uses a logarithm-based cross entropy instead of the L2 norm of the earlier example. The model is initialized, which also involves specifying an optimizer algorithm. The next two lines first perform the training and subsequently apply the trained model to the test data. As can be seen by running the script and observing the generated visual output, the classification works well.

11 Embedded Systems

This final chapter introduces the programming of electronic devices such as Arduino microcontrollers and Field Programmable Gate Arrays (FPGAs). Microncontrollers are low-powered computers that continuously run an uploaded program, and are typically used as custom controllers interfaced with external sensors. FPGAs make it possible to accelerate critical algorithms by implementing custom instructions that operate directly on the digital signal. The chapter also introduces single-board computers, electrical components, breadboard prototyping, and Verilog hardware descriptor language.

11.1 SINGLE-BOARD COMPUTERS

So far, we have only addressed the development of codes that run on "regular" computers such as desktop workstations, laptops, or cluster nodes. The lack of portability (even in the case of laptops) and high power requirements make such systems unusable for autonomous operations encountered in robotics, data processing, or edge computing. The term edge computing implies computation performed close to the source of data. This is where miniature single-board systems, microcontrollers, and FPGAs come in.

Figure 11.1(b) visualizes a collection of various miniature computational systems. Clockwise from the top left corner we have the Raspberry Pi, NVIDIA Jetson Nano, Arduino Uno, Arduino 33 BLE Sense, Arduino MKR Vidor 4000, Arduino Mega, Texas Instruments MSP430-FR5969 Launchpad, and Intel Cyclone 10LP FPGA evaluation kit. Raspberry Pi is a single-board computer developed by the Raspberry Pi Foundation. It is a complete computer shrank down to the size of a credit card. Instead of a hard drive, files are stored on an SSD card. Connectivity to external peripherals is accomplished via built-in USB and HDMI ports. The system also includes WiFi and Bluetooth connectivity. The exposed general purpose input/output (GPIO) pins can be used to drive electrical circuits and to communicate with external sensors. Another connector provides a simple interface to an external camera. By connecting a monitor and a keyboard, you gain access to a 1.4 Ghz CPU desktop for as little as $30 USD. A Raspberry Pi system can be used with a variety of operating systems. The default option is a Debian Linux variant called Raspberry Pi OS (previously called Raspbian), but other options exist including a light-weight version of Microsoft Windows. Within the O/S, you just use your favorite programming environment to develop and run the program. Since the initial release in 2012 primary as a teaching tool, Raspberry Pi has sold tens of million units. Thanks to its small size and only a 15 W power requirement, it is frequently found in robotic applications. Since 2019, NVIDIA sells a similar single-board system called Jetson Nano. While

DOI: 10.1201/9781003132233-11

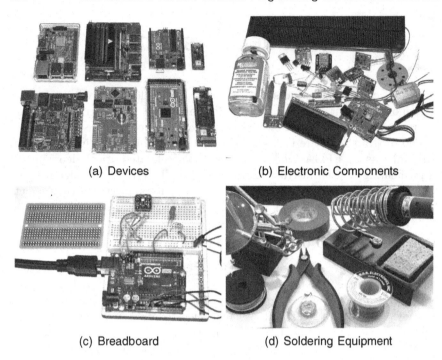

(a) Devices   (b) Electronic Components

(c) Breadboard   (d) Soldering Equipment

Figure 11.1   Collection of various electronic devices and components.

costing more than the Raspberry Pi, this system includes an integrated 128-core NVIDIA GPU for accelerating data processing and machine vision applications. It also contains GPIO and camera connectors and otherwise functions similarly to the Raspberry Pi.

These devices are essentially tiny computers running desktop-like operating systems, and as such are programmed in a manner already familiar from desktop development. However, where things get slightly novel is in interfacing with the just mentioned pins and connectors. There are numerous libraries available to simplify this task. These libraries include legacy `Wiring`, as well as the more recent `Pigpio` and `Gpiozero`. While Gpiozero is the library officially endorsed by the Raspberry Foundation, it is only available for Python. Pigpio offers bindings for C++ as well as Python. The example below illustrates how it is used to drive a GPIO pin (we discuss pins in more detail in the following section).

```
1  #include <pigpio.h>
2  ...
3  gpioInitialise()
4  gpioSetMode(3, PI_OUTPUT);      // indicate pin 3 is used for output
5  gpioDelay(1000);                // wait 1 second
6  gpioWrite(3, 1);                // drive pin 3 high (state 1)
```

This library offers numerous additional features, including the ability to communicate with I^2C components and through the serial port, pulse width modulation (PWM), and wave generation.

As tiny computers, they also inherit the ability to function as tiny webservers. When combined with these GPIO capabilities, this enables development of applications with a capacity to bridge cyber and physical boundaries. These distributed remote IO capabilities constitute a minimal enabling foundation for developing Internet of Things (IoT) capabilities. However, when fully outward-facing to the open internet, they also introduce security risks that should be properly considered.

11.2 MICROCONTROLLERS

Moving down the spatial scale, we next arrive at the microcontroller units (or MCUs) such as the Arduino or the Texas Instruments MSP430. Other vendors, such as Sparkfun or Adafruit, manufacture their own Arduino-compatible boards. Microcontrollers are not "full" computers in the sense that you cannot load an operating system onto them. Instead, microcontrollers feature a simple CPU designed to continuously run a single program that was "flashed" onto the device's Electrically Erasable Programmable Read-Only Memory (EEPROM). Low-level code that is used to control electronic devices is known as *firmware*. This program consists of two parts: 1) an initializer block that is executed whenever the system is first powered on, and 2) the main routine that runs in an infinite loop as long as the system remains powered on. Microcontrollers are typically used to *drive* electrical circuits by altering the voltage on electrical pins. A *digital pin* can have only one of two states: low and high. In the low state, the pin is connected to the ground. In the high state, the pin voltage increases to the board voltage, which tends to be either 5V or 3.3V, depending on the model. An *analog pin* can have intermediate values, although these pins are available only for input. They can be used to read voltage from analog sensors such as photoresistors or analog temperature meters. The analog signal is typically converted to an integer in the range such as $[0, 1023]$. More complex sensors, such as color detectors, magnetometers, particle counters, and so on, communicate their data over a digital interface called I^2C (Inter-Integrated Circuit). It utilizes two digital pins SCL and SDA. The SCL pin provides a *clock*. Clock is a repeating sequence of 1s and 0s that is used to delineate bits on the SDA data line. I^2C protocol typically operates at 100 kbits/s. The communication is bi-directional. Data packets are individually addressed to allow multiple I^2C devices to communicate over the same two pins.

11.2.1 ELECTRICAL COMPONENTS

Figure 11.1(b) presents a small sample of the multitude of electrical components, sensors, and other peripheral devices that can be purchased for a minimal cost from various vendors. In no particular order, we see a solar panel and a battery charging unit that could be used to power a field-deployed MCU, transistors, push and toggle buttons, various resistors, light emitting diodes (LEDs), capacitors, a speaker, a motor, a liquid crystal display (LCD) board, and sensors for measuring soil humidity, temperature, air particulate counts, and incident light color composition. These sensors are usually provided on small *breakout boards* containing the necessary electronics needed to communicate with the actual sensor. They also provide a software library that makes this communication easy to implement. We also see a bottle of silicone conformal coating. This clear coat provides protection against elements, which is needed for boards that are to be used outside. The large square "Sunny Buddy' battery charging board in the bottom right has this coating applied, although this is impossible to tell from the picture.

Some microcontrollers already come with sensors built in. Despite its miniature size, the Arduino Nano Sense 33 BLE contains an accelerometer, gyroscope, magnetometer, an RGB light sensor, proximity detector, microphone, Bluetooth connectivity, and also supports machine learning. However, microcontrollers are generally used to interface with external sensors or other electrical components. This is where the previously mentioned transistors are used. Transistors are semiconductor devices that revolutionized computing. They contain three pins. Even a small amount of current flowing through one pair of pins allows (possibly much larger) current to flow through the other pair. They thus act as electronic switches. The two outermost pins can be connected to a high-voltage circuit driving an electric motor or a solenoid valve. These devices require more power that can be delivered by the Arduino board. The middle transistor pin connects to one of Arduino's digital pins. Arduino code then drives the voltage on this pin high or low, which toggles the electrical current through the transistor to drive the motor. Transistors can also be used to implement algorithms by implementing digital logic gates. Another common component is the resistor. Resistors reduce the current flowing between two electrodes according to the familiar $I = V/R$. They come in several varieties. First, axial-lead resistors provide fixed and constant resistance according to a color code specified on their body. Variable resistors contain a dial that is controlled by the user. They effectively incorporate analog controller for, let's say, adjusting an LED brightness. The resistance of a photoresistor varies according to the ambient light. These devices can be connected to an Arduino's analog input pin to implement logic that activates on a light change. We demonstrate this use later in this chapter. Photoresistor could be the foundation of a home-made irrigation controller that activates sprinklers by using a transistor to open a solenoid valve whenever light is first detected after a sufficiently long period of darkness. They could also be used in a security system that activates an external camera when light is detected.

Another commonly utilized electrical components is the capacitor. Capacitors are devices that store electrical charge using isolated electrodes. The simplest design consists of two parallel metal plates enclosed in a ceramic body. The device's capacitance C relates the stored charge to the voltage between the electrodes via $C = Q/V$. Capacitors act as miniature batteries. They are used to smooth out noise in the input signal. Light-emitting diodes, or LEDs, represent another common component. As the name indicates, they are diodes that emit light. Being a diode, they allow current to flow in only one direction. LEDs can be single colored, in which case they contain just two electrical leads, with the longer anode connecting to the positive supply. They are also available in the RGB (red-green-blue) variety. This form uses four electrical connectors. Applying different voltage to these leads allows for the creation of an arbitrary color.

11.2.2 BREADBOARDS

Before an actual electronic device is built and deployed, it needs to be tested. This is where *breadboards* are used. A breadboard, shown in Figure 11.1(c) contains a grid of electrical connectors. It can be solderless, in which case electrical connections are made by pushing wire leads into the slots. Alternatively, it may require soldering as is the case with the board in the top left part of the figure. This solderable board allows us to see the electrical connectivity provided by the breadboard. Due to convention, columns are parallel to the long edge, while rows are parallel to the shorter side. Along both long edges are two *bus strips* in which slots are connected column-wise. Each bus strip consists of two columns. These strips are used to provide the positive and negative leads for the circuit. All slots along the entire column are connected together. By connecting any bus strip slot to a power supply positive terminal, all other slots in this same column can be subsequently used to drive the circuit. The other strip is connected in a similar manner to the power supply ground.

Terminals in the middle *bus strip* section are connected row-wise. Any two electrical leads sharing the same row are electrically connected. This connection does not cross over the midpoint. In this particular image, we can see an electrical circuit connecting the Arduino's digital pin 9 to an LED, which is then connected through a resistor to the ground. We also see a relative humidity sensor breakout board connected to the Arduino's positive (3.3 V) and negative (GND, ground) terminals via the bus strips. Note this board can supply 3.3 V and 5 V output, as well as the un-modulated input voltage (Vin) provided by the attached power supply (USB cable in this case). The two other wires connect the sensor's SCL and SDA pins to similarly labeled pins on the board. These two pins are used to provide data communication via the I²C protocol. Finally, once the breadboard design is finalized, it needs to converted to a portable version with permanent connections. These connections are made by soldering the leads together. Soldering involves using a heat source, such as a soldering iron, to melt conductive metal "solder" (typically made of tin or a tin-lead composite) for a brief amount of time to allow the liquid metal to flow around the leads being connected. The solder rapidly solidifies once the heat source is removed. Typical wire-soldering equipment is shown in Figure 11.1. This picture also shows a copper wick used to remove extra material, wire cutters, and a "third-hand" helping tool with a magnifying glass and clamps for holding the components together. Solder can also be purchased as a paste. This type is used to surface mount tiny components such as integrated circuit chips onto printed circuit boards (PCBs). The paste is melted using a special oven or, in a more DIY-fashion, a hairdryer.

11.2.3   ARDUINO PROGRAMMING

We now describe the basics of programming the Arduino class of microcontrollers. It should be noted that due to its popularity, many third-party vendors offer their own controllers that utilize the same instruction set as the Arduino. An Arduino code consists of two functions: `init` and `loop`. As the names indicate, the first function is called on startup, while the second one provides the code that runs continuously in an infinite loop. This is analogous to having a C code with the following logic:

```
1  void main() {
2      init();
3      while(true) loop();
4  }
```

Arduinos are programmed by connecting them to a computer over the USB port. Once programmed (implying the code is uploaded to the board's EEPROM), the microcontroller can be disconnected and powered by a battery.

Let's now demonstrate a simple Arduino program. Consider the wiring in Figure 11.1(c). Specifically, we are using a wire to connect Arduino's digital pin 9 to the breadboard's row 20, column a. In column e (remember that all columns are wired together across the row in this bus strip section), we place the positive long leg of a red LED. The negative cathode leg connects to row 21, column a. At another place in this row, such as column c, we place a resistor. Here we are using one rated at 220 Ω, as can be decoded from the strip colors, but any resistor will do (actually, you don't really need one, although it may lead to the LED burning out faster). The other leg of the resistor is connected to row 24. At this row, we also placed a short wire connecting to the breadboard's negative bus strip. This strip is subsequently connected with a longer wire to the board's ground, denoted by GND. This particular Arduino board contains 3 such grounding locations, and any of them will do.

We next need to power up the Arduino. This is done by connecting the board to a computer with a USB cable. Alternatively, you could run the board from a battery using the barrel plug. You will see a small indicator light turn on. The LED connected to the breadboard remains turned off. Digital pins are initially at the ground state. As such, no electricity is flowing through the circuit. We thus need to write a few lines of code to drive the pin high. The actual programming is done using an Arduino IDE (historically called Processing) that is downloaded from the `arduino.cc` website. It is available

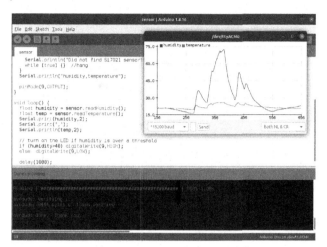

Figure 11.2 Arduino IDE running on Linux.

for a variety of operating systems including Linux. The IDE can be seen in Figure 11.2. It consists of a text editor, a console that outputs messages related to code compilation and upload, and also features a serial port monitor and plotter for communicating with the running programs over the serial port. First, check that the IDE has recognized the attached board. To do so, review the Board and Ports options under the Tools menu. Usually, the correct port and board are recognized, but if not, you can set them here manually. Next, write the following code:

```
1  void setup() {
2      pinMode(9,OUTPUT);
3      digitalWrite(9,HIGH);
4  }
5
6  void loop() {  }
```

Next, use the toolbar button containing an arrow to compile and upload the code to the Arduino. As soon as the upload finishes, you should see the LED turn on.

As you can tell, Arduino is programmed in what is essentially C++. The main difference is that we don't need to include a **main** function. We also don't need to include the **Arduino.h** header file, at least not in the main script file (for some reason, it needs to be explicitly included when developing custom libraries). This header provides prototypes for the functions used in the example. First, **pinMode** is used to make pin 9 an output pin. Arduino digital pins can be used for input or output, but not for both. We use **digitalWrite** to set this pin to the high state. This changes the voltage from 0 V to 5 V (or 3.3 V, based on the board type). It is this function that activates the electrical circuit and turns on the LED.

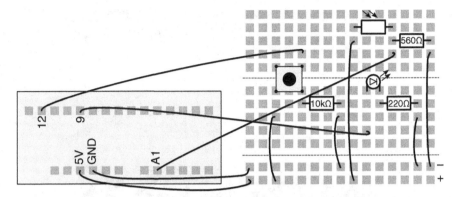

Figure 11.3 Circuit drawing for the photoresistor example.

11.2.4 PIN INTERACTIONS

Arduino Uno also contains several *analog pins*. These pins only provide read access. To illustrate, let's consider the circuit schematic in Figure 11.3. The board contains a photoresistor connected through a 560 Ω resistor to the ground with a parallel connection to Arduino's analog pin 1. We retained the LED connected to digital pin 9 and also added a push button. The button is connected to digital pin 12 via a 10 kΩ pull-down resistor. Digital pins are interrogated with `digitalRead`. This function returns 0 or 1, with the latter state indicating that the circuit is connected to the power source. In this case, the high value indicates that the button is being pressed. The `analogRead` function is used with analog pins. It returns an integer between 0 and 1024. With the selected resistor and a 5V input line, we find that in our windowless office, the readout is around 230 with the overhead lights on. With the lights off, the value drops to 30. We read around 700 under a direct flashlight illumination. We can then write the following code that automatically turns on the LED whenever the lights are switched off:

```
1   #define PIN_LED 9
2   #define PIN_BUTTON 12
3   #define PIN_PHOTORES 1
4
5   void setup() {
6       pinMode(PIN_LED,OUTPUT);
7       pinMode(PIN_BUTTON,INPUT);
8
9       Serial.begin(9600);
10      if (!Serial) {} // wait until serial port is ready
11  }
12
13  void loop() {
14      int button = digitalRead(PIN_BUTTON);  // get button state
15      int val = analogRead(PIN_PHOTORES);    // read from photores
16      Serial.println(val);       // write text to the serial port
```

```
17
18     // turn on the LED if button pressed or if the light is dim
19     if (button || val<50) digitalWrite(PIN_LED,HIGH);
20     else digitalWrite(PIN_LED,LOW);
21
22     delay(500);    // wait 0.5 seconds before checking again
23   }
```

The loop function begins by reading the button state from the digital pin 12 and the photoresistor output from analog pin 1. The analog value is also written out to the serial port, where it can be monitored using Arduino IDEs Serial Monitor to calibrate the response to the ambient environment. If the push button is pressed or if the photoresistor read-out drops below 50 (indicating darkness), the LED is activated by driving pin 9 high. Otherwise, it is turned off. We wait 0.5 seconds before querying the pins again.

11.2.5 USING LIBRARIES

We next demonstrate the use of a custom sensor such as the Adafruit Si7021 Temperature and Relative Humidity Sensor. This sensor costs $10 at the time of the writing, and ships in the form of a breakout board with 5 lead holes labeled. Vin, 3Vo, GND, SCL, and SDA. The board comes with header pins that you need to solder into these holes. Alternatively, you can just solder in wires. Vin and GND are the positive and negative terminals for the board. The 3Vo pin can be used as a 3.3 V power source for other components. We ignore it. The two other leads provide the connectivity over the I^2C interface. To use the sensor, first connect the Vin and GND connectors to the Arduino's 5V (or 3.3V) and GND pins, either directly or via a breadboard. Then connect the SCL and SDA leads to similarly labeled pins on the Arduino. These are usually found along the upper end of the digital pins closest to the USB plug.

Next, make sure you have the Adafruit Si7021 Library installed. Navigate to Tools→Manage Libraries and search for the sensor name. If not already available, install the library by clicking Install. Then, upload the following code:

```
1    #include "Adafruit_Si7021.h"
2    Adafruit_Si7021 sensor = Adafruit_Si7021();
3
4    void setup() {
5      Serial.begin(9600);
6      while (!Serial) {  delay(10);  }  // wait for the serial port
7
8      if (!sensor.begin()) {
9        Serial.println("Did not find Si7021 sensor!");
10       while (true) {}  // hang
11     }
12     Serial.println("humidity,temperature");   // write out header
            for plotter
13     pinMode(9,OUTPUT);           // mark pin 9 as output
14   }
15
```

```
16   void loop() {
17       float humidity = sensor.readHumidity();    // read data from
             the sensor
18       float temp = sensor.readTemperature();
19       Serial.print(humidity,2);
20       Serial.print(",");
21       Serial.println(temp,2);
22
23       // turn on the LED if humidity is over a threshold
24       if (humidity>50) digitalWrite(9,HIGH);
25       else    digitalWrite(9,LOW);
26
27       delay(1000);       // wait one second before querying again
28   }
```

This code assumes you still have the LED connected to the digital pin 9. Next, select Serial Monitor from the Tools menu. You will see output such as

```
44.33,23.61
44.34,23.64
```

where the first value is the relative humidity in % and the second one is temperature in °C. Alternatively, select Serial Plotter from this same menu. You should see a graph with blue and red lines mapping to these values. Now breathe onto the sensor or place your finger over it. The humidity line will increase, and the LED will light up.

11.2.6   INTERFACING WITH PYTHON

As you just saw, the Arduino IDE provides basic logging and plotting support. Similar interface can be easily added to your analysis codes using libraries that provide support for serial port communication. In Python, this can be accomplished using the **Serial** module. The listing below illustrates its use. We first open the connection by specifying the port name and bandwidth. The bandwidth needs to match the setting used in the Arduino code, otherwise the data will be corrupted. On Linux, the serial ports are given names such as /dev/ttyACM2. Since we don't know exactly which port the Arduino is attached to, this script attempts to open communication over the first five ports. Make sure to close Serial Monitor or Plotter in the Arduino IDE before running the code, since only one application can be communicating over the serial port at the same time.

```
1    import matplotlib.pyplot as plt
2    from matplotlib import animation
3    import serial
4
5    # try to find a serial port
6    port_ok = False
7    for i in range(0,5):
8        try:
9            port_name = "/dev/ttyACM%d"%i
10           print(port_name)
11           ser = serial.Serial(port_name,9600)
```

```
12          print("Opened port "+port_name)
13          port_ok = True
14          break
15      except serial.SerialException:
16          pass
17  if (not(port_ok)):
18      raise Exception("Failed to open port")
19
20  plt.close('all')
21  fig = plt.figure(1)  # use figure
22  temp_plot, = plt.plot([],[],color='black',label='Temp (C)')
23  rh_plot, = plt.plot([],[],color='gray',linestyle='-.',linewidth
           =3,label='RH%')
24  plt.xlim([0,200])
25  plt.ylim([0,100])
26  plt.legend()
27  plt.grid()
28
29  T = []
30  RH = []
31  samples = []
32  def read(n):
33      global rowj
34      global phi,ndi
35      line = str(ser.readline());  # read line from serial port
36      line = line[2:-5]    #eliminate trailing b' and \r\n
37
38      pieces = line.split(',');    # split by comma
39      RH.append(float(pieces[0]))  # append data
40      T.append(float(pieces[1]))
41      samples.append(n)
42      temp_plot.set_data(samples,T) # update plots
43      rh_plot.set_data(samples,RH)
44
45  anim=animation.FuncAnimation(fig,read,frames=200,interval=100,
           repeat=False)
46  plt.show()
```

The actual reading is accomplished with Serial.readline(). As the name suggests, this function returns a string that corresponds to a single new-line terminated line. The actual returned value will be a *string literal* like

`b'44.33,23.61\r\n`

We eliminate the leading and trailing characters and subsequently split the string by commas. The individual components are converted from a string to a float, and then appended to the end of a list. We then use line plot's `set_data` method to update the plots to generate output as illustrated in Figure 11.4.

11.2.7 CUSTOM LIBRARIES

We close the discussion on Arduino by describing steps needed to develop custom libraries. Libraries can be useful for reducing code re-use between projects. They are also necessary for commercial electronic components, as we already saw when interfacing to the humidity sensor in Section 11.2.5. Libraries are collections of C++ code placed in Arduino/Libraries folder. Each

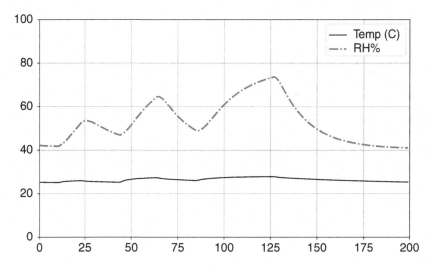

Figure 11.4 Real-time visualization of Arduino data using Python.

library resides in a directory matching the library name. Inside, you include a `src` folder that contains the source and header files. An optional `examples` folder contains sub-folders, each with a corresponding `.ino` file. These examples show up in the Arduino IDE under File → Examples. The main folder also needs to include files `keywords.txt` and `library.properties`. The first file is used to identify library keywords to include in syntax highlighting. The second file specifies library properties which are needed for the library to be considered valid. Finally, an empty `.development` file can be included to make the source files editable from the Arduino IDE. To illustrate, we now build a simple library that blinks the LED. First, we need the header file. Unlike sketch codes, libraries need to explicitly include Arduino.h.

```
1   #ifndef Blink_h
2   #define Blink_h
3
4   class Blink {
5   public:
6       // pin specifies pin connected to the LED, duration is in
              seconds
7       Blink(int pin, int duration):pin{pin},duration{duration} {
              all_ok = true; }
8       ~Blink() {}
9
10      operator bool() {return all_ok;}
11      bool operator !() {return !all_ok;}
12
13      void run();
14
15  protected:
16      int pin;
```

```
17      int duration;
18      bool all_ok = false;
19  };
20  #endif
```

Save this file as `Arduino/Libraries/Blink/Blink.h`. As you can see, this is a standard C++ code. While not demonstrated here, the code can utilize dynamic memory allocation and pointers, with the caveat that Arduino memory space is greatly limited. Arduino Uno has only 2 kB of dynamic memory, while MKR Vidor offers, comparatively, roomy 32 kB. The destructor would then be used to free the memory space. Next, save the following code as `Arduino/Libraries/Blink/Blink.cpp`:

```
1  #include "Arduino.h"
2  #include "Blink.h"
3
4  void Blink::run() {
5      digitalWrite(pin,HIGH);
6      delay(duration*1000);
7      digitalWrite(pin,LOW);
8      delay(duration*1000);
9  }
```

Then, use `Arduino/Libraries/Blink/keywords.txt` to specify the defined object names. The two types correspond to class and functions, respectively, and the space on each line needs to be a tab.

```
Blink   KEYWORD1
run     KEYWORD2
```

Finally, add `Arduino/Libraries/Blink/library.properties` file with the following fields:

```
name=Blink
version=1.0.0
author=name <email@server.com>
maintainer=name <email@server.com>
sentence=Blink Example
paragraph=Blinks a specified LED for a given number of seconds
category=Signal Input/Output
url=https://github.com/your_library_url
includes=Blink.h
dot_a_linkage=true
```

The library is included in an Arduino sketch using `#include <Blink.h>`. The IDE automatically recognizes the library, and compiles and links it during the project build. An example of usage is given below.

```
1  #include "Blink.h"
2
3  #define PIN_LED 9
4  Blink blink(PIN_LED,5);   // instantiate as a global object
5
6  void setup() {
```

```
7      while (!blink) { /* hang */ }
8  }
9
10 void loop() {
11     blink.run();
12 }
```

The above section offered only the cursory introduction to the Arduino. Some other functionality of interest may include the Wire library which provides support for communication over the I²C protocol. Instead of utilizing `digitalWrite` and `digitalRead` functions, pin states can be accessed through bit-wise operation on register variables called PORTD, PORTB, and PORTC. Companion registers DDRD, DDRB, and DDRC control the pin directionality. This direct access makes it possible to set multiple pins at once and is also faster than going through the wrapper functions.

11.3   FPGAS AND VERILOG

Arduino microcontrollers are an attractive option when computation in a small, low-power package is required. However, they lack the speed needed for high-throughput processing. The Arduino Uno CPU runs at only 16 Mhz. The more capable Vidor 4000 operates at 48 Mhz. Although 3× faster than the Uno, this clock speed is paltry compared to the 1.5 GHz Raspberry Pi, especially given that the Pi costs only one-third of the Vidor. Field Programmable Gate Arrays (FPGAs) offer an attractive synergy between the small spatial and power requirements of microcontrollers and the higher processing power of single board computers.

FPGA is essentially a programmable processor. All codes that we have encountered so far were executed on some kind of a CPU. The processor understands a multitude of basic instructions, such as "add two numbers", or "jump execution to a new address if a condition is true". The role of the compiler is to convert the textual source code to a binary sequence of these elementary instruction. The instructions are fed to the CPU one by one at the processor clock frequency. An FPGA instead offers a collection of programmable logic *gates* that control the flow of electrical signals. An AND gate receives signal on two input wires, and lets it pass only if both inputs are being driven (have current flowing through them). An OR gate passes the signal whenever at least one of the two signals is present. The gates are programmable in the sense that a new configuration of active gates can be flashed to the device as many times as needed. This is a very important characteristic to keep in mind. Even though the FPGAs can be programmed in languages that share some similarities with higher-level languages, the compiled output is not a set of instruction for some CPU. Instead, it is the description of an electrical circuit. The implemented logic operates directly on the electrical signal in the form of individual bit-wise 1s and 0s. The presence of a large number of gates implies that mulitple data bits can be processed concurrently. This makes FPGAs attractive in applications that need to process a large amount of data rapidly, such as video encoding. FPGAs and the closely related ASICs (Application Specific Integrated Circuits) are also commonly found in space instruments. Unlike an FPGA, ASICs are not re-programmable. For those of you who grew up with compact discs, FPGAs are the CD-RWs to the ASICs CD-ROMs. Producing ASICs requires a fixed initial cost to set up the tooling machinery and thus they are practical only in consumer devices requiring massive production runs.

FPGAs can be programmed using two different ways. One option is to literally draw a circuit schematic diagram in which input wires are connected through logic gates. We can only imagine that doing this for anything but a simple demo would be extremely frustrating and time consuming. The second option involves using a *hardware descriptor language* (HDL) such as VHDL or Verilog. With this approach, one begins with a textual source code. The code is subsequently synthesized into the underlying circuit description. It is important to keep the underlying hardware architecture in mind when writing HDL code. It is for example possible to write Verilog code that is syntactically correct, and can even be run using PC-based simulators, but one that cannot be *synthesized* (translated to the hardware description). The circuit and HDL approaches for programming FPGAs can be combined. It is not unusual to see Verilog or VHDL used to implement custom logic blocks, and have these blocks subsequently wired together using the graphical circuit schematics.

11.3.1 DEVELOPMENT ENVIRONMENTS

In order to program an FPGA, we first need to install a development environment capable of compiling the code, assigning pins, and programming the chip. One popular option when working with Intel FPGAs is Intel Quartus. This application, along with the supported FPGAs, was initially developed by Altera, which was subsequently acquired by Intel. For this reason, you will frequently encounter Altera labels when working in Quartus. You will of course also need an FPGA. The next sections assume that you have access to the Intel Cyclone LP Development Kit. It consists of the Intel Cyclone 10 LP FPGA integrated into a circuit board featuring several buttons, LEDs, and digital pin headers that share the footprint with Arduino Uno. This makes it possible to attach the FPGA board to the Arduino in the form of a shield. The board connects to the computer via USB.

Prior to starting up Quartus, you should make sure that your computer can communicate with the FPGA. This can be done on Linux by running jtagconfig from the Quartus installation bin directory:

```
user@system:~/intelFPGA_lite/19.1/quartus/bin$ ./jtagconfig
1) Cyclone 10 LP Evaluation Kit [1-11]
    020F30DD   10CL025(Y|Z)/EP3C25/EP4CE22
    020D10DD   VTAP10
```

If the FPGA is connected but you see a message that no JTAG hardware is available, run sudo killall -9 jtagd several times to terminate any currently running instances of the JTAG daemon. JTAG is the communication protocol used with devices such as the Arduino or the FPGA. Subsequently, run

```
sudo ./jtagd --user-start
```

Then run `jtagconfig` again, with it hopefully finding the attached hardware.

11.3.2   FPGA HELLO WORLD

Once Quartus starts, select File → New Project and click through the wizard specifying the project location. Name this project `blink`. If prompted, also specify the FPGA to use. The actual model, such as 10CLO25YU256I7G is found on the chip. Once the new project loads, once again click on File → New, but this time select Verilog HDL file. A blank text editor will appear. Type the following:

```
1  module blink (input wire button1,
2                input wire button2,
3                output wire LED);
4  // connect the buttons to the LED via an AND gate
5  assign LED = button1 | button2;
6  endmodule
```

Save this file as `blink.v`. Compile the project by clicking on the blue triangle "Start Compilation" toolbar option. This command is also found under the Processing menu.

You have just created a new module called `blink`. Think of a module as a physical chip with exposed electrical leads. These leads are wired to the rest of the circuit. Some of these leads receive input signals, while others provide output. Our module expects two input wires and provides one output. The code above specifies the logic that happens inside this "chip". Specifically, we can see that output is assigned to the result of a bitwise OR operation between the two inputs. The two input wires are labeled button1 and button2, while the output wire is labeled LED. While the names are arbitrary, they help us guess the intention of this module. The two inputs are assumed to connect to push buttons, while the output is connected to an LED. A wire carries just a single bit (0 or 1) of information. Wires can also be in undefined and high impedance states, denoted by x and z, but we ignore these in this section. Bi-directional wires are defined using `inout`.

The development board LED are wired such that they need to be driven low to turn on. In other words, the wire connecting to the LED needs to be assigned the value of 0 for the LED to produce light. Push buttons operate in a similar way in that pressing them drives them low. The OR expression evaluates to zero only when both buttons are in the low state (in other words, pressed). Thus, the above logic results in the LED turning on if, and only if, both buttons are pressed at the same time. Note that the AND operation would be used on an Arduino in order to drive the LED pin high if, and only if, both buttons are in high.

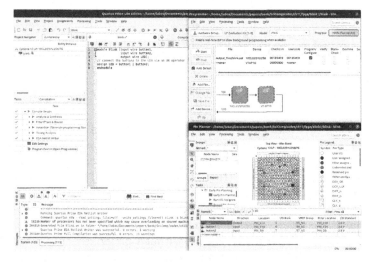

Figure 11.5 Quartus interface showing the code editor, the programmer, and pin assignment planner.

In order for this project to actually work, we need to connect the module's wires to the board's electrical connections corresponding to the buttons and the LED. This is accomplished using the Pin Planner found under the Assignments menu. You should see a diagram of the FPGA as shown in the bottom right of Figure 11.5. You may also encounter a blank screen with a message that an AUTO device is in use. This message indicates that an FPGA model was not specified during the project creation. It is easily remedied by navigating to Assignments → Device.

The bottom section of the pin planner contains 3 listed node names: LED, button1, and button2. If you don't see anything populated here, make sure to compile the project first. While the direction is set, the location fields are blank. Double-clicking the Location field will bring up a drop-down list from where you can select the FPGA *pin* to connect to the node. The FPGA is soldered onto the circuit board over many individual traces (the electrically conductive paths built into the board that look like thin wires). While you may have some luck with a direct inspection of the circuit board, a simpler option is to download the PCB schematic from Intel's website (search for Intel Cyclone 10 LP Evaluation Kit documentation). At the time of this writing, this schematic is found at `https://www.intel.com/content/dam/altera-www/global/en_US/support/boards-kits/cyclone10/c10lp-eval-a1-sch.PDF`. Page 14 of the schematic contains the wiring diagrams for the LEDs and the user push buttons. Specifically, we can note that the LEDs are given names `USER_LED0` through `USER_LED3`, while the push buttons are called `USER_PB0` through `USER_PB3`. Next, pages 4 and 5 contain wiring schematic for the FPGA pins.

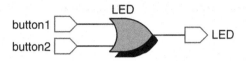

Figure 11.6 RTL diagram of our simple module.

In Bank 5, we find listings for the LEDs and find that they map to pins L14, K15, J14, and J13. Banks 6 and 7 contain the push buttons. They map to E15, F14, C11, and D9. This information is also tabulated in the User's Guide. With this in mind, we assign `PIN_L14` to the LED, and `PIN_F14` and `PIN_D9` to button 1 and button 2.

Compile the code again, and then open Programmer from the Tools menu. Under Hardware Setup, select the Cyclone 10 LP Evaluation Kit option. If this field is blank, try restarting `jtagd` as outlined previously. Then click Start. There is a possibility that the programmer fails. The error messages appear under the System tab of the Messages section of the main Quartus IDE. You may find a note that the programmer expected 1 device but found 2. There are actually two JTAG devices on the board. To remedy it, click the Auto Detect button on the Programmer. Select the correct FPGA class (10CL025Y in our case). Subsequently, click on the FPGA on this device and select Change File from the toolbar. Navigate to the `output_files` folder of the current project, and select the `.sof` file. Also make sure that the Program/Configure checkbox is selected. Clicking Start should now lead to a successful upload. Confirm that pressing push buttons PB3 and PB1 at the same time indeed leads to the LED turning on. These buttons are very tiny and thus the code was written with two non-adjacent buttons to make pressing them at the same time slightly easier.

11.3.3 SCHEMATIC VIEW

You can visualize the corresponding *netlist* (the electrical connectivity) by selecting Tools → Netlist Viewers. The RTL Viewer will generate the view shown in Figure 11.6. The part in the middle is the universal symbol of an OR gate. If you so prefer, you can design the FPGA logic by drawing these circuits directly. This is accomplished by adding a new file of Block Diagram/Schematic type.

11.3.4 EVENTS

A curious feature of the previously developed module is that there is no "loop" as may be expected based on our discussion of programming microcontrollers. Furthermore, assume that we wired in a few additional LEDs so the code reads

```
1   assign LED1 = button1 | button2;
2   assign LED2 = button1 | button2;
3   assign LED3 = button1 | button2;
```

Based on your experience in procedural programming in C++ or Python, you may expect that LED1 turns on before LED2, and LED3 turning on last. But this is not the case. These three activations happen at the same time. This is because the **assign** keyword specifies electrical connectivity. The code is analogous to physically connecting electrical wires leading from the buttons to the LEDs. As such, the LED states change continuously based on the input drivers.

Verilog does however support procedural blocks that execute in a linear fashion when some prescribed event takes place. These blocks are defined with the **always** keyword. The general expression is

```
1  always @(sensitivity list) begin
2  // some code
3  end
```

Instructions between **begin** and **end** execute in sequential order (these keywords are not needed if we have just a single instruction to run). The event is specified using a *sensitivity list*. There are essentially two forms. First, we can write

```
1  always @(input1 or input2)
```

to add code that runs whenever either input changes value. We can also write

```
1  always @(posedge input1 or negedge input2)
```

to run the code only when input1 changes from 0 to 1, or when input2 changes from 1 to 0.

Your code may, and often will, contain multiple **always** blocks. These blocks execute independently of each other. Their ordering in the source file has no impact on the final synthesized output. We need to be cognizant of race condition, which was previously covered in the discussion of multithreading in Chapter 9. In practice, it means that we need to avoid multiple blocks driving the same output. Your code may also include **initial** blocks. Code in these blocks is not synthesized and is used solely to initialize variables.

11.3.5 DATA TYPES AND ASSIGNMENTS

Verilog understands two types of data: *variables* and *nets*. Variables are analogous to their use in other languages. They are named memory blocks that store data. Nets are electrical connections. A *wire* is type of a net. The *assign* keyword is used to drive nets. The value on the left-hand side changes automatically whenever the operands on the right side change.

We already saw how to interface with individual wires. A single wire represents a single bit of information. Wires can also be grouped into a *bus*. For instance, we may write

```
1  wire [3:0] buffer;    //4-bit bus
2  wire [15:0] index;    //16-bit bus
```

The indices are listed with the least-significant bit on the right. Opening the Pin Planner, you will see that you now can (and need to) assign pins to the individual components. Brackets placed after the variable name are used to create vectors (arrays). Multi-dimensional vectors are allowed, i.e. `wire [4:0]` `buffer[2][3];`. Besides a `wire`, net data types also include `wand`, `wor`, `tri`, and `supply0` and `supply1`. `wand` and `wor` can be used to define the truth table for a net driven by multiple inputs. `tri` is synonymous with `wire`. The two supplies provide a continuous 0 or 1 states.

Variables are declared using `reg`, `integer`, and `time` keywords. The first option allows us to define a data register spanning arbitrary number of bits. Registers can also be used to declare arrays of bytes, which can subsequently store strings. The `integer` keyword declares a 32-bit register for storing unsigned whole numbers. The time format is used to store program run time for debugging purposes. The Verilog language also defines `real`, `shortreal`, and `realtime` data types, but these are not supported by Intel Quartus as they cannot be synthesized. The code below illustrates several different variable types.

```
1   reg [7:0] buffer;    // 8-bit register
2   reg [8*20:0] str;    // 21-character string
3   integer my_int;
4
5   always @(posedge clock) begin
6    buffer = 8'hFF; // some 8-bit binary value
7    buffer[3:0] = 4'b1100; // overwrite the bits 3 through 0
8    str <= "Hello World!"; // set string
9    my_int <= 123;     // some integer value
10  end
```

Note that numbers can be specified with an explicit binary width as well as base. The 8 bits of buffer are first all set to 1 by assigning the value of 255 using the hexadecimal format. We subsequently overwrite the 4 lowest bits to a new value specified in binary. Also note that we are using two types of assignment operators. `=` is considered a *blocking* operator. The first assignment completes before the second assignment is made. `<=` is a *non-blocking* assignment. It has a similar effect as the `assign` keyword. Assignments specified using this operator all take place concurrently at the completion of the always block. A particular benefit of this operator is that it allows to define a *flip-flop* using just two instructions:

```
1   a <= b;
2   b <= a;
```

In C++, this operation would require a third temporary variable to store the initial value of `a`.

11.3.6 CLOCKS

Let's now consider a modified version of our program that toggles an LED on or off after some fixed number of events. While it may be tempting to tie the event handler to a button press, pressing a button does not lead to a simple state transition from 1 to 0. Instead, there is an initial oscillatory transition during which the signal "bounces" between the two states. In the absence of a secondary algorithm to debounce the signal, the LED would activate after a seemingly arbitrary number of button clicks.

Instead, we can take advantage of a *clock*. A clock is a device (historically implemented using quartz crystals) that oscillates between a 0 and a 1 state at a fixed frequency. The Intel Cyclone board we are using includes a clock operating at 50 MHz, and another one with a programmable frequency supporting frequencies between 3 kHz and 200 MHz. The 50 MHz oscillator ties into Cyclone LP pins E1 and E2 (these correspond to the positive and negative signals). Any pin could serve as a clock if connected to an external signal generator.

Let's now consider the first 11 integers written out in the binary format:

0000	0001	0010	0011	0100	0101	0110	0111	1000	1001	1010
0	1	2	3	4	5	6	7	8	9	10

We can notice that bit 2 (the third bit from the right) flips every 4 values. We can thus implement a blink operation by defining a sufficiently large counter. We next tie in logic to the clock that increments the counter by one one each clock tick. Finally, a sufficiently high-order bit is wired to the LED to cause the LED to toggle in response to that bit changing its value. We need to utilize a bit that flips at a frequency low enough to be visually comprehensible. Bit 23 offers one such possibility, as it flips once ever 0.17 seconds given the 50 Mhz clock. Bit 26 flips once every 1.34 seconds. We can thus write the following code:

```
1   module blink (input wire clock ,
2   output wire[1:0] LED);
3
4   reg[27:0] counter;
5   initial
6     counter = {28{1'b0}};    // set all bits to 0
7
8   always @(posedge clock) begin
9   counter <= counter + 1;    // increment on tick
10  end
11
12  assign LED[0] = counter[23];    //L14
13  assign LED[1] = counter[26];    //K15
14  endmodule
```

Next, in the Pin Planner, assign pins L14 and K15 to the two LEDs, and Pin E1 to the clock. Finally, create a new Synopsis Design Constraints File, located under the Other Files section of the New File dialog box.

```
create_clock -name clock -period "50MHz" [get_ports clock]
set_false_path -from * -to [get_ports LED[0]]
set_false_path -from * -to [get_ports LED[1]]
```

Save this file as blink.sdc. This file provides additional information about the system, such as pin capacitance. It is mainly used to describe clocks for the Time Analyzer. The false path commands indicate that the LEDs do not have any timing constraints. Compile and program the FPGA. You should observe the two LEDs blinking at different frequencies.

11.3.7 CONTROL STATEMENTS

Verilog supports additional control statements that you are already familiar with from other languages, including **if-else**, **for** loops, and **case** branches. Functions can be defined using **function** and **task** keywords. The second variant is used for code blocks that return multiple values. We can also directly include logic gates using **and**, **or**, and **xor** operators. Below is a generic example illustrating these concepts:

```verilog
1  module ops(input a, input b,
2  output c, output reg [2:0] d);
3
4  reg [7:0] cnt;
5  integer i;
6  reg [3:0] x,y,z;
7
8  always @(a) begin
9      // if statement example
10     if (cnt>10) cnt = 0;
11     else cnt = cnt+1;
12
13     // for loop example
14     for (i=0;i<7;i=i+2) cnt[i] = a|cnt[i];
15
16     // case example
17     case (cnt)
18         'b00:    d = 0;
19         'b01:    d = 2;
20         default: d = 'b111;
21     endcase
22
23     // function call example
24     z = add(x,y);
25  end
26
27  // function example
28  function [3:0] add (input [3:0] x,y);
29     begin
30        add = x+y;
31     end
32  endfunction
33
34  xor(c,a,b);  // logical gate, c = a xor b
35
36  endmodule
```

11.3.8 IP BLOCKS

As we have already seen, integer math is natively supported. Unfortunately, coming from the numerical analysis world, we are accustomed to dealing with real values. In some instances, it maybe feasible to implement *fixed-point arithmetic*. This system assumes a fixed multiplier that scales the fractional numbers to integers. For example, the value 1.618 may be represented as 1618. The other option is to take advantage of built-in *Intellectual Property (IP) Cores*. IP Cores (or IP Blocks) are modules that provide some useful functionality. They serve the same purpose as the standard library functions in C++. While it is perfectly feasible to implement your own version of the exp() function with the help of Taylor series, it is much easier to just #include <math.h>.

The same is true on the FPGA. We could implement our own logic for floating point mathematics following the IEEE 754 standard, but this would lead to unnecessary effort duplication as others have already done so before us. The logic of IP Blocks, as the name suggests, tends to fall under companies' intellectual property and many firms generate revenue from licensing specific FPGA modules. Luckily for us, Intel Quartus includes a large collection of IP cores for handling math expressions, as well as providing other functionality such as Phase Locked Loops (PLLs), serial communication, and signal generation.

The IP Catalog, which is found under Tools but is also part of the standard desktop layout, contains the collection of available IP modules. To demonstrate its use, navigate to Basic/Arithmetic, and select "FP_FUNCTIONS Intel FPGA IP". You may also notice a variety of ALTFP modules, but these legacy Altera IP blocks are no longer present in recent versions of Quartus's library, despite still appearing in the catalog. You will next be prompted for a name of the custom variation of this IP core. Name it something such as mySqrt and make sure that Verilog is selected. In the Wizard, select Square Root under Roots. You will see that several files, includes mySqrt.v have appeared in your project folder. You should see the mySqrt entity appear under IP Components view of the Project Navigator. If it does not, make sure that mySqrt.qip is added under the Files view. Opening the mySqrt.v file, we can notice that it simply declares a module that in turn calls another module. The module's declaration provides us an insight into the list of arguments that need to be provided. Specifically, we see that it requires a clock, a reset bit, an input 32-bit register for the square root input, and another 32-bit register to store the output. The reset signal initializes the module on start-up.

Using the module in the rest of our code is simple. The main thing to remember is that Verilog is a hardware descriptor language. As such, we do not "call" a module in procedural **always** block the way you may be used to calling functions in C++. Instead, the module is a physical entity that is permanently added to the circuitry. The square root logic continues to be evaluated continuously at the input clock frequency, although it is possible to add an optional enable port to bypass the calculation. In the rest of the code, we need to decide when to actually use the square root module output. Specifically, let's assume that a top-level module is already wired to a data bus providing the 32-bit IEEE 754 data stream, and to another that expects the output. The most significant bit in this format indicates a negative number.

Hence we may write the following code:

```
1   module ip ( input clock ,
2   input reset ,
3   input wire [31:0] in ,
4   output wire [31:0] out );
5
6   reg  [31:0] calc_out;
7   myFP myFP_inst (clock , reset , in , calc_out );   // instantiate the
        module
8
9   // pass output only if input is positive
10  assign out = ( in [31]==0) ? calc_out : 32'b0;
11  endmodule
```

The above module instantiation uses positional parameters. It is also possible to directly specify the target for each argument, as in

```
1   myFP myFP_inst (.clk (clock ) , .areast (reset ) , .a( in ) , .q(calc_out ));
```

11.3.9 SIMULATION

As the FPGA code grows more complex, it becomes more likely to encounter a situation in which the observed behavior deviates from the expectation. Attempting to debug an FPGA algorithm by illuminating the LEDs when some condition is met is clearly a possibility but not the ideal one. Luckily, Verilog code can be run on a PC. This is known as a *simulation*.

To illustrate, consider the following code

```
1   'timescale 1ns / 10ps   // set time unit size and time tolerance
2   module blink (input wire clock , input wire disp , output wire [1:0]
        LED);
3       reg [27:0] counter ;
4       initial
5           counter = {28{1'b0}};   // clear all bits
6
7       always @(posedge clock ) begin
8           counter <= counter + 1;   // increment on tick
9       end
10
11      always @(disp) begin        // display counter on disp signal
            change
12          $display ("time:%6d, counter:%b" , $time , counter );
13      end
14
15      assign LED[0] = counter [18];
16      assign LED[1] = counter [22];
17  endmodule
18
19  // test bench
20  module blink_tb () ;
21      reg clock ;
22      reg disp ;
23      wire [1:0] LEDs ;
24      parameter MAX_TICKS = 1000000;
25
```

```
26    //instantiate the module
27    blink blink_inst(.clock(clock), .disp(disp),.LED(LEDs));
28
29    initial begin
30        clock = 1'b0;
31        disp = 1'b0;
32        forever #1 clock = ~clock;    // tick clock every time unit
33    end
34
35    initial begin
36        #200 disp = ~disp;       // flip disp after 200 time units
37        #800 disp = ~disp;
38        #1000 disp = ~disp;
39    end
40
41    initial begin
42        $monitor("time: %3d, LED0:%b, LED1:%b",$time,LEDs[0],LEDs
              [1]);
43    end
44
45    initial begin
46        $dumpfile("blink_tb.vcd");   // open output file
47        $dumpvars(0, blink_tb);      // save all data from blink_tb
              module
48        #(MAX_TICKS)                 // wait MAX_TICKS
49        $display("Done!");           // show message and terminate
              simulation
50        $finish;
51    end
52    endmodule
```

This code contains two modules: **blink** and **blink_tb**. The first one is a slightly modified version from the example blink algorithm from page 429. The LEDs are associated with lower bits in order to reduce the number of clock ticks needed to see a change. We also added a new wire called **disp** and added an **always** block that listens to this wire. Verilog is unique among programming languages in that only a small subset of valid language constructs actually form syntax that can be synthesized into hardware. Many commands and keywords are only valid in simulations running on a PC. One example is the system function $disp. It works similarly to C **printf**, with the output directed to the console. This function is used by this **always** block to visualize the counter variable in binary format. Other examples use the already mentioned **real** variable type as well as various system functions for performing mathematical calculations.

The second module defines the test bench. It is customary to name test benches by appending **_tb** to the name of the module. Note that this test bench module does not have any inputs or outputs. Its primary role is to instantiate the module being tested. The test bench also drives all signals that the tested module expects. You can notice that we defined a one-bit variable called clock. In the first initial block we use the **forever** keyword to continuously invert (or tick) this clock. We could have alternatively written

```
1  always begin
```

```
2  #1 clock = ~clock;
3  end
```

This syntax of an always block without a sensitivity list is valid only in simulation codes (in other words, it cannot be synthesized). The # statement delays the execution of the next statement by the specified number of time units. always blocks run in parallel, so pausing the execution of one has no impact on the timing in another block. The actual time associated with a single time unit is given by the first value in the 'timescale statement on line 1. The second value specifies the smallest time delay that can be specified. In this example, the smallest allowable delay is #0.01 or 10 ps.

We also define another initial block that first waits for 200 time units and then flips the disp register. This causes the always block in the blink module to trigger and display the current state of the counter variable. We repeat this operation two more times, after another 800 time units (simulation time 1000) and another 1000 ticks (simulation time 2000). Remember that while this is happening, the clock specified in the first initial block continues ticking. The third initial block uses the $monitor statement to generate display automatically whenever any of the monitored variables change. Finally, the last block uses $dumpfile and $dumpvars commands to output to the disk time history of all module data. The 0 parameter specifies the output level, while the second one lists the module to track. As you can imagine, this file can grow fairly massive! As such, it is best to automatically finish the simulation after some specified number of clock cycles.

Since this code is not going to be synthesized to a particular FPGA, we do not need to use Intel Quartus to run it. Instead, we utilize the Icarus Verilog compiler, iverilog, which also comes with a run-time engine called vvp. Compile and run this code using the following commands

```
testbench$ iverilog blink_tb.v -o blink_tb
testbench$ vvp ./blink_tb
VCD info: dumpfile blink_tb.vcd opened for output.
time:        0, counter:00000000000000000000000000000000
time:        0, LED0:0, LED1:0
time:      200, counter:00000000000000000000000001100100
time:     1000, counter:00000000000000000000000111110100
time:     2000, counter:00000000000000000001111101000
time: 524287, LED0:1, LED1:0
Done!
```

We obtain a print out of the counter register at the specified time points. Also, after some time has elapsed, the monitor catches the LED[0] changing state. The working directory will contain a large (23 Mb on our system) .vcd file. This file can be examined using another useful tool called **gtkwave**. Executing

```
testbench$ gtkwave blink_tb.vcd
```

will bring up a GUI where you can plot time histories of the clock, counter, LED and disp registers. A typical output is shown in Figure 11.7.

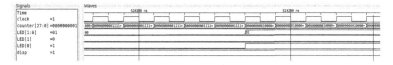

Figure 11.7 Visualization of data registers using GTKWave.

11.3.10 FPGA SYNERGISTIC USE WITH AN ARDUINO

The previously mentioned MKR Vidor 4000 is a special breed of the Arduino microcontroller since it contains an integrated Intel Cyclone 10 LP FPGA. Thus conceptually this device could be used for applications, such as image processing, which require processing more data that can be handled by the Arduino CPU. The purpose of this section is to illustrate how to use a custom FPGA code on this microcontroller. At least as of this writing, the Arduino IDE does not support programming the FPGA directly. There is however a step-by-step guide available at `https://docs.arduino.cc/tutorials/mkr-vidor-4000/vidor-gsvhdl`. The companion files are available at `https://github.com/wd5gnr/VidorFPGA`. This repo contains a Quartus project under `viderdemo/projects/MKRVIDOR4000_template`, as well as a sample Arduino sketch found under `viderdemo/blink-sketch`. The repo also includes a small C program for converting the compiled output from Quartus to an ASCII format compatible with Arduino's JTAG programmer.

In order to get started, we recommend that you copy the Quartus project files and the two Verilog codes from the template folder into the top-level projects directory containing the **constraints** and **ip** folders. Then, edit the .qsf file to update file paths, specifically by removing the `../../` parts. Here you can also update references to the included files in case they were renamed. Then open the project in Quartus. As the first step, verify that pins are assigned in the pin planner. If not, load the assignments from the constraints folder. Next, let's review the top Verilog module. It species a large list of input, output, as well as bidirectional wires. These nets are organized by their use. For instance, there is a group of wires that can be used for the FPGA to gain access to the system RAM. There are also pins used to communicate over the Arduino's Mini PCI-express as well as HDMI connectors. However, of interest to us are the pins that interface to the digital and analog pins. The digital pins are accessed through a 15-bit **bMKR_D** bus, while the analog pins are found in a 7-bit bus called **bMKR_A**. After instantiating timing modules, the module includes the **user.v** file. This is where your custom code should go. We can use it to write a version of the blink algorithm:

```
1  reg [27:0] counter;
2  reg [3:0] active_bit;
3  reg out_val;
4
5  //D6 connects to the LED, D5 is an input on/off signal
6  assign bMKR_D[6] = bMKR_D[5]?counter[22+active_bit]:0; // send 0
```

```
              if D5 is LOW
 7
 8  initial begin
 9     active_bit = 4'd5;
10  end
11
12  always @(posedge bMKR_D[4]) begin
13     if (active_bit >0) active_bit = active_bit -1;
14     else active_bit = 5;
15  end
16
17  always @(posedge wOSC_CLK)
18  begin
19     if (!rRESETCNT[5])    // wait until boot up
20     begin
21        counter <= 0;      // use 28'hfffffff to turn on LED on start
22     end
23     else
24     begin
25        counter <= counter+1;
26     end
27  end
```

This is a modification of the code that comes with the project. In the original version, the D5 pin is used select the pin connected to the LED. Here we use it to toggle the LED blinking on and off. Instead of hardcoding the bit used to drive the LED, we compute the assignment using the `active_bit` variable. This variable is initialized to 5, which makes the LED to initially use bit 27 for its blinking rate. Digital pin 4 transition from 0 to 1 is used as an indicator that the active_bit variable should be decremented. Once it reaches 0, it is rewound back to the initial value of 5. As such, D4 can be "clicked" 5 times to speed up the LED flicker rate. The subsequent click returns the rate to the slowest setting. The final code section is retained from the example. The main `top.v` code contains a section that increments the reset counter on each clock tick. The algorithm shown here waits for 32 clock cycles before our custom increment logic starts. This gives the FPGA time to wrap up any other possible initialization.

Next, use Quartus to compile the project. In the `output_files` directory you will find a file called `vidordemo.ttf`. This file needs to be converted to an ASCII representation utilizing the included script. For example

```
vidor/vidorcvt$ g++ vidorcvt.c -o vidorcvt
vidor/vidorcvt$ cd ../
vidor$ vidorcvt/vidorcvt < output_files/vidordemo.ttf > app.h
```

The resulting `app.h` is just a collection of binary data written out as ASCII integers, such as

```
2,2,2,64,64,64,64,0,0,0,0,132,132,132,132,196,
196,196,196,0,0,0,0,0,0,0,0,0,0,0,0,4,
```

Next copy this file to the folder containing the sample Arduino sketch. There are several versions of this sketch available on the internet. Here we use a variant in which the entire FPGA initialization was moved to setup_fpga() function found in defines.h. This file starts off by constructing a memory buffer containing the packet header followed by the custom bit stream from app.h. The setup function then uploads the buffer to the FPGA using JTAG functions.

```
1   #include <wiring_private.h>
2   #include "jtag.h"
3   #include "defines.h"
4
5   int blink = LOW;
6   int bit = LOW;
7
8   #define SET_PIN 5
9   #define LED_PIN 6
10  #define BIT_PIN 4
11
12  void setup() {
13      setup_fpga();
14      Serial.begin(9600);
15
16      pinMode(SET_PIN, OUTPUT);
17      pinMode(BIT_PIN, OUTPUT);
18      pinMode(LED_PIN, INPUT);
19      digitalWrite(SET_PIN, blink);
20      digitalWrite(BIT_PIN, bit);
21
22      Serial.println("Select No Line Ending, enter 'go', 'stop', or
            'bit' to control LED");
23  }
24
25  void loop() {
26      static int oldstate = -1;
27      static int num = 0;
28      static int oldbit =0;
29      int state;
30
31      // check if we have any input from the serial port
32      if (Serial.available() > 0)  {
33          String msg = Serial.readString();
34          if (msg.equals("go")) blink=HIGH;   else if (msg.equals("
                stop")) blink=LOW;
35          digitalWrite(SET_PIN, blink);
36          if (msg.equals("bit")) {
37              digitalWrite(BIT_PIN,LOW); delay(1);
38              digitalWrite(BIT_PIN,HIGH);
39          }
40      }
41
42      // read the LED pin value that is set by the FPGA
43      state = digitalRead(LED_PIN);
44      if (state != oldstate)  {
45          Serial.print(state);
46          if (++num == 40)  {
47              Serial.println();
48              num = 0;
```

```
49            }
50
51            oldstate = state;
52        }
53  }
```

Finally, connect the Arduino's D9 pin to an LED anode, with the cathode line connected to the ground through a small resistor. Then open the Serial Monitor. Make sure to change the line endings option to "No Line Ending". Failing to do so, will include the \n character in the string received on the Arduino, which will then cause the comparison algorithm to fail. Typing in 'go' or 'stop' and pressing enter will toggle the LED. Typing in 'bit' causes the Arduino code to send a positive edge signal on D9 by first driving the bit low, waiting a short time, and then driving it high. This will cause the always block to trigger, and run the active_bit decrement logic.

Appendix A: Fortran 77 Syntax

This appendix provides a brief summary of some commonly used Fortran 77 syntax, and a brief comparison of common programming constructs. Please note that this section is *not* intended to be a Fortran 77 tutorial. The readers are referred to the Fortran 77 language manual for a complete reference guide. Instead, this appendix is targeted at engineers who are tasked with modifying or converting a legacy Fortran 77 code but are otherwise not familiar with the language syntax.

FIXED LINE FORMAT

Legacy Fortran 77 codes are written using a fixed format, in which line lengths are limited to 72 characters. The first five columns must be left blank or can contain a numeric label (line number). Line continuation is indicated by a non-zero character in column 6. Comments are generated by writing c or * in column 1. Inline comments are generated using !, which ignores the rest of the line. Fortran 77 is not case-sensitive.

MAIN PROGRAM AND SUBROUTINES

The syntax for a main program "Main" which further calls subroutines subroutineA, and subroutineB(x,y,z) is

```
c234567
      program main
      . . . .
      call subroutineA
      . . . .
      call subroutineB(x,y,z)
      . . . .
      stop
      end

      subroutine subroutineA
      . . . .
      return
      end

      subroutine subroutineB(x1,y1,z1)
      . . . .
      return
      end
```

Popular compilers include **gfortran** from the gcc environment and the **ifort** Intel Fortran compiler.

DOI: 10.1201/9781003132233-A

439

DATA TYPE, ARRAY, PARAMETER, COMMON BLOCK

By default, any variable starting with i,j,k,l,m or n are assumed to be of integer type, while those starting with a-h or o-z are of a real data type. Variables can also be declared in the declaration block. **real*4** declares real data type in single-precision (4 bytes). **real*8** declares real data in double-precision (8 bytes). **complex** and **complex*8** is used to declare a complex data type consisting of two **real*4** elements. **complex*16** declares complex data type consisting fo two double precision elements. Integer data types are declared using **integer*2**, **integer*4**, and **integer*8**. Variables declared using **integer** default to **integer*4**. Character strings are declared using **character*n**, where n is the desired string length. Logical types are specified using **logical**.

The **parameter** command is the convenient way to define values of parameters (constant expressions) used throughout the program. For example,

    parameter (imax=50, jmax=100)

defines the dimension of the array X(imax, jmax) as X(50,100).

dimension is used to define an array. For example

    dimension X(imax,jmax)

defines a 2-dimensional array with bonds at 1 and imax, and 1 to jmax. **dimension** X(imin:imax, jmin:jmax) defines a 2-dimensional array with bonds at imin and imax, and jmin to jmax. It is important to remember that Fortran arrays are one-indexed, instead of the zero-indexing used in Python and C++.

An array can also be defined using the common block, instead of using **dimension**. Any variables and array can be defined and stored using a common block. The syntax for defining a common block that contains variables is **common /blockname/ variablename**. The data in a common block can be passed along to subroutines. The following example illustrates defining the arrays and transferring data from the main program to the subroutine using the common block:

```
program main
implicit real*8 (a-h, o-z)
parameter (imax=50,jmax=100)
common /domain/ x(0:imax,0:jmax), y(0:imax,0:jmax)
common /constant/ dx, dy
. . . .
call subroutineA
. . . .
stop
end

subroutine subroutineA
implicit real*8 (a-h, o-z)
parameter (imax=50,jmax=100)
common /domain/ x(0:imax,0:jmax), y(0:imax,0:jmax)
common /constant/ dx, dy
. . . .
```

```
    return
    end
```

In this example, the double precision arrays x(0:50, 0:100) and y(0:50, 0:100) are defined in the common block. These arrays and the double precision variables `alpha` are transferred to the subroutine through the use of common block.

CONDITIONAL STATEMENTS

The syntax for the "if" statement include:

```
if  (....)  then
....
endif

if  (...)
....
else
....
end if

if  (...)
....
else if  (....)
....
end if

if  (...)  go to  ...
```

Conditional expressions use .eq., .ne., .lt., .le., .gt., and .ge. for equality, $\neq$, $<$, $\leq$, $>$, and $\geq$, respectively. For example,

```
if  (A.ge.10)  then
....
endif
```

ITERATIONS

Iterative operations are carried out using a "do loop". The syntax for a do loop is:

```
do  i=1, imax
....
enddo
```

or

```
do  100  while  ( A .eq. B )
....
100    continue
```

INPUT AND OUTPUT

Data is read in using **read(i,j)**, and the output is written using **write(i,j)**, where i defines the file or "unit" for the I/O and j defines the format of I/O. **read**(*,*) or **read**(5,*) denote input from the keyboard, with no specified format. **write**(*,*) or **write**(6,*) denotes denotes output on the screen, with no specified format.

The format specifier is used to generate a fixed width output. This item can also reference a line which specifies the format. For example

```
      real*8 A
      integer*4 B
      write(*,20) A, B
20    format(F12.3, I10)
```

will write out the real variable A as padded 12 characters, with 3 decimal points, followed by the value of the integer variable B written out using 10 characters.

For data output, one typically uses the **open** command to open up a file for writing and the **close** command to close the file after output.

EXAMPLE

The following example illustrates the described concepts. The main code generates a 2-dimensional mesh and outputs the mesh. The mesh is written to file f100.dat. The variables are output with 4 decimal points.

```
c23456————————————
      program main
      implicit real*8 (a-h, o-z)
      parameter (imax=50,jmax=100)
      common /domain/ x(0:imax,0:jmax), y(0:imax,0:jmax)
      common /constant/ dx, dy, x0, y0
      call ReadInput

c———generate mesh
      do i=1, imax
      do j=1, jmax
        x(i,j) =x0 + float(i)*dx
        y(i,j) =y0 + float(j)*dy
      call WriteOutput
      stop
      end


      subroutine ReadInput
      implicit real*8 (a-h, o-z)
      parameter (imax=50,jmax=100)
      common /domain/ x(0:imax,0:jmax), y(0:imax,0:jmax)
      common /constant/ dx, dy, x0, y0
      read(5,*) x0, y0
      read(5,*) dx, dy
      return
      end


      subroutine WriteOutput
      implicit real*8 (a-h, o-z)
      parameter (imax=50,jmax=100)
      common /domain/ x(0:imax,0:jmax), y(0:imax,0:jmax)
      open(unit=100, form='formatted", status='unknown')
```

```
do i=0, imax
do j=0, jmax
write(100,200) x(i,j), y(i,j) \\
enddo
enddo
200 format('x=', f10.4, 'y=', f10.4) \\
close(unit=100)
return
end
```

COMPARISON OF COMMON CONSTRUCTS

Fortran	C++	Python
2D Arrays		
!array **dimension** X(ni,nj) *!array in common* *block* **common** /block/ X(ni, nj)	/* *static* */ **double** X[ni][nj]; /* *dynamic* */ **double** **x = **new** **double***[ni]; **for** (**int** i=0;i<ni;i ++) x[i] = **new double**[nj];	**import** numpy as np np = np.zeros(ni,nj)
Subroutines / Functions		

! using a variable
subroutineA (X,Y)
dimension X(ni,nj)
...
return
end
! with a common block
subroutineA
common/**block**/X(ni,nj)
...
return
end
	int x(**float** f) {
	...
	return A;
	}
		def X(f):
		...
		return A

Iteration		

do i=1,imax
...
end do
! with labels
do 100 i=1,imax
...
100 **continue**
	for (**int** i=0;i<imax;i
	++) {
	...
	}
	
	// *while loop*
	while (test) {
	...
	}
		for i **in range**(imax):
		...
		while test:
		...

Conditionals		

if (X.**ge**.A)
...
endif
!with goto
if (X.**lt**.A) **goto** 100
...
100 **continue**
	if (test) {
	...
	} **else if** (test) {
	... }
	else { ... }
	
	switch (val) {
	case X: ...; **break**;
	default: ...;
	}
		if test:
		...
		elif test:
		...
		else:
		...

Input and Output		
read(∗,∗) **read**(∗,100) 100 **format** (...) **write**(∗,∗) **write**(∗,100) 100 **format** (...)	// *C style* scanf("%d",&val); printf("%d, %.2g\n", ival,fval); // *C++ style* cin>>val; cout<<val<<", "<< setprecision(2)<<fval <<endl;	val = **input**() **print**("%d, %.2g"%(ival,fval))

References

Anderson, Dale A. et al. (2021). *Computational Fluid Dynamics and Heat Transfer*. 4th ed. CRC Press.

Anderson, John D. Jr. (1990). *Modern Compressible Flow*. McGraw Hill.

Anton, Howard and Chris Rorres (2000). *Elementary Linear Algebra*. 8th ed. John Wiley & Sons, Inc.

Aris, Rutherford (1962). *Vectors, Tensors, and the Basic Equations of Fluid Mechanics*. Dover.

Balestriero, Randall and Richard G. Baraniuk (2021). "Mad Max: Affine Spline Insights Into Deep Learning". In: *Proceedings of the IEEE* 109.5, pp. 704–727. DOI: 10.1109/JPROC.2020.3042100.

Bewley, Thomas R. (2018). *Numerical Renaissance: Simulation, Optimization, & Control*. Renaissance Press. ISBN: 978-0-9818359-0-7. URL: http://NumericalRenaissance.com.

Bird, G. A. (1994). *Molecular Gas Dynamics and the Direct Simulation of Gas Flows*. Oxford University Press.

Birdsall, C. K. and A. B. Langdon (1991). *Plasma Physics via Computer Simulation*. CRC Press.

Birdsall, Charles K (1991). "Particle-in-cell charged-particle simulations, plus Monte Carlo collisions with neutral atoms, PIC-MCC". In: *IEEE Transactions on Plasma Science* 19.2, pp. 65–85.

Boyce, William E. and Richard C. DiPrima (1997). *Elementary Differential Equations and Boundary Value Problems*. John Wiley & Sons.

Boyd, Iain D. and Thomas E. Schwartzentruber (2017). *Nonequilibrium Gas Dynamics and Molecular Simulation*. Cambridge University Press.

Brieda, Lubos (2019). *Plasma Simulations by Example*. CRC Press.

Briggs, William L., Van Emden Henson, and Steve F. McCormick (2000). *A Multigrid Tutorial*. 2nd ed. SIAM.

Brunton, Steven L. and J. Nathan Kutz (2022). *Data-Driven Science and Engineering*. Cambridge University Press.

Burden, Richard L. and J. Douglas Faires (2001). *Numerical Analysis*. 7th. Brooks / Cole Publishing Company.

Chan, Tony F and Jianhong Shen (2005). *Image Processing and Analysis: Variational, PDE, Wavelet, and Stochastic Methods*. SIAM.

Cheng, Chio-Zong and Georg Knorr (1976). "The integration of the Vlasov Equation in Configuration Space". In: *Journal of Computational Physics* 22.3, pp. 330–351.

Ferziger, Joel H. and Milovan Perić (2002). *Computational Methods for Fluid Dynamics*. 3rd. Springer.

Fife, John Michael (1998). "Hybrid-PIC modeling and electrostatic probe survey of Hall thrusters". PhD thesis. Massachusetts Institute of Technology.

Fitzgerald, Scott and Michael Shiloh, eds. (2013). *Arduino Projects Book*. Arduino LLC.

Fukunaga, Keinosuke (1990). "Chapter 6 - Nonparametric Density Estimation". In: *Introduction to Statistical Pattern Recognition (Second Edition)*. 2nd. Boston: Academic Press, pp. 254–299. DOI: https://doi.org/10.1016/B978-0-08-047865-4.50012-0.

Giles, Michael B. (Jan. 2016). "Algorithm 955: Approximation of the Inverse Poisson Cumulative Distribution Function". In: *ACM Trans. Math. Softw.* 42.1. ISSN: 0098-3500. DOI: 10.1145/2699466.

Gonoskov, A. (2022). "Agnostic conservative down-sampling for optimizing statistical representations and PIC simulations". In: *Computer Physics Communications* 271.108200.

Goodfellow, Ian, Yoshua Bengio, and Aaron Courville (2016). *Deep learning*. MIT press.

Hastie, Trevor, Robert Tibshirani, and Jerome H Friedman (2009). *The Elements of Statistical Learning: Data Mining, Inference, and Prediction*. Vol. 2. Springer.

Hastings, Daniel and Henry Garrett (1996). *Spacecraft-Environment Interactions*. Cambridge University Press.

Hockney, R.W. and J. W. Eastwood (1988). *Computer Simulations Using Particles*. IOP Publishing.

Hughes, Thomas J. R. (2000). *The Finite Element Method*. Dover Publications.

Jackson, John David (1992). *Classical Electrodynamics*. 3rd. John Wiley & Sons.

Jardin, Stephen (2010). *Computational Methods in Plasma Physics*. CRC Press.

Kays, William, Michael Crawford, and Bernhard Weigand (2005). *Convective Heat and Mass Transfer*. McGraw-Hill.

Kitware Inc. (2021). "The VTK User's Guide". In: 11th. Kitware, Inc. Chap. 19.3.

Kutz, J. Nathan (2013). *Data-Driven Modeling & Scientific Computation*. Oxford University Press.

Leveque, Randall J. (2002). *Finite Volume Methods for Hyperbolic Problems*. Cambridge University Press.

MathWorld, Wolfram (n.d.). *Sphere Point Picking*. https://mathworld.wolfram.com/SpherePointPicking.html. accessed Nov. 2023.

Matsuda, Kouichi and Rodger Lea (2013). *WebGL Programming Guide*. Addison-Wesley.

Nesterov, Yurii and Vladimir Spokoiny (Apr. 2017). "Random Gradient-Free Minimization of Convex Functions". In: *Foundations of Computational Mathematics* 17.2, pp. 527–566. ISSN: 1615-3383. DOI: 10.1007/s10208-015-9296-2.

O'Neil, Peter V. (1995). *Advanced Engineering Mathematics*. 4th. Brooks / Cole Publishing Company.

Oberkampf, William L. and Christopher J. Roy (2010). *Verification and Validation in Scientific Computation.* Cambridge University Press.

Panton, Ronald L. (2005). *Incompressible Flow.* 3rd. John Wiley & Sons.

Powers, David L. (2009). *Boundary Value Probems.* 6th ed. Harcourt Academic Press.

Press, William H (2007). *Numerical recipes: The Art of Scientific Computing.* 3rd. Cambridge University Press.

Robbins, Jennifer Niederst (2012). *Learning Web Design.* O'Reilly.

Robey, Robert and Yuliana Zamora (2021). *Parallel and High Performance Computing.* Manning Publications.

Russell, Stuart J and Peter Norvig (2010). *Artificial Intelligence: a Modern Approach.* London.

Salih, A (2013). "Streamfunction-vorticity formulation". In: *Department of Aerospace Engineering Indian Institute of Space Science and Technology, Thiruvananthapuram-Mach*, p. 10.

Sanders, Jason and Edward Kandrot (2011). *CUDA by Example.* Addison-Wesley.

Schroeder, Will, Ken Martin, and Bill Lorensen (2006). *The Visualization Toolkit.* 4th ed. Kitware, Inc.

Seidelin, Jacob (2012). *HTML5 Games.* John Wiley and Sons.

Sterling, Thomas Lawrence (2002). *Beowulf Cluster Computing with Linux.* MIT Press.

Stroustrup, Bjarne (2018). *A Tour of C++.* Addison-Wesley.

Tribble, Alan C. (2003). *The Space Environment.* Princeton University Press.

Tu, Jiyuanand, Guan-Heng Yeoh, and Chaoqun Liu (2013). *Computational Fluid Dynamics, a Practical Approach.* Elsevier.

Vincenti, Walter G. and Charles H. Kruger (2002). *Introduction to Physical Gas Dynamics.* Krieger Publishing Company.

Vranic, Marija et al. (2015). "Particle merging algorithm for PIC codes". In: *Computer Physics Communications* 191, pp. 65–73.

Wilt, Nicholas (2013). *The CUDA Handbook.* Addison-Wesley.

Yee, Kane (1966). "Numerical Solution of Initial Boundary Value Problems Involving Maxwell's Equations in Isotropic Media". In: *IEEE Transactions on Antennas and Propagation* 14.3, pp. 302–307.

Zhang, Zhuomin M. (2007). *Nano/Microscale Heat Transfer.* McGraw-Hill.

Index

Printed in the United States
by Baker & Taylor Publisher Services